Experimental, Numerical and Field Approaches to Scour Research

Experimental, Numerical and Field Approaches to Scour Research

Editors

Yee-Meng Chiew
Jihn-Sung Lai
Oscar Link

MDPI • Basel • Beijing • Wuhan • Barcelona • Belgrade • Manchester • Tokyo • Cluj • Tianjin

Editors
Yee-Meng Chiew
Nanyang Technological University
Singapore

Jihn-Sung Lai
National Taiwan University
Taiwan

Oscar Link
Universidad de Concepción
Chile

Editorial Office
MDPI
St. Alban-Anlage 66 4052 Basel,
Switzerland

This is a reprint of articles from the Special Issue published online in the open access journal *Water* (ISSN 2073-4441) (available at: https://www.mdpi.com/journal/water/special_issues/scour).

For citation purposes, cite each article independently as indicated on the article page online and as indicated below:

LastName, A.A.; LastName, B.B.; LastName, C.C. Article Title. *Journal Name* **Year**, *Volume Number*, Page Range.

ISBN 978-3-0365-2195-4 (Hbk)
ISBN 978-3-0365-2196-1 (PDF)

Cover image courtesy of Chiew Yee Meng.

Contents

About the Editors

Yee-Meng Chiew

Dr. Chiew Yee Meng, who is currently is a Professor in at the School of Civil and Environmental Engineering at the Nanyang Technological University (NTU), has forty years of research and consulting experience in the field of sediment transport, scour and erosion, turbulence, hydraulics and coastal/offshore engineering. He particularly is interested in the scour at of hydraulic structures, such as bridge piers, submarine pipelines, monopiles, ship propellers, culverts, bridge abutments, and jets, etc. and how such erosion affects the integrity of the surrounding structures and sediment bed. His research interests also include how turbulence affects fluvial and coastal hydraulics and their overall engineering impact, and the effect of seepage on the turbulence characteristics in both open channel and closed-conduit flows. Dr Chiew has close to two hundred archival publications and serves as an Associate Editor of the Journal of Hydraulic Engineering, (ASCE) and the International Journal for Sediment Research (IJSRC). He was the Chairman of the 2nd International Conference on Scour and Erosion, held in Singapore in 2004. Throughout his career, Dr Chiew has acted as a specialist consultant to numerous international and local companies as well as governmental organizations, offering his expert advice to solve challenging engineering projects.

Jihn-Sung Lai

Dr. Jihn-Sung Lai is currently a research fellow of the Hydrotech Research Institute at National Taiwan University (NTU). He is also the Chief Executive Officer of the Center for Weather Climate and Disaster Research, NTU, and the R&D Center for Hydrometry, Hydrotech Research Institute, NTU. His research expertise includes river mechanics, sediment transport, hydraulic structure, hydraulic physical and numerical modeling, reservoir sediment desilting techniques, flood inundation simulation, and disaster prevention and response. Dr. Lai has authored over 70 peer-reviewed SCI journal paper publications in his areas of expertise. In addition, he served as an associate editor of several journals and the co-chairman of the 3rd International Workshop on Sediment Bypass Tunnels, held in Taipei, Taiwan, in 2019. In addition, he has received 25 patents in the development of flow and sediment monitoring and measurement devices.

Oscar Link

Dr. Oscar Link is a Professor in the Civil Engineering Department of the Faculty of Engineering at Universidad de Concepción in Chile. He is interested in the broad field of river engineering, focusing on the development of practical solutions to complex problems, such as bridge scour, fish passages through hydropower dams, and flood risk management in unregulated rivers. In his twenty-years career, Dr. Link has authored 50 peer-reviewed SCI journal paper publications in his areas of expertise and acted as a specialist consultant to numerous international and local companies as well as governmental organizations, including the Inter-American Development Bank and the World Bank. He is co-author of design guidelines for the estimation of bridge scour recently published by the German Association for Water, Wastewater and Waste, DWA. Currently, Dr. Link is an Associate Editor of the Hydrological Sciences Journal.

Preface to "Experimental, Numerical and Field Approaches to Scour Research"

Whether it occurs in the fluvial, coastal, or offshore environment, scour can endanger the integrity of adjacent engineering structures, leading not only to economic loss but also to the loss of human lives. The interdependent interaction between the structure-induced flow field and sediment motion poses a highly complex physical mechanism, which comprises both temporal and spatial variations in flow property and has baffled researchers and engineers for a long time. Intensive research likely began in the mid-20th century, when extensive infrastructural development, including the construction of numerous bridges, ports, and offshore structures, was necessary due to the ravages of the Second World War. The objective was to provide cost-effective engineering design for these structures. Despite such research efforts, a full and complete understanding of the subject remains elusive. Additionally, new structures are now required in more extreme environments, presenting new challenges to researchers. This book presents fourteen research papers and one editorial based on the results of innovative research studies (experimental, numerical and field-based) conducted around the world. The editors hope that this will collectively provide new insights for researchers to explore the highly intriguing and challenging subject of scour.

Yee-Meng Chiew, Jihn-Sung Lai, Oscar Link
Editors

Editorial

Experimental, Numerical and Field Approaches to Scour Research

Yee-Meng Chiew [1,*], Jihn-Sung Lai [2] and Oscar Link [3]

[1] School of Civil and Environmental Engineering, Nanyang Technological University, Singapore 639798, Singapore
[2] Hydrotech Research Institute/Department of Bioenvironmental Systems Engineering, National Taiwan University, Taipei 10617, Taiwan; jslai525@ntu.edu.tw
[3] Department of Civil Engineering, Faculty of Engineering, Universidad de Concepción, Concepción 403000, Chile; olink@udec.cl
* Correspondence: cymchiew@ntu.edu.sg

Received: 11 June 2020; Accepted: 17 June 2020; Published: 19 June 2020

Abstract: Scour, which represents a fascinating and complex engineering problem involving a number of physical mechanisms and interactions, has motivated a vast amount of research following different but complementary methodological approaches such as experimental, numerical, and field methods. Far from being completely understood, scour remains one of the main hazards for many structures such as bridges, submarine pipelines, offshore wind turbines, etc. Thus, scour is currently a very active field of research with important open questions that are unanswered and practical challenges to be tackled. This Special Issue aims to bridge knowledge gaps by collecting fourteen papers to provide a wide view of scour types and different research approaches, with innovative ideas and inferences.

Keywords: scour; sediment transport; research approaches

1. Introduction

Scour, which represents an intriguing and complex engineering problem involving many physical mechanisms and interactions, has motivated a vast amount of research following different but complementary methodological approaches such as experimental, numerical, and field methods. Notwithstanding the many research findings that can be found in published literature, the subject is far from being completely understood. Scour continues to be one of the main hazards for many structures such as bridges, submarine pipelines, offshore wind turbines, etc., as is shown in the catastrophic failure of Houfeng Bridge in September 2008 (Figure 1) and the serious scour problems downstream of the grade-control structure leading to the demise of Zhongzheng Bridge in July 2013 (Figure 2). Consequently, this subject remains a very active area of research with important unanswered questions and practical challenges that need to be resolved.

Published evidence reinforce the need for more research in other scour types besides pier and abutment scour, particularly considering the importance of the use of alternative sources of energy other than hydrocarbon, with the aim to alleviate the threat of carbon dioxide emission and climate change. With the number of wind farms built in recent years or to be built in the near future by European and Asian countries (particularly China), obtaining a comprehensive understanding of the mechanism of scour around monopiles and finding the most cost-effective scour countermeasure are particularly needed. Despite the important works done in Europe on monopile scour, e.g., Whitehouse et al. [1], the authors did not incorporate the effect of vibrations on scour. Such limitations were highlighted in a recent review paper by Fredsøe [2], who lamented such a lack of consideration in past researches. However, more recent researchers, e.g., Guan et al. [3], have begun to include this concern in monopile scour studies. The effect of vibrations on scour is presently still in its infancy stage, and more effort is clearly needed in this area, whether experimentally or numerically. The effect of vibrations on scour is not limited to monopiles but also submarine pipelines. Scour around submarine pipelines started with researchers working in the 2-dimensional case, in which laboratory tests were conducted in narrow flumes. The early empirical 2-dimensional pipeline scour researches [4,5] were extended to numerical studies [6,7]; and later to 3-dimensional scour; with the latter better epitomizing the prototype condition. Some of the earliest 3-dimensional works were empirically conducted by [8–10], which were supplemented by numerical studies, although some of the earlier numerical studies have not been vigorously verified due to the lack of experimental data [11]. The effect of vibration on pipelines is equally an important consideration, which also was ignored by these researchers. This effect was only seriously examined in the work of Li et al. [12], who showed how vibrations significantly amplify the scour formation process. Later works include the experimental works of [13–16].

Figure 1. Failure of Houfeng Bridge in Dajia River, Taiwan due to pier-scour in September 2008 (courtesy of Yee-Meng Chiew).

(a) (b)

Figure 2. Scour downstream of grade-control structure leading to failure of Zhongzheng Bridge in Touqian River, Taiwan (**a**) scour downstream of grade-control structure in 2009 (courtesy of Yee-Meng Chiew); (**b**) Scour leading to bridge failure in 2013 (courtesy of Dr. Hong JH).

In comparison with pier- or even pipeline-induced scour, propeller scour receives notably less attention. Most of the earlier works were carried out in Europe, especially those conducted in Queen's University Belfast, with the first doctoral dissertation by Hamill [17] and subsequent doctoral studies. A second group of propeller scour works was conducted in Singapore, with very meticulous flow field measurements collected using the particle image velocimetry technique [18], giving comprehensive descriptions of the propeller-induced flow field and providing new insights into the scouring mechanism associated with the development of the scour hole. While weirs often are built along rivers, sometimes for flow regulation or as a water intake structure (as a side weir), their effects on scour have not been extensively studied. Interestingly, when the weir is very high, it essentially becomes a dam. On the other hand, if it is low, it acts similarly to a sill and eventually resembles a grade-control structure.

Reservoirs that are caused by dam construction act to impound water in rivers, mainly for water supply and flow regulation, invariably modifying the channel bed slope. Accordingly, their formation has a tendency to induce sediment deposition, which is a kind of negative scour [19,20], while a grade-control structure can have a profound influence on local scour [21]. Reservoirs also interrupt the course of sediment transport through river systems, causing decreased sediment loads because of trapping by upstream dams [22,23]. The interference of sediment trapping by reservoir exerts considerable morphological effects on downstream river channels, including riverbed incisions, riverbank instabilities, scour-induced damages to infrastructure (e.g., bridges, pipeline-crossings, embankments, and levees), channel width variations, and coastal erosion. Many reservoir management strategies and effective countermeasures have been investigated to reduce sediment deposition in reservoirs and increase sediment supply for downstream river systems [24,25]. To remove trapped sediment deposits from reservoirs, mechanical dredging is one of the commonly used measures. However, the disposal of dredged sediment is costly.

From another viewpoint, dredged sediments may be considered as a resource that provides effective environmental benefits. By adding sediment to a river, an approach termed sediment replenishment or sediment augmentation can be applied to compensate for the lack of sediment loads downstream of dams. In field practice, it can be planned and practiced in specific hydrological and geographic areas. Okano et al. [26] investigated reservoir sedimentation management by depositing coarse sediment (mainly sand and gravel sizes) for downstream replenishment. The conceptual idea of the replenishment method is to place sediments on the downstream floodplain before the arrival of floods. During the flood season, the replenished sediment will be scoured and distributed downstream. The applied replenishment method requires a sufficient amount of discharge in order to scour the replenished sediment. In practice, sediment deposited in a reservoir can be periodically dredged and strategically placed on the floodplain downstream of the dam. Most commonly, the sediments added are gravel and sand obtained from gravel quarries in the floodplain or other sources. Additionally, sediments also may be dredged from the reservoir delta deposits [27,28]. Ock et al. [28] reviewed methods in the context of sediment replenishment and compared implementation activities undertaken in Japan and the USA. According to sediment placement, sediment replenishment methods are implemented with mechanical rehabilitation for re-creating gravel or sand bar features through fluvial processes. This comparative study provides useful information for engineers to adopt proper methods corresponding to river-specific high-flow and sediment regimes. Laboratory experiments and field investigations have been conducted to improve the understanding of the transport processes of sediment replenishment on the river basin scale [29,30]. Notwithstanding these research, additional works clearly are needed to ensure the most cost-effective solution to address the dire problems of reservoir storage and downstream bed load reduction that are prevalent worldwide presently.

The editors of this Special Issue have carefully selected fourteen papers to provide a wide view of scour types using different research approaches. We hope that the contents of these papers, which do not originate from them, will provide guidelines for practicing engineers and researchers in their works.

2. Contributions

This Special Issue comprises fourteen papers, the majority of which examine the scour topic from the empirical or experimental approaches (64%); the others are founded on the numerical and field approaches, and a review paper. The types of structure involved are piers, pipelines, propellers, turbines, weirs, reservoirs, and grade-control structures (GCS), with piers leading the group with five papers or 36%. The review paper summarizes research works to-date relating to bridge foundation, namely piers and abutments, in supercritical flows. The number of papers published in terms of the research methodology or structure type, namely experimental approach and pier scour, respectively, is representative of the number of papers most commonly found within scour research in archival journals and conference proceedings. Notwithstanding the number of published research works in a particular area, this does not negate the importance of scour research in the other areas and research approaches. The limited number of scour research using field data in this Issue highlights the need for more efforts in the field approach. The papers of this Special Issue, which are listed in alphabetical order of the lead author, are summarized as follows:

Chavan et al. Turbulent Flow Structures and Scour Hole Characteristics around Circular Bridge Piers over Non-Uniform Sand Bed Channels with Downward Seepage

This paper [31] presents the experimental results of the scour hole characteristics forming around single vertical pier sets on a non-uniform sand bed with and without downward seepage. Empirical equations for the evaluation of the scour hole characteristics such as the length, width, area, and volume, including the downward seepage parameter, are proposed and experimentally tested. Predictions give reasonably good agreement with the experimental data.

Cui et al. Scour Induced by Single and Twin Propeller Jets

The authors conduct laboratory experiments to investigate scour induced by single and twin propeller jets [32]. Propellers with three and six blades typically used in British ports and harbors are scaled according to the similitude theory and reduced propeller models were created with a 3D printer using biodegradable the polylactic acid filament, PLA, material. A laser rangefinder is used to measure the scour depth during the experiments. Empirical relations for the computation of the scour depth along the propeller axis are provided.

Guan et al. Scour Evolution Downstream of Submerged Weirs in Clear Water Scour Conditions

The authors propose exponential equations to estimate the temporal variation in the scour depth and scour length downstream of a submerged weir [33]. The study is useful in the development of models capable of estimating scour depth downstream of weirs in rivers or coastal areas, for which the overtopping conditions are very common. The proposed equations for scour hole dimensions, profiles, and sizes will be good tools for hydraulic engineers in the design of scour countermeasures.

Lee et al. Scouring of Replenished Sediment through Reservoir Flood Discharge Affects Suspended Sediment Concentrations at Downstream River Water Intake

This paper describes a comprehensive study on the transport of replenished sediment following two methodological approaches, namely numerical modelling and physical scale modelling [34]. Valuable field data are used for the models' calibrations and validations. The study makes an important contribution to the use of dredged reservoir sediments focusing on fine (cohesive) sediments for sediment replenishment, which in turn contributes to the morphological stability of the river downstream a dam. In this way, an alternative for the disposal of dredged fine sediments taking advantage of scour is provided.

Lin et al. Visible Light Communication System for Offshore Wind Turbine Foundation Scour Early Warning Monitoring

This paper proposes a scour monitoring system for offshore wind turbine installations using visible light communication (VLC) modules [35]. The feasibility of the proposed system is investigated through laboratory scour experiments. The novel technique is a promising low-cost alternative for field scour monitoring.

Link et al. Scour at Bridge Foundations in Supercritical Flows: An Analysis of Knowledge Gaps

Scour at bridge foundations, viz., piers and abutments caused by supercritical flows, is reviewed and knowledge gaps are identified and discussed [36]. Based on published data, the review paper finds that the scour depth caused by supercritical flows is unexpectedly much lower than that found in subcritical flows, and argues that the reasons of this behavior remain uncertain. It surmises that interactions of the horseshoe vortex with the detached hydraulic jump and the wall jet flow observed in supercritical flows could be a reason. Finally, it provides recommendations for future research in this topic.

Ma et al. Case Study: Model Test on the Effects of Grade Control Datum Drop on the Upstream Bed Morphology in Shiting River

This case study describes the construction of a 1:80 undistorted physical model of a 1.3 km-long river reach in Shiting River, China, to investigate the impact of the grade-control datum (GCD, which is defined as the crest elevation of the grade-control structure) on the upstream bed morphology [37]. It provides a plausible designed consideration for the solution of scour countermeasure against excessive erosion around an existing bridge on the river.

Mathieu et al. Two-Phase Flow Simulation of Tunnel and Lee-Wake Erosion of Scour below a Submarine Pipeline

This paper presents a numerical investigation of the scour phenomenon around a submarine pipeline [38]. Numerical simulations are performed using a two-phase flow model for sediment transport implemented in an open-source computational fluid dynamics (CFD) toolbox. The paper focuses on the sensitivity of the granular stress model and the turbulence model with respect to the predictive capability of the two-phase flow model. The numerical results show no sensitivity to the granular stress model. However, the results are strongly dependent on the choice of turbulence model.

Quezada et al. Numerical Study of the Hydrodynamics of Waves and Currents and Their Effects in Pier Scouring

The authors use the Reynolds averaged Navier–Stokes (RANS) equations via REEF3D numerical modeling to simulate the hydrodynamic field and estimate the scour around cylindrical piles due to co-directional and opposite waves and currents [39]. The relative velocity of a current is defined as an indicator that adequately measures the interactions between currents and waves under the condition of co-directional flows. It concludes that the dimensionless scour depth will be less when waves and currents come from opposite directions.

Sun et al. Temporal Evolution of Seabed Scour Induced by Darrieus-Type Tidal Current Turbine

This paper presents an experimental study on the temporal evolution of seabed scour around the monopile foundation of a Darrieus-type tidal current turbine [40]. Their results show that the scour depth is inversely correlated with the tip–bed clearance between the turbine and seabed. Empirical equations are also proposed for the prediction of the temporal scour depth around the turbine, although this has not been extensively verified.

Wang et al. Experimental Investigation of Local Scour Protection for Cylindrical Bridge Piers Using Anti-Scour Collars

The authors conduct live bed pier scour experiments to investigate the effects of collars as a scour countermeasure [41]. Three variables, namely the collar installation height, collar external diameter, and collar protection range, are tested. Important design guidelines such as the recommended collar shape, installation height, and collar diameter are provided.

Wei et al. Vortex Evolution within Propeller Induced Scour Hole around a Vertical Quay Wall

This study advances the technique of oblique particle image velocimetry (OPIV) to conduct a detailed experimental study on the flow field in a scour hole induced by a propeller, including the effect of a vertical quay wall [42]. Concurrent measurements of scour and flow during the temporal development of the scour hole are performed successfully. The evolution of the vortex system and its interaction with the developing scour hole are examined in detail, elucidating the distinct influence of different quay types on the formation of flow patterns and the resulting scour profiles. The near-bed flow characteristics that are indicative of the erosive flow mechanisms are discussed; this provides new insights into the scour mechanism associated with the development of the scour hole.

Williams et al. Examination of Blockage effects on the Progression of Local Scour around a Circular Cylinder

This paper describes investigations of the effect of the blockage ratio, which is defined as the ratio of pier diameter and flume width, on scour around a circular cylinder [43]. Unlike past researchers who stated how such effects are negligible when the blockage ratio exceeds 0.1 and 0.125 for clear-water and live-bed conditions, respectively, they concluded that the effect also is related to the relative coarseness, which is defined as the ratio of pier diameter and median grain size of the bed sediment. It suggests that further experiments are needed for the range of relative coarseness ratios exceeding 100 in order to establish the role of blockage ratio effects to refine existing scour estimation methods.

Yang et al. Local Scour at Complex Bridge Piers in Close Proximity under Clear-Water and Live-Bed Flow Regime

This experimental study investigates the characteristics of scour at complex bridge piers in close proximity under both clear-water and live-bed flow regimes [44]. The results are compared with those for a single complex pier. A different scour pattern is observed when more than one complex pier with variable arrangements is present. Moreover, the resulting scour pattern also is different for clear-water and live-bed conditions, with an important influence of migrating bed forms on scour under the latter condition.

3. Conclusions

This Special Issue presents up to date research on scour following experimental, numerical, and field approaches covering the interaction of the flow and sediment bed around different structures such as bridge piers, monopiles, propellers, turbines, weirs, grade-control structures, and pipelines. The majority of the papers examined the subject from the empirical or experimental approach (64%); the others are founded on the numerical and field approaches, and a review paper. The leading group of papers was that devoted to the pier scour, with five papers or 36%. The review paper summarizes research works to-date relating to bridge foundations, namely piers and abutments, in supercritical flows. The number of papers published in terms of the research methodology or structure type, namely the experimental approach and pier scour, respectively, are representative of the number of papers most commonly found within scour research in archival journals and conference proceedings. Notwithstanding the number of published research works in a particular area, this does not negate the importance of scour research in the other areas and research approaches.

Author Contributions: All three writers contributed equally in the preparation and writing of the paper. All authors have read and agreed to the published version of the manuscript.

Funding: This research received no external funding.

Conflicts of Interest: The authors declare no conflict of interest.

References

1. Whitehouse, R.J.S.; Harris, J.M.; Sutherland, J.; Rees, J. The nature of scour development and scour protection at offshore windfarm foundations. *Mar. Pollut. Bull.* **2011**, *62*, 73–88. [CrossRef] [PubMed]
2. Fredsøe, J. Pipeline-seabed interaction. *J. Wtrw. Port Coast. Ocean Eng. ASCE* **2016**, *142*, 03116002. [CrossRef]
3. Guan, D.; Chiew, Y.M.; Melville, B.W.; Zheng, J. Current-induced scour at monopile foundations subjected to lateral vibrations. *Coast. Eng.* **2019**, *144*, 15–21. [CrossRef]
4. Sumer, B.; Fredsøe, J. Scour below pipelines in waves. *J. Wtrw. Port Coast. Ocean Eng. ASCE* **1990**, *116*, 307–323. [CrossRef]
5. Chiew, Y.M. Mechanics of local scour around submarine pipelines. *J. Hydraul. Eng. ASCE* **1990**, *116*, 515–529. [CrossRef]
6. Liang, D.; Cheng, L. A numerical model for wave-induced scour below a submarine pipeline. *J. Wtrw. Port Coast. Ocean Eng. ASCE* **2005**, *131*, 193–202. [CrossRef]
7. Fuhrman, D.R.; Baykal, C.; Sumer, B.M.; Jacobsen, N.G.; Fredsøe, J. Numerical simulation of wave-induced scour and backfilling processes beneath submarine pipelines. *Coast. Eng.* **2014**, *94*, 10–22. [CrossRef]
8. Fredsøe, J.; Hansen, E.A.; Mao, Y.; Sumer, B.M. Three dimensional scour below pipelines. *J. Offshore Mech. Arct. Eng.* **1988**, *110*, 373–379. [CrossRef]

9. Cheng, L.; Yeow, K.; Zhang, Z.; Teng, B. Three-dimensional scour below pipelines in steady currents. *Coast. Eng.* **2009**, *56*, 577–590. [CrossRef]
10. Wu, Y.S.; Chiew, Y.M. Three-dimensional scour at submarine pipelines. *J. Hydraul. Eng. ASCE* **2012**, *138*, 788–795. [CrossRef]
11. Chen, B.; Cheng, L. Numerical investigations of three-dimensional flow and bed shear stress distribution around the span shoulder of pipeline. *J. Hydrodyn.* **2004**, *16*, 687–694.
12. Li, F.Z.; Dwivedi, A.; Low, Y.M.; Hong, J.H.; Chiew, Y.M. Experimental investigation on scour under a vibrating catenary pipe. *J. Eng. Mech. ASCE* **2013**, *139*, 868–878. [CrossRef]
13. Hsieh, S.C.; Low, Y.M.; Chiew, Y.M. Flow characteristics around a circular cylinder subjected to vortex-induced vibration near a plane boundary. *J. Fluids Struct.* **2016**, *65*, 257–277. [CrossRef]
14. Guan, D.; Hsieh, S.C.; Chiew, Y.M.; Low, Y.M. Experimental study of scour around a forced vibrating pipeline in quiescent water. *Coast. Eng.* **2019**, *143*, 1–11. [CrossRef]
15. Guan, D.; Hsieh, S.C.; Chiew, Y.M.; Low, Y.M.; Wei, M. Local scour and flow characteristics around pipeline subjected to vortex-induced vibrations. *J. Hydraul. Eng. ASCE* **2020**, *146*, 04019048. [CrossRef]
16. Tofany, N.; Low, Y.M.; Lee, C.-H.; Chiew, Y.M. Two-phase flow simulation of scour beneath a vibrating pipeline during the tunnel erosion stage. *Phys. Fluids* **2019**, *31*, 113302. [CrossRef]
17. Hamill, G.A. Characteristics of the Screw Wash of a Manoeuvring Ship and the Resulting Bed Scour. Ph.D. Thesis, Queen's University Belfast (United Kingdom), Belfast, UK, 1987.
18. Wei, M.; Chiew, Y.M. Characteristics of propeller jet flow within developing scour holes around an open quay. *J. Hydraul. Eng.* **2018**, *144*, 04018040. [CrossRef]
19. Lai, J.S.; Chang, F.J. Physical modeling of hydraulic desiltation in Tapu Reservoir. *Int. J. Sediment Res.* **2001**, *16*, 363–379.
20. Wang, H.W.; Kondolf, G.M.; Desiree, T.; Kuo, W.C. Sediment management in Taiwan's reservoirs and barriers to implementation. *Water* **2018**, *10*, 1034. [CrossRef]
21. Bormann, N.E.; Julien, P.Y. Scour downstream of grade-control structures. *J. Hydraul. Eng. ASCE* **1991**, *117*, 579–594. [CrossRef]
22. Shen, H.W.; Lai, J.S. Sustain reservoir useful life by flushing sediment. *Int. J. Sediment Res.* **1996**, *11*, 10–17.
23. Kondolf, G.M.; Gao, Y.X.; Annandale, G.W.; Morris, G.L.; Jiang, E.H.; Zhang, J.H.; Cao, Y.T.; Carling, P.; Fu, K.D.; Guo, Q.C.; et al. Sustainable sediment management in reservoirs and regulated rivers: Experiences from five continents. *Earth's Future* **2014**, *2*, 256–280. [CrossRef]
24. Annandale, G.W.; Morris, G.L.; Karki, P. *Extending the Life of Reservoirs: Sustainable Sediment Management for Dams and Run-of-River Hydropower*; World Bank Group: Washington, DC, USA, 2016.
25. Stähly, S.; Franca, M.J.; Robinson, C.T.; Schleiss, A.J. Sediment replenishment combined with an artificial flood improves river habitats downstream of a dam. *Nat. Sci. Rep.* **2019**, *9*, 5176. [CrossRef] [PubMed]
26. Okano, M.; Kikui, M.; Ishida, H.; Sumi, T. Reservoir sedimentation management by coarse sediment replenishment below dams. In Proceedings of the Ninth International Symposium on River Sedimentation, Yichang, China, 18–21 October 2004; pp. 1070–1078.
27. Sumi, T.; Kantoush, S.A. Sediment replenishing measures for revitalization of Japanese rivers below dams. In Proceedings of the 34th IAHR World Congress, Brisbane, Australia, 26 June–1 July 2011; pp. 2838–2846.
28. Ock, G.; Sumi, T.; Takemon, Y. Sediment replenishment to downstream reaches below dams: Implementation perspectives. *Hydrol. Res. Lett.* **2003**, *7*, 54–59. [CrossRef]
29. Battisacco, E.; Franca, M.J.; Schleiss, A.J. Sediment replenishment: Influence of the geometrical configuration on the morphological evolution of channel-bed. *Water Resour. Res.* **2016**, *52*, 8879–8894. [CrossRef]
30. Kondolf, G.M. Sediment management at the river-basin scale: Challenges and opportunities. In Proceedings of the 3rd International Workshop on Sediment Bypass Tunnels, Taipei, Taiwan, 9–12 April 2019.
31. Chavan, R.; Gualtieri, P.; Kumar, B. Turbulent flow structures and scour hole characteristics around circular bridge piers over non-uniform sand bed channels with downward seepage. *Water* **2019**, *11*, 1580. [CrossRef]
32. Cui, Y.; Lam, W.H.; Zhang, T.; Sun, C.; Hamill, G. Scour induced by single and twin propeller jets. *Water* **2019**, *11*, 1097. [CrossRef]
33. Guan, D.; Liu, J.; Chiew, Y.-M.; Zhou, Y. Scour evolution downstream of submerged weirs in clear water scour conditions. *Water* **2019**, *11*, 1746. [CrossRef]
34. Lee, F.-Z.; Lai, J.-S.; Guo, W.-D.; Sumi, T. Scouring of replenished sediment through reservoir flood discharge affects suspended sediment concentrations at downstream river water intake. *Water* **2019**, *11*, 1998. [CrossRef]

35. Lin, Y.-B.; Lin, T.-K.; Chang, C.-C.; Huang, C.-W.; Chen, B.-T.; Lai, J.-S.; Chang, K.-C. Visible light communication system for offshore wind turbine foundation scour early warning monitoring. *Water* **2019**, *11*, 1486. [CrossRef]
36. Link, O.; Mignot, E.; Roux, S.; Camenen, B.; Escauriaza, C.; Chauchat, J.; Brevis, W.; Manfreda, S. Scour at bridge foundations in supercritical flows: An analysis of knowledge gaps. *Water* **2019**, *11*, 1656. [CrossRef]
37. Ma, X.; Wang, L.; Nie, R.; Yang, K.; Liu, X. Case study: Model Test on the effects of grade control datum drop on the upstream bed morphology in Shiting River. *Water* **2019**, *11*, 1898. [CrossRef]
38. Mathieu, A.; Chauchat, J.; Bonamy, C.; Nagel, T. Two-phase flow simulation of tunnel and lee-wake erosion of scour below a submarine pipeline. *Water* **2019**, *11*, 1727. [CrossRef]
39. Quezada, M.; Tamburrino, A.; Niño, Y. Numerical study of the hydrodynamics of waves and currents and their effects in pier scouring. *Water* **2019**, *11*, 2256. [CrossRef]
40. Sun, C.; Lam, W.H.; Lam, S.S.; Dai, M.; Hamill, G. Temporal evolution of seabed scour induced by darrieus-type tidal current turbine. *Water* **2019**, *11*, 896. [CrossRef]
41. Wang, S.; Wei, K.; Shen, Z.; Xiang, Q. Experimental investigation of local scour protection for cylindrical bridge piers using anti-scour collars. *Water* **2019**, *11*, 1515. [CrossRef]
42. Wei, M.; Cheng, N.-S.; Chiew, Y.-M.; Yang, F. Vortex evolution within propeller induced scour hole around a vertical quay wall. *Water* **2019**, *11*, 1538. [CrossRef]
43. Williams, P.; Balachandar, R.; Bolisetti, T. Examination of blockage effects on the progression of local scour around a circular cylinder. *Water* **2019**, *11*, 2631. [CrossRef]
44. Yang, Y.; Melville, B.W.; Macky, G.H.; Shamseldin, A.Y. Local scour at complex bridge piers in close proximity under clear-water and live-bed flow regime. *Water* **2019**, *11*, 1530. [CrossRef]

Article

Local Scour at Complex Bridge Piers in Close Proximity under Clear-Water and Live-Bed Flow Regime

Yifan Yang *, Bruce W. Melville, Graham H. Macky and Asaad Y. Shamseldin

Department of Civil and Environmental Engineering, The University of Auckland, Private Bag 92019, Auckland 1142, New Zealand
* Correspondence: yyan749@aucklanduni.ac.nz

Received: 24 June 2019; Accepted: 19 July 2019; Published: 24 July 2019

Abstract: In this study, we investigated the characteristics of scour at complex bridge piers in close proximity. The experiments were performed under both clear-water and live-bed flow regimes. We compare our results with those for a single complex pier. Further, the performance of existing predictors is discussed. In this study, four typical pier arrangements were adopted, including side-by-side with aligned or 30° skewed flow, staggered, and tandem. The results show that the skew angle for a side-by-side arrangement significantly accelerates the clear-water scour development at all the vertical piles as well as between the piers, and the most scoured pile shifts from the upstream end to the downstream end of the upstream pier flank. The staggered and tandem pier arrangement show significant protection to the downstream pier for both the developing rate and the equilibrium scour depth. When the flow velocity exceeds the threshold for general bed motion, the clear-water scour pattern for all the pier arrangements may be altered significantly due to the upstream sediment supply, the weakened protection effect, and the enhanced flow contraction. The bed-forms migrate via the bridge opening and are damped gradually by the flow, and thus the response of the bed morphology under live-bed conditions is quite unsteady.

Keywords: scour; complex bridge pier; pier-pier proximity; temporal evolution; equilibrium scour pattern; bed-form migration; scour predictor evaluation

---

1. Introduction

Scour is one of the most common damages to bridges in fluvial environments. The approaching flow diverted by a pier leads to a series of large-scale turbulence coherent structures, e.g., horseshoe vortex, surface roller, and wave vortices, which consequently increase the kinetic energy and the eroding capacity of the flow. A scour hole forms around the pier when the surrounding erodible bed materials are significantly entrained and eroded away by flow features. At the same time, the pier itself is further exposed. The existence of a scour hole significantly alters the inherent frequency, stress, or bearing conditions of the pier and leads to either a settling, bending, or aggravated wobbling effect. The failure of a bridge foundation damages not only the engineering purposes of the bridge itself but also possibly the downstream ecological environment, e.g., spawning beds (Melville and Coleman [1]).

Bridge piers are usually built with complex geometry due to mechanical, geotechnical, and structural considerations. Thus, complex pier forms are more widely adopted around the world than the single pier form with uniform cross-sections. A typical complex bridge pier often consists of a wall-like column supporting the bridge deck and superstructures, a pile-cap under the column, and a group of piles supporting the pile-cap. Complex bridge piers are usually built in close proximity to each other for large bridges with multiple lanes, parallel companion bridges (especially highway-railway bridges combination), and bridges with continuous spans on flood plains. The remnants of previously existing bridges that have been demolished or destroyed may also influence adjacent newly built ones. Figure 1 show the typical cases of complex piers in close proximity. The flow field and sediment transport mechanism are significantly altered at a complex pier with another one nearby. Thus, the existing scour predictors may not be applied to such situations directly and need further adjustment with caution.

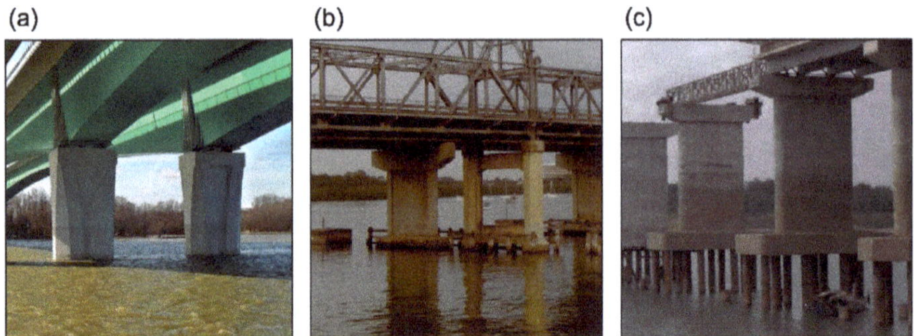

Figure 1. Typical cases of complex bridge piers in close proximity: (**a**) Maria Skłodowska-Curie Bridge in Warsaw, Poland; (**b**) Ryde Bridge in Sydney, Australia; (**c**) Unknown bridge under construction. (Photos from internet).

Previous studies have paid much attention to current-induced scour at multiple piles (Hannah [2]; Elliott and Baker [3]; Zhao and Sheppard [4]; Sumer et al. [5,6]; Ataie-Ashtiani and Beheshti [7]; Amini et al. [8]; Liang et al. [9]; Lança et al. [10]; Das and Maxumdar [11]; Wang et al. [12]; Khaple et al. [13]; and Kim et al. [14]) and scour at complex piers (Jones and Sheppard [15]; Coleman [16]; Sheppard and Glasser [17]; Ataie-Ashtiani et al. [18]; Grimaldi and Cardoso [19]; Beheshti & Ataie-Ashtiani [20,21]; Moreno et al. [22–24]; Ferraro et al. [25]; Amini et al. [26]; Baghbadorani et al. [27]; and Yang et al. [28]), respectively. However, there is still no solution for dealing with scenarios with multiple complex piers, indicating an obvious gap of the existing knowledge. This gap is the impetus of the present study. As a result of the insufficient knowledge, people usually tend to design with large safety redundancy when dealing with real engineering practices, leading to excessive cost and push the original budget much higher. In addition, it is also worth mentioning that the interaction between bridge piers and other bridge foundation components, e.g., the abutments, has also been studied by Oben-Nyarko and Ettema [29] and Sturm et al. [30,31]. They found that the extent and depth of the abutment's scour hole is much greater than that of a nearby pier and thus the scour aggravation caused by pier proximity is usually negligible.

This paper is aimed at providing an insight into the scour pattern at complex bridge piers in close proximity and determining the quantitative and qualitative scour features under both clear-water and live-bed flow regime. The morphological response of the erodible bed at the adjacent piers may vary significantly with or without general sediment transport and bed-form migration, which are common fluvial factors that should not be neglected.

2. Experimental Setup

The experiments in the present study were performed in a 2.4-m wide, 0.6-m deep, and 25-m long sediment-recirculating flume in the Fluid Mechanics Laboratory at the University of Auckland. The flume slope is adjustable up to a 1% slope. The flume is equipped with two water pumps, allowing a maximum of 1100 liters per second discharge and one sand pump. A 4.0-m long, 0.6-m deep sediment recess box, i.e., the test section for scour experiments, is located 11-m downstream from the water outlet section. The sediment used in the present study was uniform quartz sand with median particle diameter d_{50} = 0.84-mm and geometrical standard deviation σ_g = 1.30. This sediment grain size is coarse enough to be considered as non-cohesive and non-ripple-forming.

The complex pier model used in the present study was made from solid materials and represents a form typical of the bridge piers that have failed in the past several decades in New Zealand, as described by Melville and Coleman [1]. These failures include the Bulls Road Bridge on SH1, the Blackmount Road Bridge, the Oreti River Road Bridge, and the Whakatane River Road Bridge. The model consists of a rectangular wall-like column (310-mm long and 30-mm wide), a rectangular pile-cap (362-mm long, 120-mm wide, and 60-mm thick) and a 2 × 4 pile group (25-mm in diameter for each vertical pile). The model has a geometric scale of 3:50 and is similar to the primary model used by Yang et al. [28]. Figure 2 shows the schematic drawings of the configuration of the complex pier model.

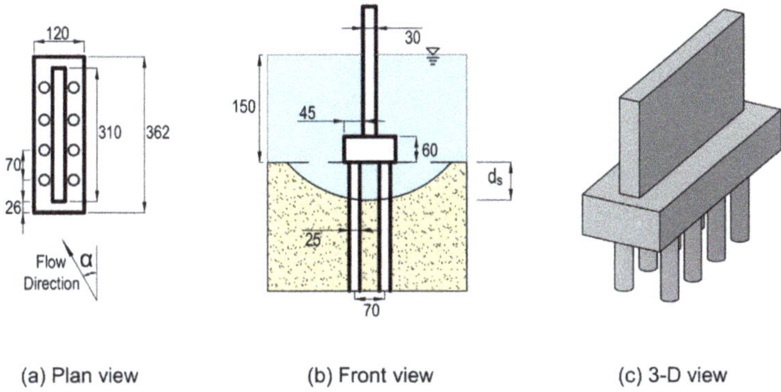

Figure 2. Model shape and dimensions (unit: mm): (**a**) plan view; (**b**) front view; and (**c**) 3-D view.

Depth-averaged velocities were measured by a Nortek Vectrino+ ADV made in Norway. A relationship was found between the measured flow velocity, the reading of the electromagnetic flow meter, and the working frequency of the pump controller. The scour depth was measured using a SeaTek multiple transducer array (MTA) made in USA; detailed specification can be referred in [32]. Each transducer has a cylindrical shape and is 12.7-mm in diameter and 25.4-mm in length. The transducers were fixed onto a transversal frame across the flume and in holes drilled in the pile cap, to measure instantaneous scour depth around the pier and at each pile throughout the duration of the tests. The transducers emit ultrasound pulses with a frequency up to 20-Hz and thereby measure the distance by analyzing the reflected signal.

Four different pier arrangements were used to represent typical scenarios of complex pier proximity in the field, including: (a) two adjacent side-by-side complex piers that are both aligned to the approaching flow; (b) two adjacent side-by-side complex piers are that are both skewed to the approaching flow with $\alpha = 30°$; (c) two aligned complex piers that are staggered with 30° angle between the flow direction and the line connecting the pier centres; and (d) two aligned complex piers with tandem arrangement. Besides the side-by-side arrangements, two adjacent complex piers can also support different bridge decks, which are common for companion bridges with multiple lanes. If so, the position relationship between two adjacent piers can be either staggered or tandem. The distance between two adjacent piers is based on typical real cases, while a closer distance is rarely adopted in engineering practices. Figure 3 shows the schematic drawings and photos of the experimental setups before filling the flume. The red circles mark the locations where the MTA transducers were placed to monitor real-time bed level data. Further transducers were also set upstream of the pier to monitor undisturbed bed-forms but are not located in the figure. Figure 4 provides more detailed setup information of the MTA transducers at one of the piers, as well as the numbering scheme for the transducers. Previous studies, including Zhao and Sheppard [4], Lança et al. [10], and Yang et al. [28]

have shown that, for scour at a group of piles or a stand-alone skewed complex pier, the maximum scour depth usually occurs at the downstream end of the upstream pile column, i.e., A3 or A4.

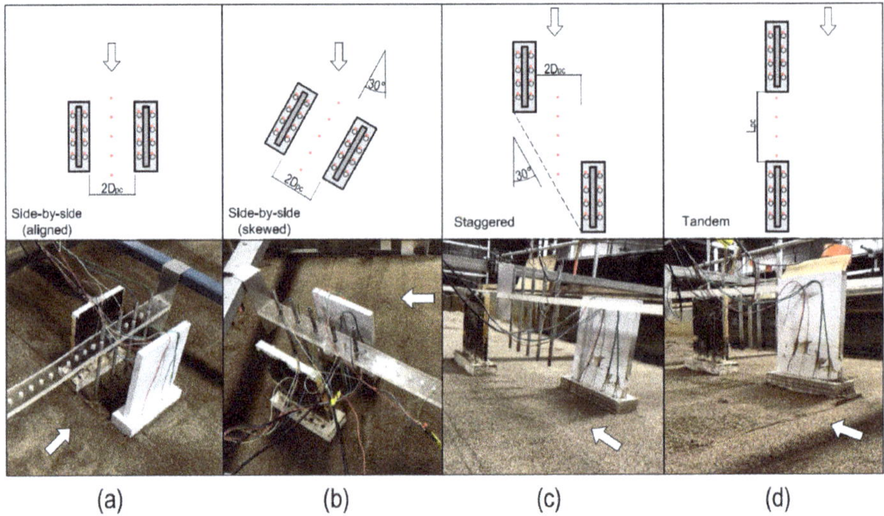

Figure 3. Setup of four different pier arrangement: (**a**) side-by-side arrangement with aligned flow; (**b**) side-by-side arrangement with 30° skew angle; (**c**) staggered arrangement; (**d**) tandem arrangement. The small red circles represent the locations where the multiple transducer array (MTA) transducers were set and real-time bed level data are available.

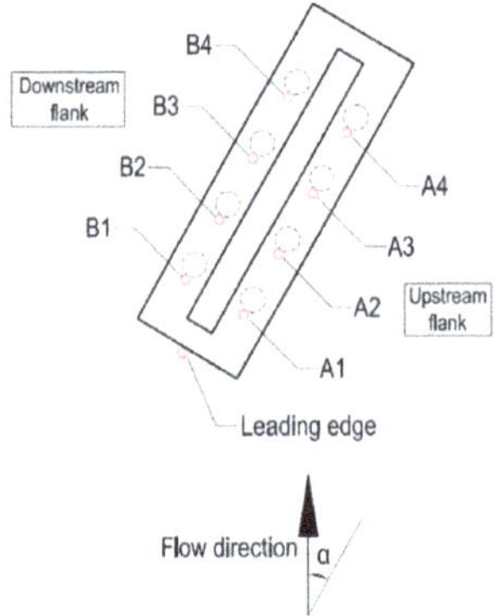

Figure 4. Definition and numbering scheme of the MTA transducers.

The experimental design in the present study in shown in Table 1. The flow depth y_0 was fixed at 0.15 m. The pile-cap elevation was fixed at H_c = 0.06 m and the base of the pile-cap was at the original bed level. This position is typical and leads to enlarged scour hole, as previous studies (Ataie-Ashtiani et al. [18], Moreno et al. [22], and Yang et al. [28]) have shown that the scour depth is maximized when the pile-cap is near the original bed level. For each pier arrangement, a wide range of flow intensity ratios (U/U_c) were tested, from 0.9 to 4.0–5.2, to investigate scour features under both clear-water ($U/U_c \leq 1$) and live-bed ($U/U_c > 1$) flow regimes. When $U/U_c > 1$, the sediment particles over the entire channel bed start to move downstream and lead to consecutive migrating dunes with wavy patterns.

Table 1. Experimental design (D_{pc} = pile-cap width, L_{pc} = pile-cap length).

Series	Pier Arrangement	Pier Alignment	Pier Distance	Supporting the Same Bridge Deck?
1	Side-by-side	Aligned	$2D_{pc}$	Yes
2	Side-by-side	30° skewed	$2D_{pc}$	Yes
3	Staggered	Aligned	$2D_{pc}$	No
4	Tandem	Aligned	L_{pc}	No

3. Clear-Water Scour Experiments

3.1. Temporal Evolution

The temporal evolutions of the experiments in the present study are shown in Figure 5, displaying the development with time of scour depth at the transducers. Due to the many measuring locations, the data are displayed with increased transparency towards downstream for a clearer distinction. To reduce the potential distraction caused by excessive lines, part of the data of minor significance are deleted from the graph, while the general trends can still be shown clearly. It should also be noted that the transducers are not functional when they are buried or the distance to the bed is smaller than 0.03-m. Thus, part of the scour data at the beginning stage are not available, such that some of the curves start at a non-zero value.

Figure 5a shows the scour temporal evolution for the side-by-side arrangement with aligned flow. Data from just one of the complex piers are displayed, as the scour patterns are symmetric. Scour at two pile columns of any one of the adjacent piers are very close, and the scour depth at each column decreases towards downstream. Scour between the two complex piers, where the flow is usually contracted and accelerated, is minor during the first 300,000 s (≈3.5 days), followed by comparatively rapid scour when the scour holes produced by the two piers expand and merge. It indicates that the current level of proximity for this arrangement will not lead to significant contraction scour between two piers under clear-water flow regime. The maximum in-between scour depth occurs at the upstream end of the space and decreases towards downstream, which is similar to the natural profile of the merged scour holes.

When there is a 30° skew angle (which is common for river bends or braided channels), as shown in Figure 5b, the scour temporal evolutions at all the measuring locations suggest consistent trends with rapid deepening during the first 100,000 s followed by comparatively slow development of the scour holes to reach equilibrium. Similar to the scour features at stand-alone skewed complex piers (Yang et al. [28]) and pile groups (Zhao and Sheppard [4]; Lança et al. [10]), the maximum scour depths at both of the adjacent piers occur at the downstream ends of the upstream pier flanks (pile A4). Thus, only the data of the pile column at the upstream pier flank (pile A1–A4) of each pier are shown. Generally, the scour depth at each pile column increases towards downstream, in contrast to the aligned arrangement. The scour between the two piers initiates rapidly at the beginning of the experiment without significant stagnation and the maximum equilibrium scour depth occurs at the middle section of the in-between space.

Figure 5c shows the scour temporal evolution at two staggered complex piers. The downstream pier shows an "ascend-descend-equilibrium" trend, which is conjectured to be due to the incoming sediment eroded from the upstream pier filling the downstream scour hole. A similar trend can also be observed for the bed-level data between the piers. The sediment particles entrained and eroded away from the scour hole at the upstream pier are transported towards downstream by wake vortices with a certain skew angle to the pier's centerline; it is observed that the inner path of transport may partially overlap with the downstream pier. As a result, the scour depths at the downstream complex pier are significantly smaller than the upstream pier. The maximum scour depth at each pier still occurs at the first-row piles. However, for the downstream pier, the scour at the inner side (pile A1) is reduced by the upstream sediment supply.

For the tandem pier arrangement, as shown in Figure 5d, the temporal trend is much simpler than the staggered arrangement. The scour at the downstream pier is significantly mitigated by the upstream pier, which leads to strong vortex shedding observed and thereby the weakened scouring capacity in the lee-wake area.

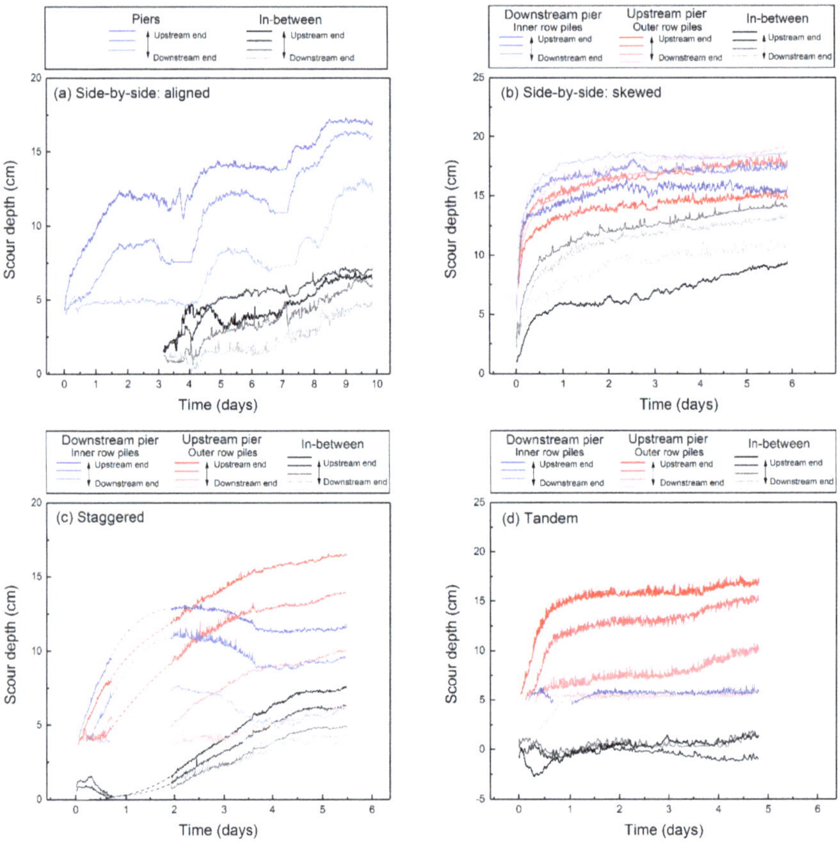

Figure 5. Temporal evolution of clear-water scour depths ($U/U_c = 0.9$) measured with: (**a**) side-by-side arrangement without skew angle; (**b**) side-by-side arrangement with 30° skew angle; (**c**) staggered arrangement; and (**d**) tandem arrangement.

3.2. Equilibrium Bed Level

The equilibrium scour depths are calculated using the temporal evolution curves shown in Figure 5 and the extrapolation method validated by Yang et al. [28]. Figure 6 shows the equilibrium bed level for each pier arrangement in the present study with a viewing direction perpendicular to the approaching flow. This figure acts as a supplement to the temporal evolution data to clearly illustrate the maximum range of scouring excavation that can possible happen under clear-water flow regime. The scour at inner and outer pile columns are generally symmetric for aligned complex piers, while an oblique alignment leads to more complex scour patterns at and between the two piers. The scour between piers under clear-water flow regime is found to be caused by the expansion and merging of the scour holes rather than contraction scour due to the symmetric scour pattern observed in Figure 6a, and thus the spatial distribution of in-between scour generally follows the natural profile of the scour holes. For the staggered and tandem pier arrangement tested, the greatest scour occurred at the leading side (pile A1 and B1) of the upstream pier. The protection effect of the upstream pier is similar to the performance of sacrificial piles as stated by Melville and Hadfield [33]. This protection is due to the increased sediment supply for the downstream pier and, as also mentioned above and observed during the experiments, the lee-wake turbulence and the vortex shedding path caused by the upstream pier overlapped with the downstream pier.

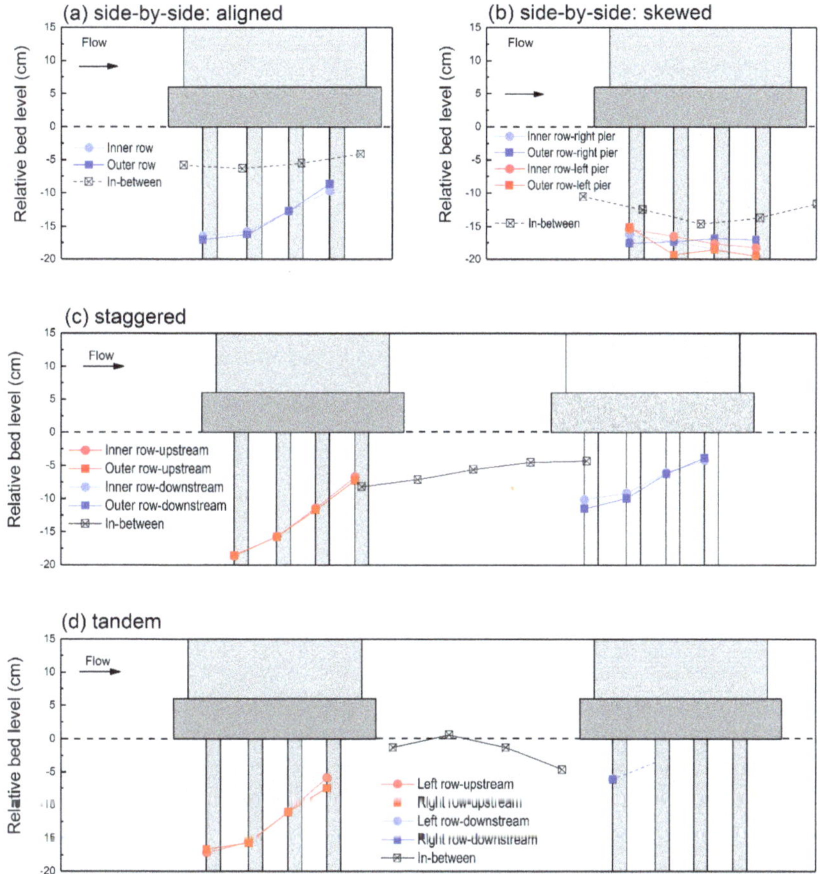

Figure 6. Equilibrium bed-levels: (**a**) side-by-side arrangement without skew angle; (**b**) side-by-side arrangement with 30° skew angle; (**c**) staggered arrangement; and (**d**) tandem arrangement. The measuring locations can also be referred in Figure 3 for more detailed information.

4. Live-Bed Scour Experiments

4.1. Mean Scour Depth at Complex Piers

A live-bed flow regime occurs when the mean flow velocity exceeds the threshold of sediment incipient motion. The general sediment motion on the bed surface leads to the formation of bed-forms and there is general transport of sediment downstream. Under a live-bed flow regime, the scouring process fluctuates due to the alternate arrival of the crests and troughs of the bed-forms and thus the instantaneous bed level data are time-averaged to obtain a mean scour depth. A dynamic equilibrium may be reached with mean scour depth constant over time. The equilibrium mean scour depth is dependent on the flow intensity ratio (U/U_c) and may show a "descend-ascend" trend with increasing U/U_c, as reported by Chee [34], Chiew [35], and Melville [36].

Figure 7 shows the mean scour depths as functions of flow intensity ratio U/U_c at selected measuring locations, including the most scoured pile of each complex pier and the upstream, downstream, and middle sections of the space between piers. It should also be noted that the bed-forms observed in the present study were dunes or washed-out dunes (for large flow intensity) migrating towards downstream. Other bed-form patterns, including ripples and anti-dunes, were not observed due to the characteristics of non-ripple-forming coarse sand (usually with $d_{50} > 0.6$-mm).

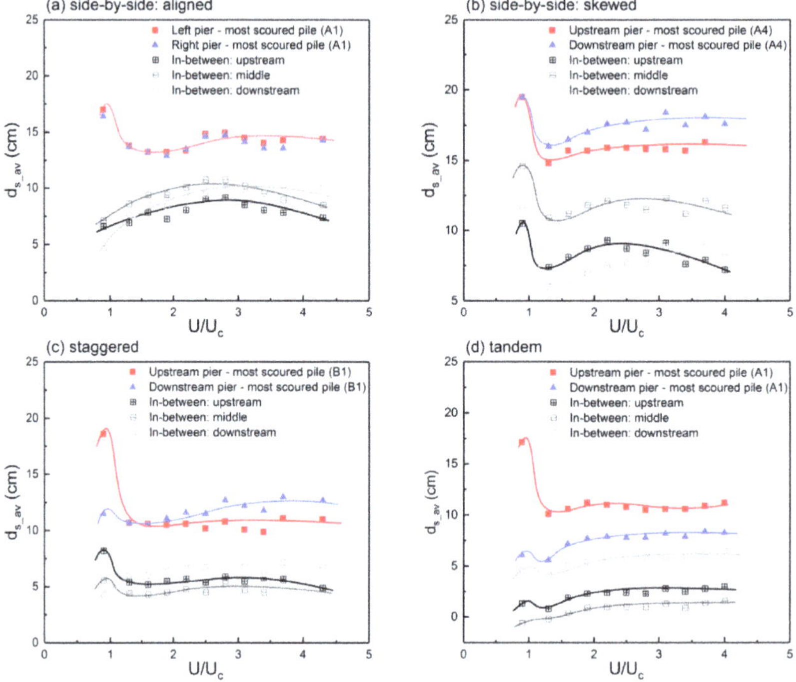

Figure 7. Relationship between equilibrium mean scour depth and flow intensity ratio at various locations. (**a**) side-by-side: aligned; (**b**) side-by-side: skewed; (**c**) staggered; and (**d**) tandem.

Figure 7a shows the scour variation of a side-by-side pier arrangement with aligned flow. The maximum scour depth at the two adjacent complex piers stays reasonably constant with increasing flow intensity, while the peak value at the transitional flat-bed stage ($U/U_c = 4 \sim 5$) is smaller than the peak at the clear-water threshold, which is in accord with the conclusions of Chee [34], Chiew [35], and Melville [36] for non-ripple-forming coarse sediment. The typical "descend-ascend" trend with flow intensity is absent for scour between the two piers, and the scour peak occurs for $U/U_c = 2 \sim 3$, when the approaching bed-form size is usually maximized. The location of maximum scour between the piers moves to the downstream section with flow intensity increasing above 3.5.

When there is a 30° skew angle between the piers and the flow direction, the mean live-bed scour depth at the downstream pier exceeds that at the upstream pier (Figure 7b). It is conjectured that the incoming sediment rapidly erodes away without further accumulation between the piers due to the strong wake vortices and the contracted flow. Thus, the scour at the inner side of the downstream pier (upstream pier flank) will be subjected to less influence of the approaching bed-forms, while, in contrast, a stronger filling effect exists at the outer side of the upstream pier. Furthermore, the "descend-ascend" trend is present for scour between the two piers, which indicates that, compared with Figure 7a, this scour more resembles that in a unified scour hole, i.e., the two scour holes have merged to form a single extensive scour hole.

The scour trend of the staggered pier arrangement is shown in Figure 7c. Under a clear-water flow regime, the scour at the downstream pier is significantly reduced by the lee effect and by the sediment supply from the upstream pier. However, the scour depth at the upstream pier drops sharply with increasing flow after entering the live-bed flow regime due to the approaching dunes, while, in contrast, the downstream pier is much less subjected to the influence of bed-forms with a live-bed peak larger than that at the clear-water threshold. The live-bed scour depths at the two piers are generally close but the downstream pier is slightly more scoured, which is in contrast to the trend with clear-water scour. A possible explanation is that with high flow intensity the lee protection of the upstream pier disappears while the diverted flow acting directly on the downstream pier with a certain skew angle actually strengthens the downstream scouring capacity. This is also in accordance with the observed diverted flow path behind the upstream pier. In addition, the attenuation effect of downstream vortex shedding becomes negligible due to the stronger flow.

In Figure 7d, the scour trend at the upstream pier in the tandem arrangement is similar to the staggered arrangement, which indicates that, for both arrangements, the upstream piers are not significantly affected by the downstream pier and still maintain the same scour features as a stand-alone pier. In contrast, the scour at the downstream pier is greatly attenuated. The downstream pier, as well as the measuring locations between the piers, suggest similar and uniform trends of slightly more scour with increased flow intensity and show a live-bed scour peak greater than that at the clear-water threshold. This feature is similar to the downstream pier in staggered arrangement and presumably tends to disappear with increasing distance between the two piers. Generally, for staggered or tandem arrangements with significant upstream/downstream distinction, the most hazardous scour overall may occur at the leading piles of the upstream pier under a clear-water flow regime.

4.2. Bed-Form Migration between Adjacent Piers

Besides the live-bed scour at each complex pier, it is also important and useful to understand the scour pattern and the varying trend within the space between two bridges, where the altered hydrodynamic and morphological features may affect human activity or natural processes, e.g., debris drifting and accumulating, boating, etc. Figure 8 shows both the clear-water and live-bed scour features between two piers. The shaded areas represent the varying range of bed level under a live-bed flow regime, and the data with $\frac{U}{U_c} = 2.2$ are plotted specifically because the bed-form size is usually maximized at this flow intensity with a medium celerity, and thus the features are usually most typical. In addition, the mean magnitude of bed level fluctuation at each measuring location is also shown in the small embedded graphs within Figure 8.

In Figure 8a, for side-by-side arrangement with aligned approaching flow, the range of bed level between two piers under live-bed condition is always lower than that at the clear-water threshold and the most significant additional scour occurs at the downstream section.

An opposite relationship can be observed in Figure 8b when a 30° skew angle exists, where the scour depth at the clear-water threshold stays larger than live-bed scour for any flow intensity and the difference is also maximized at the downstream section. A feature shared by the side-by-side arrangements with or without pier skewness is that the maximum live-bed in-between scour between the piers tends to occur at the central measuring location.

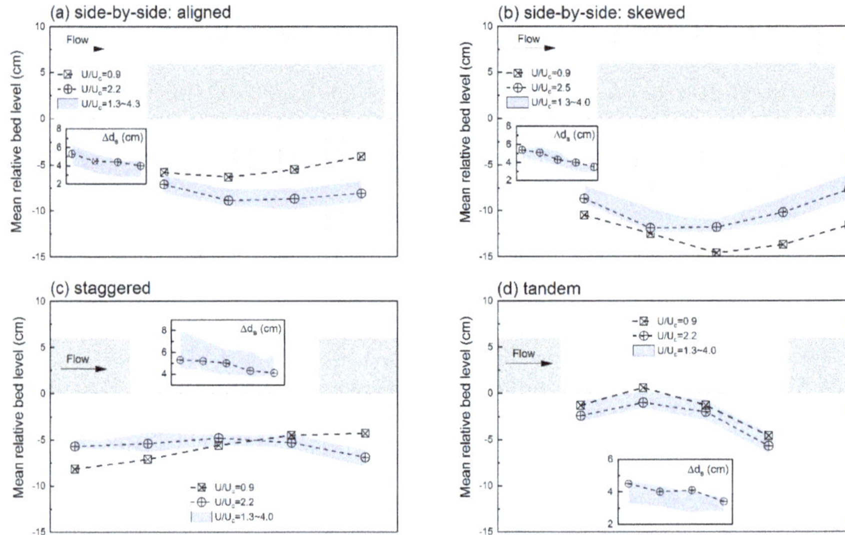

Figure 8. Live-bed scour range between adjacent piers. The embedded subplots show the trend of mean scour fluctuation at each measuring location. (**a**) side-by-side: aligned; (**b**) side-by-side: skewed; (**c**) staggered; and (**d**) tandem.

In Figure 8c, for the staggered arrangement, the increased flow intensity, and the corresponding approaching bed-forms attenuate the scour at the upstream end (adjacent to the rear of the upstream pier) but, in contrast, aggravate the scour at the downstream end (adjacent to the face of the downstream pier). In accordance with the conjecture in the last section, this phenomenon is also the outcome of the weakened lee-wake protection to the downstream pier with large flow intensity.

In Figure 8d, the scour depths under clear-water and live-bed flow regimes are similar. Only very minor scour occurs between the piers except close to the face of the downstream pier as part of the natural profile of the downstream scour hole.

For all the pier arrangements in the present study, the mean magnitude of the bed level fluctuation between the piers always decreases from upstream to downstream, regardless of the varying trend of the mean bed level, as shown by all the embedded graphs in Figure 8. This indicates that the bed-forms are damped when migrating downstream between the two adjacent complex piers. This damping effect can be influenced by other factors besides the bed level profile. This effect is discussed further below.

Cross-correlation analysis is used to determine the level of similarity of two data series. With time series (such as instantaneous bed-level data in the present study) a time lag between two series can be identified by a cross-correlation analysis performed after displacing one series relative to the other. Guan et al. [37] used cross-correlation analysis to calculate the migrating celerity of bed-forms between two locations, which can be determined by dividing the known distance between the locations by the calculated time lag of the correlation peak. In the present study, correlation analysis has been performed on data from pairs of adjacent measuring locations between the piers. Data with $U/U_c = 2.2$ (a typical value as mentioned above) are shown in Figure 9 and the trends with other flow intensity ratios are similar. In the figure, a higher correlation peak represents a greater similarity between the time series (i.e., less deformation of the bed-forms) and a smaller time lag represents a faster migrating celerity between the two adjacent measuring locations. The correlation coefficients are normalized by the largest peak value in each subplot.

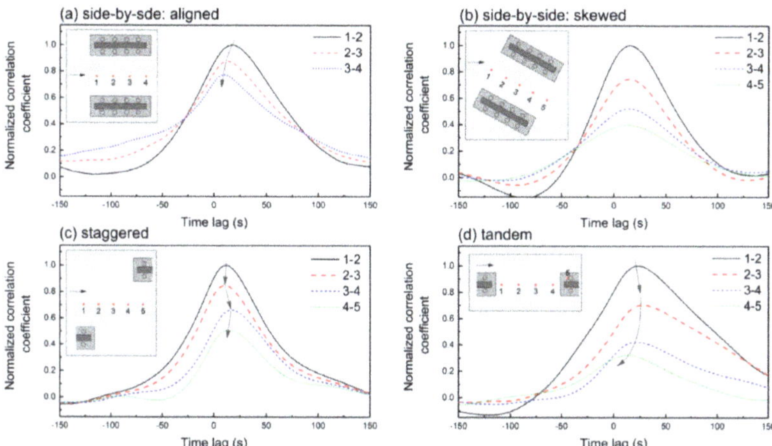

Figure 9. Cross-correlation analysis for the four different pier arrangements in the present study ($U/U_c = 2.2$). (**a**) side-by-side: aligned; (**b**) side-by-side: skewed; (**c**) staggered; and (**d**) tandem.

In Figure 9a, for a side-by-side pier arrangement with aligned flow, bed-forms keep accelerating and are damped continuously when passing between the piers, although the bed level does not vary significantly as shown in Figure 8a. The acceleration is probably due to the flow contraction that is magnified with large flow intensity under a live-bed flow regime. In contrast, with a 30° skew angle (Figure 9b), bed-forms tend to pass between the piers with a constant celerity but are damped to a much greater extent than with the aligned arrangement. The enhanced damping effect can be attributed to the larger vertical (and also horizontal) extent of the scour hole, in which the approaching bed-forms may collapse and lose the original shape features. This is also in accord with the findings of our previous unpublished study on the relationship between mean scour fluctuation and scour depth.

In Figure 9c for the staggered pier arrangement, a process of repeated acceleration and deceleration can be observed. A possible explanation is that the migrating bed-forms firstly leave the region dominated by the upstream pier and then enter the region that is significantly affected by the downstream pier. The flow tends to be more contracted and accelerated in the region closer to either one of the piers and therefore the celerity of the migrating bed-forms also tends to behave in the same way.

For the tandem arrangement, as shown in Figure 9d, the bed-forms tend to decelerate behind the upstream pier due to the lee-wake protection and then accelerate after entering the downstream scour hole before collapsing to the bottom of the leading piles. Damping of the bedforms is marked.

Generally, the bed-forms tend to accelerate in contracted flow near the piers regardless of the pier arrangement, and deceleration may occur when the lee-wake protection effect prevails. Furthermore, the magnitude of the bed-forms will always be damped during migration downstream while being eroded by the turbulence in the scour hole, which is also in accord with the decreasing mean fluctuation magnitudes shown in Figure 8.

5. Discussion

Predicting scour depth at complex piers has always been a problem due to the difficulty of analyzing the complexity of flow field and pier geometry, as reviewed extensively by Ettema et al. [38]. As a result of the limitation of contemporary computing capability, numerical modelling has not been practical and efficient enough to provide reliable scour simulation. Thus, engineers still rely heavily on empirical or semi-empirical equations based on dimensional analysis to calculate the equilibrium scour depth at piers, although errors may be significant and modifications are usually needed for unusual conditions. To date, a few prediction methods have been well developed and widely adopted around the world for complex piers, including the Coleman method (Coleman [16]), the FDOT (Florida Department of Transportation) method (Sheppard and Renna [39]), HEC-18 method (Arneson et al. [40]), and the Sheppard-Melville (S-M) method (Sheppard et al. [41]). The S-M method is a meld of the equations of Sheppard and Miller [42] and Melville and Chiew [43]. Ettema et al. [38] has noted that the S-M method is more readily adapted than the other methods as further information becomes available. However, the S-M method was originally developed for single piers and does not deal with geometric complexity. Yang et al. [28] proposed a modification for clear-water scour at complex piers by replacing the pier diameter in the original S-M method with the equilibrium pier width calculated by other equations. Yang et al. [44] extended the modification mentioned above to the live-bed flow range. The newly extended method can be expressed as below.

When $U/U_c \leq 1$:

$$\frac{d_{se}}{D_e} = 3.17 f_1 f_2 f_3 \tag{1}$$

$$f_1 = tanh\left[\left(\frac{y_0}{D_e}\right)^{0.65}\right] \tag{2}$$

$$f_2 = 1 - 1.2\left[ln\left(\frac{U}{U_c}\right)\right]^2 \tag{3}$$

$$f_3 = \frac{D_e/d_{50}}{69.25(\ D_e/d_{50})^{-0.34} + 0.14(\ D_e/d_{50})^{1.41}} \tag{4}$$

When $1 < U/U_c \leq U_{lp}/U_c$:

$$\frac{d_{se}}{D_e} = f_1\left\{2.3\left[1 - \left(\frac{\frac{U}{U_c} - \frac{U_{lp}}{U_c}}{\frac{U_{lp}}{U_c} - 1}\right)\right]^{1.1} + 3.17 f_3\left(\frac{\frac{U}{U_c} - \frac{U_{lp}}{U_c}}{\frac{U_{lp}}{U_c} - 1}\right)^{1.1}\right\} \tag{5}$$

When $U/U_c > U_{lp}/U_c$:

$$\frac{d_{se}}{D_e} = 2.3 f_1 \tag{6}$$

$$U_{lp} = max\left\{5U_c, 0.6\sqrt{gy_0}\right\} \tag{7}$$

Specifically, U_{lp} represents the flow velocity where the bed-forms are washed out, i.e., the transitional flat-bed stage. The equivalent pier width D_e is calculated by the method of Coleman [16] for aligned complex piers and that of Sheppard and Renna [38] for skewed complex piers.

The performance of Equations (1)–(7) is compared with the HEC-18 and FDOT methods in Figure 10. Supplementary experiments were also performed to investigate the scour at stand-alone piers and the results are included in the figure for comparison. In Figure 10a, for the side-by-side arrangement with aligned flow the live-bed scour is aggravated by the proximity of another pier, while no significant difference exists under the clear-water flow regime. In contrast, in Figure 10b with a 30° skew angle, the scour depths at both the piers are reduced by the presence of another pier, although the horizontal extent of the merged scour hole is typically somewhat larger than the hole produced by any individual complex pier. It is counter-intuitive that a larger flow blockage usually leads to greater scour depth. A possible explanation is that the presence of a downstream pier causes a greater flow separation that starts from a longer distance upstream of the pier site, then the attacking angle of the flow is actually reduced due to enhanced flow diversion. More analysis, both physical and numerical, is needed to better describe the mechanism of scour attenuation in similar situations.

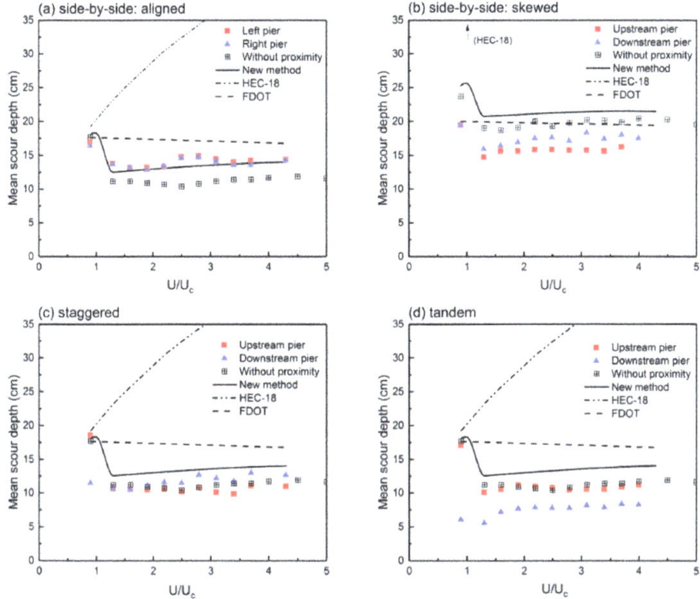

Figure 10. Scour comparison with single complex pier without proximity and performance of predictors. (**a**) side-by-side: aligned; (**b**) side-by-side: skewed; (**c**) staggered; and (**d**) tandem.

For staggered and tandem pier arrangements, in Figure 10c,d, the upstream pier is not significantly affected by the downstream pier. The modified and extended S-M method then provides an adequate but not excessive safety margin even if the scour depth is aggravated due to an adjacent pier, and the natural "descend-ascend" trend with increasing flow velocity in the live-bed regime can also be captured with good accuracy. In contrast, although the FDOT method and the HEC-18 method show fair accuracy for clear-water scour, they both tend to over-simplify the scour trend under the live-bed flow regime and may lead to significant errors. The HEC-18 method uses Froude number (rather than U/U_c as used in other popular methods) to account for the influence of flow velocity and this may be the reason for the excessive overestimation that can be observed in Figure 10. Therefore, it is suggested, for engineers and researchers, to use the modified and extended S-M method expressed by Equations (1)–(7) to predict the scour depth at the upstream pier when aligned to the approaching flow. Attention should also be paid to special conditions that may lead to scour aggravation due to the skew angle as shown in Figure 10a. In the meantime, the information provided in Figure 8 can be used to evaluate the regional bed topography, the potential location for severe scour damage, or bed level stability during bed-form migration.

It should be noted that the scour pattern at bridge sites can also be significantly influenced by other artificial structures or natural topography, e.g., weirs, sluice gates or shoals, which may even be far enough away to be neglected in error sometimes. Wang et al. [45] studies the scour aggravation caused by an adjacent submerged weir set downstream of the pier under live-bed conditions. The scour pattern for a staggered or tandem pier arrangement will be more prone to be affected by the downstream weir. In addition, the existence of the weir leads to upstream bed aggradation and consequently changes the relative pile-cap elevation to the original bed level. The relative pile-cap elevation has been considered as a key factor affecting the scour development and equilibrium scour depth at complex piers (Coleman [16]; Ataie-Ashtiani et al. [18]; Moreno et al. [22]; Yang et al. [28]). Similar effects can also be caused by general bed degradation that usually occurs downstream of sluice gates.

Furthermore, natural topography, e.g., shoals, bends, and sand bars, may cause oblique flow that much complicate the scenario for scour depth analysis.

Generally, the analysis of scour at complex bridge piers is complicated. A reliable prediction can be obtained only when the influence of pier proximity is accurately integrated with the influences of many other factors that are not addressed by this paper. The current study is still limited and only investigates the general scour pattern with typical pier arrangements and a fixed pile-cap elevation. More research needs in the future include testing extensively the influence of pier complexity, sediment type, gradation, large flow blockage due to consecutive piers, and unconventional pier arrangements, which may help creating an updated scour prediction framework that is not discussed by the current study.

6. Conclusions

The present study investigates the clear-water and live-bed scour characteristics at complex bridge piers in close proximity to each other with four different pier arrangements: side-by-side arrangement with aligned flow, side-by-side arrangement with 30° skew angle, a staggered arrangement, and a tandem arrangement. The main conclusions drawn from the paper include:

1. The pier arrangement has a significant effect on the temporal evolution of scour at individual vertical piles of each pier and also in between the two piers. Compared with the side-by-side pier arrangement with aligned flow, a flow skew angle leads to more rapid but asymmetric scour development. The extended scour holes merge between the piers and aggravate the scour development in this region. The staggered and tandem pier arrangements provide significant protection to the downstream pier by reducing the scour rate at the downstream scour hole and facilitating the filling process there.
2. The equilibrium clear-water scour pattern is also heavily dependent on the pier arrangement. For side-by-side, staggered. and tandem arrangements with aligned flow, the scour patterns at the upstream piers (or either one of the piers for side-by-side scheme) are similar to that at a stand-alone complex pier under the same conditions. The upstream-to-downstream protection effect is maximized for a tandem arrangement because the downstream pier is entirely in the lee-wake zone, and meanwhile the flow blockage is also the minimum.
3. With live-bed flow, the original scour patterns for all the pier arrangements under clear-water conditions may be altered significantly. The live-bed scour depths for side-by-side arrangements are much larger than the others due to the enhanced flow contraction between the two piers with large flow intensity. For staggered and tandem arrangements, the mean scour depth at the upstream pier drops markedly after entering the live-bed range. In contrast, at the downstream piers that are well protected under clear-water flow regime, the live-bed scour may significantly exceed the clear-water scour and even exceed scour at the upstream pier. It is conjectured that the lee-wake protection disappears with more intense flow and general sediment transport.
4. Under a live-bed flow regime, bed-forms migrate downstream via the bridge opening and are damped continuously. Thus, the bed profile and the morphological response at the erodible bed between the two piers are different to those under clear-water conditions. The relationship of the difference between the clear-water and live-bed scour depths also varies for different pier arrangements. Between two piers the mean bed level fluctuation decreases continuously with distance downstream, regardless of the bed profile. The flow near either one of the piers is diverted and accelerated, thereby increasing the bed-form celerity.
5. Compared with a single complex pier, the live-bed scour at side-by-side piers with aligned flow is greater. On the contrary, with a 30° flow skew angle, live-bed scour at both the adjacent piers is less than a stand-alone pier. For both the staggered and tandem arrangements, scour at the upstream pier is barely affected by the downstream pier. The modified and extended Sheppard-Melville method provides an adequate safety margin for most of the scenarios, while

attention should still be paid to the side-by-side piers with aligned flow, for which there is a risk with high flow intensity of higher scour than the method predicts.

6. Due to the complexity of predicting scour depth at complex piers, the conclusions drawn in the present study should be integrated with other factors that may also influence the scour pattern, e.g., the proximity to other artificial structures or natural topography, general bed level variation and channel bends.

Author Contributions: Conceptualization, Y.Y. and B.W.M.; Formal analysis, Y.Y. and G.H.M.; Investigation, Y.Y.; Methodology, Y.Y.; Supervision, B.W.M.; Validation, Y.Y.; Visualization, Y.Y.; Writing–original draft, Y.Y.; Writing–review & editing, Y.Y., B.W.M., G.H.M., and A.Y.S.

Funding: This research was funded by China Scholarship Council (CSC).

Acknowledgments: The authors would like to thank the technical support provided by the Fluid Mechanics Laboratory of The University of Auckland as well as the staff, including Geoffrey Kirby, Dan Fray, and Oane Galama. In particular, the first author would like to thank Yu Huang for her precious help during the past years.

Conflicts of Interest: The authors declare no conflict of interest.

References

1. Melville, B.W.; Coleman, S.E. *Bridge Scour*; Water Resources Publications: Highlands Ranch, CO, USA, 2000.
2. Hannah, C. Scour at Pile Groups. Master's Thesis, University of Canterbury, Christchurch, New Zealand, 1978.
3. Elliott, K.R.; Baker, C.J. Effect of pier spacing on scour around bridge piers. *J. Hydraul. Eng.* **1985**, *111*, 1105–1109. [CrossRef]
4. Zhao, G.; Sheppard, D.M. The effect of flow skew angle on sediment scour near pile groups. In *Proceedings of the Stream Stability and Scour at Highway Bridges, Compilation of Conference Papers*; American Society of Civil Engineers (ASCE): Reston, VA, USA, 1999; pp. 377–391.
5. Sumner, D.; Price, S.J.; Paidoussis, M.P. Flow-pattern identification for two staggered circular cylinders in cross-flow. *J. Fluid Mech.* **2000**, *411*, 263–303. [CrossRef]
6. Sumer, B.M.; Fredsøe, J.; Bundgaard, K. Global and local scour at pile groups. In Proceedings of the Fifteenth International Offshore and Polar Engineering Conference, Seoul, Korea, 19–24 June 2005.
7. Ataie-Ashtiani, B.; Beheshti, A.A. Experimental investigation of clear-water local scour at pile groups. *J. Hydraul. Eng.* **2006**, *132*, 1100–1104. [CrossRef]
8. Amini, A.; Melville, B.W.; Ali, T.M.; Ghazali, A.H. Clear-water local scour around pile groups in shallow-water flow. *J. Hydraul. Eng.* **2011**, *138*, 177–185. [CrossRef]
9. Liang, D.; Gotoh, H.; Scott, N.; Tang, H. Experimental study of local scour around twin piles in oscillatory flows. *J. Waterw. Port Coast. Ocean Eng.* **2012**, *139*, 404–412. [CrossRef]
10. Lança, R.; Fael, C.; Maia, R.; Pêgo, J.P.; Cardoso, A.H. Clear-water scour at pile groups. *J. Hydraul. Eng.* **2013**, *139*, 1089–1098. [CrossRef]
11. Das, S.; Mazumdar, A. Turbulence flow field around two eccentric circular piers in scour hole. *Int. J. River Basin Manag.* **2015**, *13*, 343–361. [CrossRef]
12. Wang, H.; Tang, H.; Liu, Q.; Wang, Y. Local scouring around twin bridge piers in open-channel flows. *J. Hydraul. Eng.* **2016**, *142*, 06016008. [CrossRef]
13. Khaple, S.; Hanmaiahgari, P.R.; Gaudio, R.; Dey, S. Interference of an upstream pier on local scour at downstream piers. *Acta Geophys.* **2017**, *65*, 29–46. [CrossRef]
14. Kim, H.S.; Roh, M.; Nabi, M. Computational Modeling of Flow and Scour around Two Cylinders in Staggered Array. *Water* **2017**, *9*, 654. [CrossRef]
15. Jones, J.S.; Sheppard, D.M. Local scour at complex pier geometries. In Proceedings of the Building Partnerships—2000 Joint Conference on Water Resource Engineering and Water Resources Planning and Management, American Society of Civil Engineers (ASCE), Minneapolis, MN, USA, 30 July–2 August 2000.
16. Coleman, S.E. Clearwater local scour at complex piers. *J. Hydraul. Eng.* **2005**, *131*, 330–334. [CrossRef]
17. Sheppard, D.M.; Glasser, T. Local scour at bridge piers with complex geometries. In *Contemporary Topics in Situ Testing, Analysis, and Reliability of Foundations*; International Foundation Congress and Equipment Expo; American Society of Civil Engineers (ASCE): Orlando, FL, USA, 2009; pp. 506–513.

18. Ataie-Ashtiani, B.; Baratian-Ghorghi, Z.; Beheshti, A.A. Experimental investigation of clear-water local scour of compound piers. *J. Hydraul. Eng.* **2010**, *136*, 343–351. [CrossRef]
19. Grimaldi, C.; Cardoso, A.H. Methods for local scour depth estimation at complex bridge piers. In Proceedings of the 1st IAHR European Division Congress, Heriot-Watt University, Edinburgh, UK, 4–6 May 2010.
20. Beheshti, A.A.; Ataie-Ashtiani, B. Experimental study of three-dimensional flow field around a complex bridge pier. *J. Eng. Mech.* **2010**, *136*, 143–154. [CrossRef]
21. Beheshti, A.; Ataie-Ashtiani, B. Scour hole influence on turbulent flow field around complex bridge pier. *Flow Turbul. Combust.* **2016**, *97*, 451–474. [CrossRef]
22. Moreno, M.; Maia, R.; Couto, L. Effects of relative column width and pile-cap elevation on local scour depth around complex piers. *J. Hydraul. Eng.* **2015**, *142*, 04015051. [CrossRef]
23. Moreno, M.; Maia, R.; Couto, L. Prediction of equilibrium local scour depth at complex bridge piers. *J. Hydraul. Eng.* **2016**, *142*, 04016045. [CrossRef]
24. Moreno, M.; Maia, R.; Couto, L.; Cardoso, A. Subtraction approach to experimentally assess the contribution of the complex pier components to the local scour depth. *J. Hydraul. Eng.* **2016**, *143*, 06016030. [CrossRef]
25. Ferraro, D.; Tafarojnoruz, A.; Gaudio, R.; Cardoso, A.H. Effects of pile cap thickness on the maximum scour depth at a complex pier. *J. Hydraul. Eng.* **2013**, *139*, 482–491. [CrossRef]
26. Amini, A.; Melville, B.W.; Ali, T.M. Local scour at piled bridge piers including an examination of the superposition method. *Can. J. Civ. Eng.* **2014**, *41*, 461–471. [CrossRef]
27. Baghbadorani, D.A.; Ataie-Ashtiani, B.; Beheshti, A.; Hadjzaman, M.; Jamali, M. Prediction of current-induced local scour around complex piers: Review, revisit, and integration. *Coast. Eng.* **2018**, *133*, 43–58. [CrossRef]
28. Yang, Y.; Melville, B.W.; Sheppard, D.M.; Shamseldin, A.Y. Clear-water local scour at skewed complex bridge piers. *J. Hydraul. Eng.* **2018**, *144*, 04108019. [CrossRef]
29. Oben-Nyarko, K.; Ettema, R. Pier and abutment scour interaction. *J. Hydraul. Eng.* **2011**, *137*, 1598–1605. [CrossRef]
30. Sturm, T.; Abid, I.; Melville, B.; Xiong, X.; Stoesser, T.; Bugallo, B.F.; Hong, S. *Combining Individual Scour Components to Determine Total Scour*; National Cooperative Highway Research Program Reports; Transportation Research Board: Washington, DC, USA, 2017; pp. 24–37.
31. Sturm, T.W.; Ettema, R.; Melville, B.W. *Evaluation of Bridge-Scour Research: Abutment and Contraction Scour Processes and Prediction*; National Cooperative Highway Research Program Reports; Transportation Research Board: Washington, DC, USA, 2011; pp. 24–27.
32. SeaTek Instrumentation. Available online: www.seatek.com (accessed on 23 July 2019).
33. Melville, B.W.; Hadfield, A.C. Use of sacrificial piles as pier scour countermeasures. *J. Hydraul. Eng.* **1999**, *125*, 1221–1224. [CrossRef]
34. Chee, R.K.W. Live-bed Scour at Bridge Piers. Master's Thesis, University of Auckland, Auckland, New Zealand, 1982.
35. Chiew, Y.M. Local scour at bridge piers. Ph.D. Thesis, University of Auckland, Auckland, New Zealand, 1984.
36. Melville, B.W. Live-bed scour at bridge piers. *J. Hydraul. Eng.* **1984**, *110*, 1234–1247. [CrossRef]
37. Guan, D.; Melville, B.W.; Friedrich, H. Live-bed scour at submerged weirs. *J. Hydraul. Eng.* **2015**, *141*, 04014071. [CrossRef]
38. Ettema, R.; Constantinescu, G.; Melville, B. Flow-Field Complexity and Design Estimation of Pier-Scour Depth: Sixty Years since Laursen and Toch. *J. Hydraul. Eng.* **2017**, *143*, 03117006. [CrossRef]
39. Sheppard, D.M.; Renna, R. *Bridge Scour Manual*; Florida Department of Transportation: Tallahassee, FL, USA, 2010.
40. Arneson, L.A.; Zevenbergen, L.W.; Lagasse, P.F.; Clopper, P.E. *Evaluating Scour at Bridges*, 5th ed.; FHWA-HIF-12-003, Hydraulic Engineering Circular No. 18 (HEC-18); Federal Highway Administration: Washington, DC, USA, 2012.
41. Sheppard, D.M.; Demir, H.; Melville, B. *Scour at Wide Piers and Long Skewed Piers*; National Cooperative Highway Research Program Reports 682; Transportation Research Board: Washington, DC, USA, 2011.
42. Sheppard, D.M.; Miller, W. Live-bed local pier scour experiments. *J. Hydraul. Eng.* **2006**, *132*, 635–642. [CrossRef]
43. Melville, B.W.; Chiew, Y.M. Time scale for local scour at bridge piers. *J. Hydraul. Eng.* **1999**, *125*, 59–65. [CrossRef]

44. Yang, Y.; Melville, B.W.; Macky, G.H.; Shamseldin, A.Y. Experimental study on live-bed scour at complex bridge pier with bed-form migration. *J. Hydraul. Res.*. manuscript submitted and under review.
45. Wang, L.; Melville, B.W.; Guan, D. Effects of upstream weir slope on local scour at submerged weirs. *J. Hydraul. Eng.* **2018**, *144*, 04018002. [CrossRef]

Review

Scour at Bridge Foundations in Supercritical Flows: An Analysis of Knowledge Gaps

Oscar Link [1], Emmanuel Mignot [2,*], Sebastien Roux [3], Benoit Camenen [4], Cristián Escauriaza [5], Julien Chauchat [6], Wernher Brevis [7] and Salvatore Manfreda [8]

1 Department of Civil Engineering, University of Concepción, Edmundo Larenas 215, Concepción 4030000, Chile
2 University of Lyon, INSA de Lyon, LMFA, 20 avenue Einstein, 69621 Villeurbanne CEDEX, France
3 Centre d'Analyse Comportementale des Ouvrages Hydrauliques (CACOH), Compagnie Nationale du Rhône, 69007 Lyon, France
4 Irstea, UR RiverLy, centre de Lyon-Villeurbanne, 5 Rue de la Doua, CS 20244, F-69625 Villeurbanne CEDEX, France
5 Departamento de Ingeniería Hidráulica y Ambiental, Pontificia Universidad Católica de Chile, Santiago 3580000, Chile
6 Institute of Engineering, University Grenoble Alpes, CNRS, Grenoble INP, LEGI, 38000 Grenoble, France
7 Departamento de Ingeniería Hidráulica y Ambiental and Departamento de Minería. Pontificia Universidad Católica de Chile, Santiago 3580000, Chile
8 Department of European and Mediterranean Cultures, University of Basilicata, via Lanera 20, 75100 Matera, Italy
* Correspondence: Emmanuel.mignot@insa-lyon.fr

Received: 17 June 2019; Accepted: 8 August 2019; Published: 10 August 2019

Abstract: The scour at bridge foundations caused by supercritical flows is reviewed and knowledge gaps are analyzed focusing on the flow and scour patterns, available measuring techniques for the laboratory and field, and physical and advanced numerical modeling techniques. Evidence suggests that the scour depth caused by supercritical flows is much smaller than expected, by an order of magnitude compared to that found in subcritical flows, although the reasons for this behavior remain still unclear. Important questions on the interaction of the horseshoe vortex with the detached hydraulic-jump and the wall-jet flow observed in supercritical flows arise, e.g., does the interaction between the flow structures enhance or debilitate the bed shear stresses caused by the horseshoe vortex? What is the effect of the Froude number of the incoming flow on the flow structures around the foundation and on the scour process? Recommendations are provided to develop and adapt research methods used in the subcritical flow regime for the study of more challenging supercritical flow cases.

Keywords: bridge foundations; scour; supercritical flows; sediment hydraulics

---

1. Introduction

A vast amount of research on scour at bridge piers and abutments (referred here to as bridge foundations) has been conducted in the past, mostly focusing on results obtained from flume experiments with subcritical flow conditions, i.e., with a Froude number smaller than 1 and sand as a bed material. Even though some issues are still unresolved, the current knowledge has enabled the development of a number of guidelines for bridge design in different countries, e.g., HEC-18 in the US [1], in New Zealand [2], DWA-M529 in Germany [3], and the Ministry of Public Works in Chile [4], among others. However, there is an important lack of knowledge in transferring these methodologies and theories to bridge foundation design when they are placed in rivers with supercritical flow conditions.

Large-scale supercritical free-surface flows can occur in different environments. Some examples can be found in cases of flooded urban streets, fish-ways, tsunami inland flows, coastal channels, and mountain rivers. This paper focuses on the flow and scouring patterns at bridge foundations in rivers with supercritical conditions. The occurrence of supercritical flows in rivers is defined by high longitudinal slopes (>1%) and/or rapid flood waves. Commonly, steep rivers present gravel beds or mixtures of fine and coarse sediments, containing all possible sizes, from clay and silt up to boulders tens of centimeters in size. In dentritic networks, streams with a low Strahler's order (i.e., <3) are steep and produce flash floods but normally possess a small cross-sectional width. Therefore, deck bridges without foundations in these riverbeds are usually selected. At piedmont, however, rivers widen and it is common to observe cross sections with widths over 50 m, where bridge foundations may have to be included. Salient examples of such configurations are often encountered in steep watersheds subjected to heavy rains, such as on the Panamericana Route along Perú and Chile, La Réunion Island (Indian Ocean) in Taiwan or Japan, and also in a few European Alpine piedmont rivers (Figure 1). The examples in Figure 1 clearly highlight that supercritical flows are associated with a significant amount of energy for scouring and dynamic loading of the superstructure. Wood debris can also enhance the risk of pier stability. Such flows thus produce among the worst hydraulic conditions for bridge design. A recent bridge collapse due to scour in a supercritical flow occurred at the Rivière Saint Etienne in the La Réunion island due to cyclone Gamède. This bridge, which connected a road with traffic of 65,000 vehicles per day, collapsed and modified the terrestrial transport route for a long time (Figure 1f,g), thereby producing large economic losses. This event motivated, in France, the funding of specific experimental studies on flow and scour patterns around bridge foundations in supercritical flows, thus opening a new line of research. This event also evidenced an important lack of knowledge, with implications for the hydraulic design of bridge foundations in many regions of the world where supercritical conditions occur. Indeed, high flow velocities, along with high sediment transport and turbidity, rapid changes in the local morphology, and air entrainment make it complex and sometimes impossible to perform flow and scour measurements in the field [5] or even in laboratory facilities [6].

In this paper, we identify the knowledge gaps in scour at bridge foundations in rivers with supercritical flows. These problem areas can be summarized as knowledge gaps in flow dynamics, past obstacles in flat and scoured beds, and scour patterns and mechanisms. This paper also reviews the applicability and limitations of the existing methodological approaches typically used in subcritical flows and sand beds, including field and laboratory measuring techniques for flow and scour, as well as physical and numerical modeling techniques. The link between experimental or numerical work at the local process scale and the long-term river dynamics that finally determine bridge failures is also highlighted. In most cases, we are forced to start with a well-studied case of scour in sand caused by a subcritical flow to provide a referential basis. The paper concludes with final remarks on the results of our analysis.

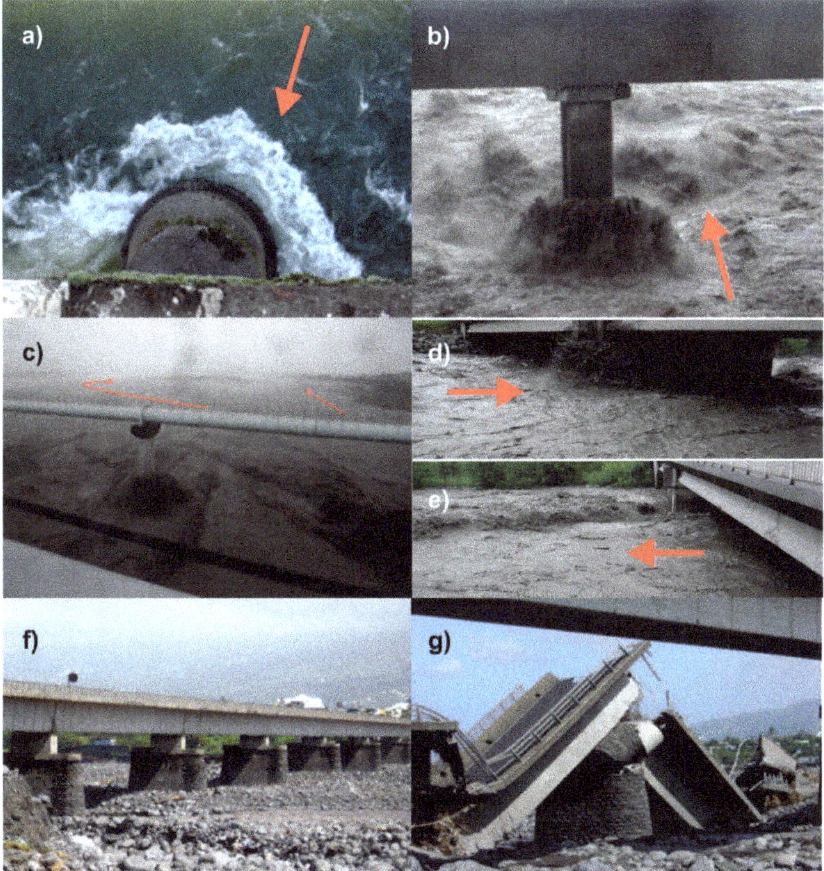

Figure 1. Photographs of supercritical flows at bridge piers in the Biobío River, Chile (**a**), La Rivière des Galets, La Réunion island (**b**), The Choshui River, Taiwan ((**c**) [7]), and the Arc-en-Maurienne River, France ((**d**) and (**e**) [8]). Photographs of the bridge at Rivière St-Etienne (La Réunion island) before (**f**) and after (**g**) the 2007 collapse ((**g**) [9]). Red arrows indicate the flow direction.

2. Flow Patterns around Bridge Foundations

Scour at bridge foundations involves complex interactions between the three-dimensional, unsteady, and turbulent flow, and the movable riverbed. The flow patterns around foundations have been described for cylinders mounted on plane beds, as well as for cylinders in scoured holes in both sub and supercritical flows. Important differences in the flow field around bridge foundations can be highlighted between these two flow regimes.

2.1. Surface Mounted Emerging Obstacles on Flat Beds

2.1.1. Subcritical Flows

Because of the adverse pressure gradient induced by an obstacle mounted on a flat surface, the incoming flow separates at the upstream junction of the obstacle and the bed, reorganizing into a complex large-scale dynamically-rich coherent structure, known as a horseshoe vortex (HSV) system, which wraps around the front and the flanks of the obstacle [10,11]. The turbulent HSV system presents a bi-modality of the probability density functions of velocity and pressure fluctuations with two dominant modes called the backflow and the zero–flow modes [11,12]. The study in [13] documented the presence of a third mode, the intermediate mode, which is described as a mode close to the zero-flow mode, but with less intensity in the vertical velocity component of the near-wall jet [14–16]. The main HSV interacts with larger structures, impinging the separated region from upstream, as well as with hairpin vortices that develop underneath the horseshoe vortices [12,17,18]. The stagnation pressure causes an additional bow wave at the upstream free surface, adjacent to the obstacle, which, in the upstream symmetry plane, rotates in a direction opposite to the HSV. The stagnation pressure also causes a sideward acceleration of the flow at the sides of the cylinder [19].

2.1.2. Supercritical Flows

The study in [20] recently showed that the flow pattern around an obstacle in a supercritical flow varies as a function of the obstacle width (D) to flow depth (h) ratio, distinguishing the following patterns:

The detached hydraulic jump pattern: at high ratios of obstacle width to flow depth, i.e., D/h > 0.5–2.0 a bow-wave like, detached, hydraulic jump takes place in front of the obstacle [21]. From up to downstream, the flow first slows down to pass from the supercritical to subcritical regime and eventually stops at the upstream face of the obstacle. This transition from the super- to sub-critical regime takes place through the detached hydraulic jump wrapping around the obstacle (Figure 2a). The foot of the hydraulic jump then follows a hyperbolic curve in the horizontal plane. Near the obstacle, the flow is in the subcritical regime, so a horseshoe vortex occurs in the near-bed region at the foot of the foundation. The foot of the horseshoe vortex then follows an elliptic curve [21]. Moreover, the separation distance between the obstacle and both the detached jump and the horseshoe vortex appears to increase with the non-dimensional flow depth (h/D) and decreases with the Froude number of the incoming flow [22].

The wall jet pattern: at small ratios of foundation width to flow depth (i.e., D/h < 0.5–2.0) a so-called wall-jet-like bow wave develops (Figure 2b; [20]). The flow remains in the supercritical regime (unaffected by the presence of the obstacle) until reaching the foot of the obstacle. There, it deviates and goes up along the upstream face of the obstacle and slightly towards its sides, where it is evacuated and falls down in the flow further downstream. Part of the up-going flow rolls backward and falls down at the foot of the obstacle in periodic reverse spillage.

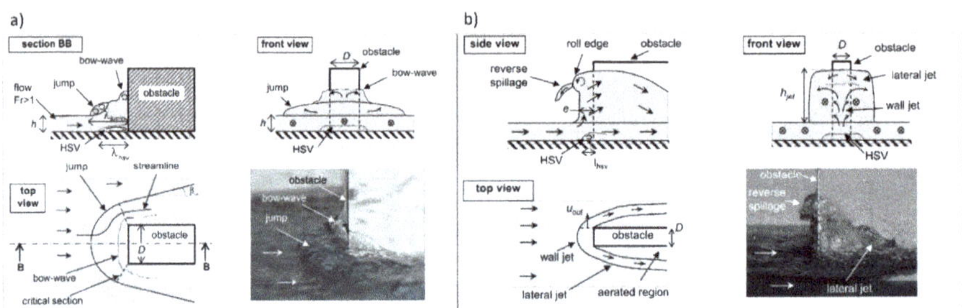

Figure 2. Flow patterns upstream of a rectangular emerging foundation in supercritical flow: (**a**) detached hydraulic jump and (**b**) wall jet pattern (adapted from [20]). HSV, horseshoe vortex.

2.2. Foundations with Scour Hole

2.2.1. Subcritical Flows

The HSV system around a scoured foundation presents similar patterns to systems around a cylinder on a plane bed, namely bimodal near wall velocity distributions, sizess, and frequencies that scale with the Reynolds number [18,23,24]. Turbulence properties are also reported by [25,26].

2.2.2. Supercritical Flows

To the best of our knowledge, no velocity measurements have been carried out in the scour hole of a foundation in a supercritical flow. The only available information are flow depths measured under laboratory conditions by [27]. These measurements were obtained using a clear-water turbulent inflow with mobile bed conditions and $D/h \approx 10$. In the presence of a scour hole, a detached hydraulic jump wrapping around the abutment was observed, as in the plane bed case, but with the detached hydraulic jump closer to the obstacle. Additionally, two parallel bow waves smaller elevations were also visible downstream from the obstacle (Figure 3).

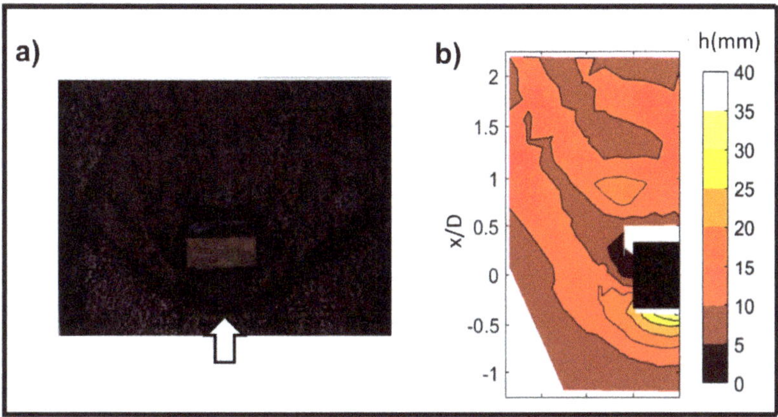

Figure 3. 2D field of water level around a scoured rectangular abutment in a clear-water supercritical flow by [27]. (**a**) Photograph at initial stage of scour, and (**b**) measured flow depth field at the quasi equilibrium stage of scour.

3. Scour around Bridge Foundations

Comprehensive studies on the variables controlling the maximum scour depth have been conducted since the state-of-the-art paper by [28]. The available information mostly corresponds to the case of scour in sand, with few exceptions for cases of scour in fine sediments [29–32] and gravel bed rivers [33–38]). Further, researchers have concentrated on the dynamics of scour [39–42]. A detailed description of the functional relationships between scour and its controlling variables [43,44] are out of the scope of the present paper. In this section, we focus on scour patterns.

3.1. Subcritical Flows

The researchers in [45] conducted an experimental study of clear water scouring around a circular cylinder, identifying the HSV as the main scour mechanism, due to the enhanced bed shear stress under the horseshoe vortices. The authors in [46] used numerical simulations of the flow field in the scour hole around a cylinder to show the importance of side slides in the scouring process. The study in [47] showed differences in the scour mechanism, depending on the flow intensity (I), i.e., the ratio between the flow velocity and the critical velocity for the incipient motion of sediment particles. From the minimum scour formation threshold at $I \approx 0.4$ to 0.6, to the incipient motion condition at $I = 1$, clear-water conditions dominate, and the HSV and downflow produce sediment entrainment and landslides, which enlarge the scour hole over time. For intensities of $1.0 < I < \approx 4.0$, a bedload occurs, and fluctuations of the scour depth in time can be observed due to the bedforms entering the scour hole. For $I > 4.0$, entrainment into suspension occurs at the undisturbed bed, and refilling of the scour hole due to deposition is expected during the falling stage of floods. The study in [48] described the morphological evolution of dune-like bed forms downstream of bridge piers and abutments that are generated by local scour. Recently, the authors in [49] conducted experiments that clearly distinguished two different scour modes: one, at the onset of erosion arising at the base of the cylinder and usually ascribed to the wrapping horseshoe vortex, which was determined and rationalized by a flow contraction effect, and another one, visible downstream of the cylinder, which consists of two side-by-side elongated holes. This pattern is observed for flow regimes close to the horseshoe scour onset, whose growth usually inhibits its spatio-temporal development.

3.2. Supercritical Flows

The study in [50] measured the maximum scour depths at piers in live-bed experiments for several flow configurations, including seven flows in the supercritical regime (Figure 4). For these flow conditions, no major increase in the maximum scour depth was reported as the Froude number exceeds 1.

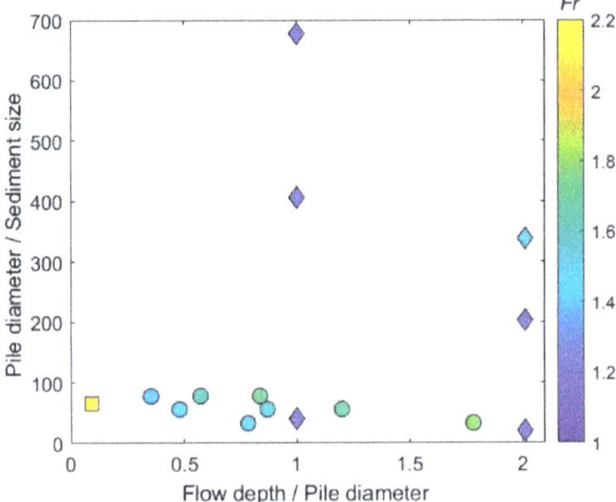

Figure 4. Measured maximum scour depth for flows in the supercritical regime: available configurations in the literature, with *Fr* as the Froude number.

The authors in [27] reported on scour at an abutment ($h/D = 0.095$) in a supercritical flow with clear-water conditions. The flow pattern was of the detached hydraulic jump type. The scour hole extended far downstream from the abutment sides (Figure 5a). Two symmetric deposition zones were observed downstream, separated by a streamwise valley of almost zero deposition elevation, with a magnitude of maximum deposition elevation about half the maximum erosion measured in the scour hole. The maximum scour depth at equilibrium was observed along the nose and the upstream part of the lateral faces of the abutment (see Figure 5b). Interestingly, the temporal evolution of the maximum scour depth agreed well with the scour formula in [51] for subcritical flows, when considering the conjugate depth and velocity downstream the straight hydraulic jump (see Figure 5c) as input flow parameters. More experimental evidence for the extrapolation of the results in [27] is needed, as the researchers in [27] investigated only a single flow configuration, with a low Reynolds number (Figure 4).

The study in [6] investigated scour at a pier or abutment in a supercritical flow using a 1:50 physical model of the projected bridge on the Rivière des Galets, located in the CNR (Compagnie Nationale du Rhône) laboratory (Figure 5d) to assess the shape and depth of the scour hole for different foundation diameters and approaching flow conditions. The ratio between the flow depth and pile diameter ranged between 0.4 and 1.8 (Figure 4), resulting in a wall-jet or detached jump flow patterns. Incoming velocities ranged from 1 to 2 m/s under a steady flow regime. The model sediment was scaled geometrically and consisted of a non-uniform mixture of sand and gravel, with a mean diameter of 1.8 mm. Around 20 tons of sand were supplied during running experiments to keep the sediment bed in equilibrium. Unfortunately, experimental conditions (turbidity of water, bed load displacement, high flow velocity) did not allow the researchers to measure the hydrodynamics in the scour hole. The measurements of the scour depth during the test were carried out using three rows of metal rods in a comb arrangement (see Figure 5f), while the equilibrium scour (see Figure 5g) was measured using an automatic tacheometer after drying the bed (Figure 5h). Figure 5i shows the maximum scour depths for a range of pier diameters and approaching flow velocities, with a flow intensity I up to 4. The maximum scour depth after the experiments appears to vary between 0.9 and 2.5 times the pier diameter, demonstrating that the scour is not as high as expected from the extrapolation of the results from subcritical flows, even though the flow velocities are much higher.

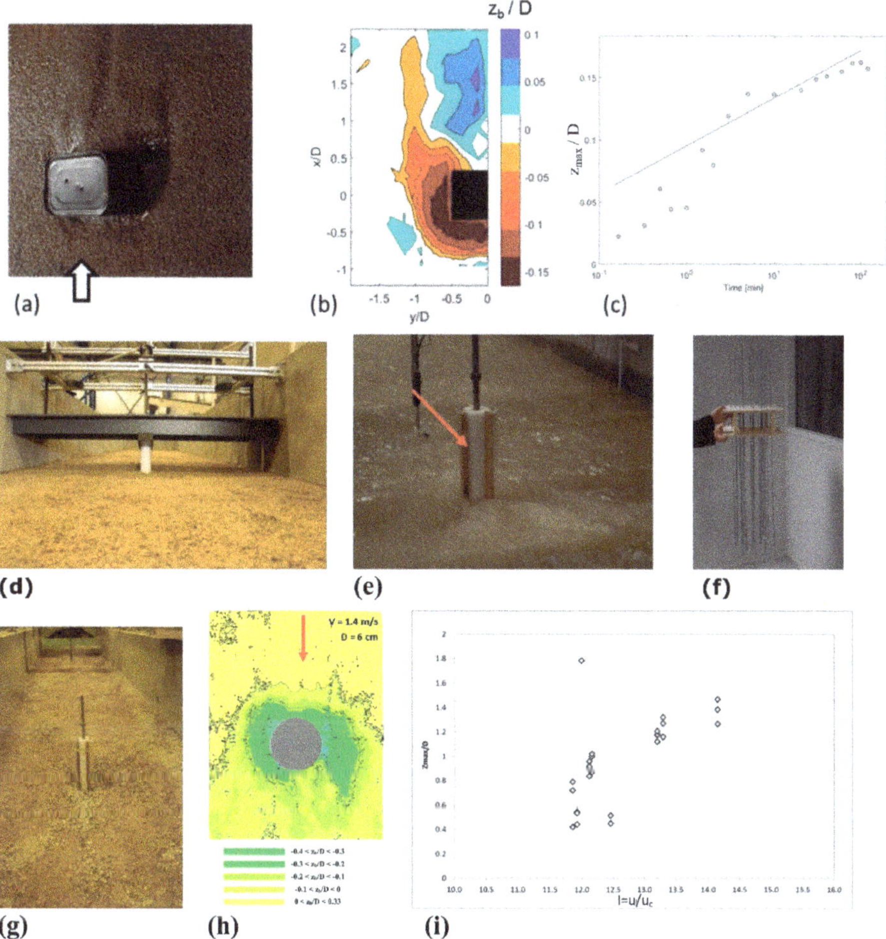

Figure 5. Clear-water scour at bridge foundations in experiments with a supercritical flow. Photograph of the clear water scour from [27] (**a**), with the corresponding normalized topography (z_b being the bed elevation) at equilibrium (**b**) and the measured time evolution of normalized maximum scour depth z_{max} (symbols) along with the prediction (plain line) obtained with the scour formula for subcritical flows [51] (**c**); live bed scour experiments from [6] (**d**), the flow pattern seen from downstream the pile (**e**), the metal rod comb used to measure the scour hole during the experiment (**f**), the scour shape at quasi-equilibrium after 2 hours at a pile of 6 cm in diameter in a 1.4 m/s flow condition (**g**, **h**), and the maximum normalized scour depth measured at the foot of the piles as a function of the flow intensity, I (**i**).

4. Recommended Methods for the Study of Flow and Scour in Supercritical Flows

Future research should be devoted to better understanding the flow processes responsible for scour at bridge foundations in watercourses with a supercritical regime along with the quantification of the maximum scour depths. Recommendations on methods to develop research on scour in supercritical flows are given below for laboratory and field measurements, as well as physical and numerical modeling.

4.1. Laboratory Techniques for Flow and Scour Dynamics

4.1.1. Standard Velocimetry

The use of micro high-speed propellers is appropriate when the quantification of local time average velocities is the measurement objective, such as in discharge measurements. However, the use of such propellers can be constrained by the presence of sediment due to the potential damage caused in the mechanical system. A modern approach for pointwise velocimetry is the use of Acoustic Doppler Velocimeters. Even though this instrument has been employed to measure the time averages and turbulent quantities in a broad range of flows, and even though the instruments have reduced dimensions, in cases of supercritical conditions, their intrusiveness may create a flow pattern similar to those caused by obstacles (Figure 2). Another consequence is the generation of oblique surface waves affecting the whole cross-section of the channel downstream.

4.1.2. Advanced Measurement of Hydrodynamic Processes

The use of advanced optical or acoustic measurement techniques for laboratory research in the supercritical regime presents many technical restrictions compared to subcritical cases. Supercritical flows, and particularly the flows developed around obstacles, show important unsteady free surface deformations, which make any attempt to access them from the surface impractical.

In the case of a flat smooth bed, measurements along the vertical planes can be performed through the bottom wall. One possibility is to use an acoustic profiler [52], located within the bed, which emits a vertical acoustic signal towards the free surface (Figure 6d). This instrument provides access to a time resolved vertical profile of three velocity components measured with a vertical resolution precision of approximately 1 mm. A second alternative is an optical technique using sourced illumination and non-orthogonal cameras located at the bottom of the channel. The most common optical technique is two-dimensional Particle Image Velocimetry (2D PIV) or 2D particle Tracking Velocimetry [53]. In this case, the reduction of optical distortions is achieved through the use of Scheimpflug adapters and optical prisms attached to the channel bottom-wall (as in standard stereoscopic or tomographic setups). A high image resolution and sampling frequency for the system becomes crucial for measuring the dynamics of small-scale structures around obstacles. A double pulse laser with a short-time interval between images is recommended to produce images with a high signal to noise ratio. An additional issue is air entrainment through the hydraulic jump or reverse spillage. In these cases, it would be necessary to consider the use of tracer particles, e.g., fluorescent particles, able to reflect the light source in a different wavelength than the light scattered by the air bubbles. This and additional band-pass filters in the camera give access to an independent quantification of particle and bubble dynamics. Finally, other alternatives available, using similar optical arrangements from the bottom, are the use of Stereo- and Tomographic-PIV ([54], Figure 5b), as well as the use of pointwise measurements performed with Laser Doppler Velocimetry [55].

The use of a movable bed constrains the use of optical measurements from the bottom. An alternative is to use refractive index-matching between the sediment and the fluid, as presented in [56], which would facilitate a similar technique to the one described above for the plane bed condition (Figure 6a).

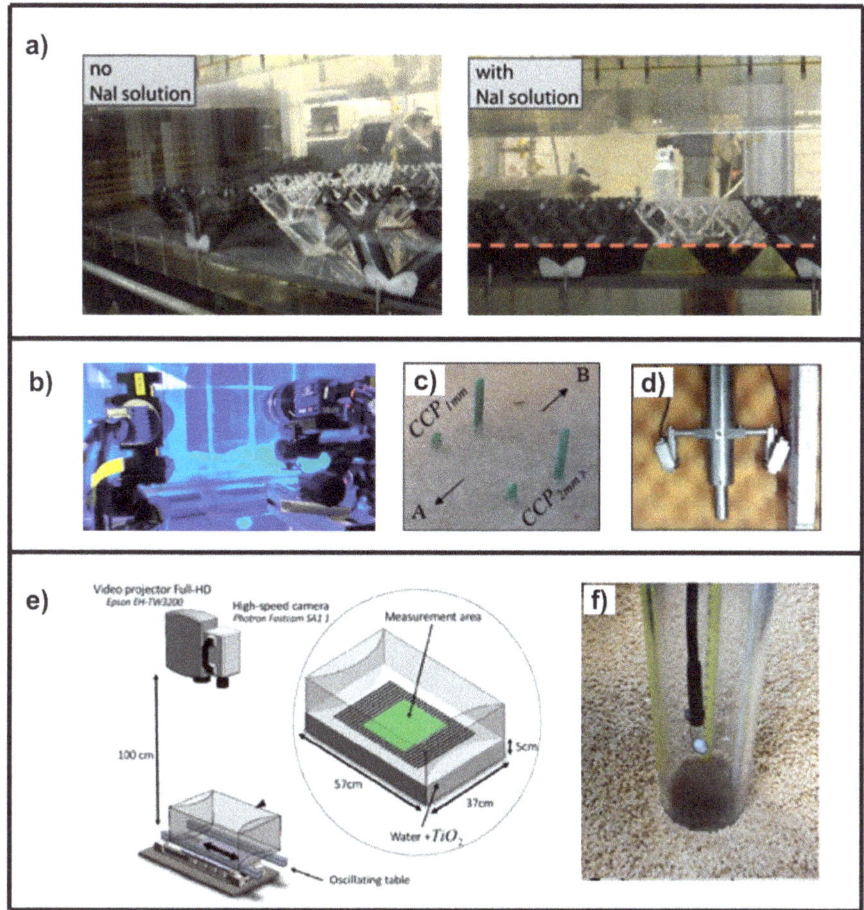

Figure 6. Innovative experimental techniques to be applied to supercritical hydrodynamics and scour processes in the laboratory: (**a**) an example of refractive-index and density matching from [57], (**b**) a view of a tomographic system, (**c**) CCP (Conductivity Concentration Profiler), (**d**) and ACVP (Acoustic Concentration and Velocity Profiler) from [58], (**e**) the application of Fourier transform profilometry to free-surface deformation, as per [59], (**f**) and an endoscopic camera placed inside a transparent pier for measuring scour evolution (from [47]).

4.1.3. Image-Based Reconstruction of a Free Surface

The time-resolved measurement of free-surface deformation is commonly performed using point-wise sensors, such as ultrasonic sensors (from above) or electrical resistive probes (within the water column). The study in [20] showed that these techniques apply to both supercritical flow patterns: the detached hydraulic jump and the wall-jet like bow wave. On the other hand, access to instantaneous 2D free-surface deformation around the pier/abutment requires projection techniques, such as the RGB-D sensor method [60], the projected grid method [27] (with a limited spatial resolution), or the Fourier transform profilometry method [61], with a higher spatial resolution (Figure 6e). These techniques require high quantum efficiency for the CCD and a high frequency sampling camera to get access to the dynamics of the surface deformation.

4.1.4. Distance Sensors for Scour Measurement

Laser distance sensors (LDS) are used to measure the scour-hole radius at different depths, thus quantifying the scour hole's geometry. In [62,63], researchers developed a measuring system composed of an LDS to measure the scour radius with an accuracy of ±0.4 mm. This LDS is placed inside a Plexiglas pier and aligned in a horizontal and radial direction, such that no refraction on the cylinder wall is observed during the measurements. The sensor is driven in the vertical direction by a step-motor with a precision of ±1/50 mm, and in the azimuthal direction, a vertical positioning system is driven by a second step-motor with an accuracy of ±1/100°, allowing the distance sensor to turn around in the scour-hole, taking various vertical profiles in different azimuthal half-planes. In this way, the geometry of the scour-hole below the original flat bed is automatically measured. The sensor performed well in tests with live bed conditions, having a flow intensity of $I = 2$. The same technique was applied to measurements in gravel [37] and sand-clay mixtures [30]. The application of LDS is, however, restricted to conditions close to clear-water and with low turbidity. It is, therefore, expected that LDS will not work properly in a number of supercritical cases where fine sediment is suspended.

Conductivity concentration profilers (see Figure 6c), originally developed in [64], were used to measure concentration profiles under sheet flow conditions in [58] using lightweight PMMA particles (1 and 3 mm). This technique is based on the inversion of the linear relationship between the sediment concentration and the conductivity of the medium. Using a grid of conductivity probes stuck on the pile, it would be possible to detect, at about 10 Hz, the position of the fixed bed interface at the pile position. Compared with LDS, the obtained information would be restricted to the bed elevation at the pile location, and no information on the 2D geometry of the scour hole would be provided. However, the advantage of such a technique is that it would be applicable even under intense live-bed conditions encountered in supercritical flow conditions for which no optical access is possible.

4.1.5. Image-Based Reconstruction of Scour during Running Experiments

The study in [65] presented a stereovision-based technique for continuous measurement of the bed morphology. This technique is capable of reconstructing instantaneous surface representations of the evolving bed with high spatial resolution during scour experiments. Two calibrated cameras must be partially submerged in the flow and record videos of the evolving bed geometry. This technique considers the texture of sediment beds and does not require the use of targets or structured light. A set of computer-vision and image-processing algorithms were developed for accurately reconstructing the surface of the bed. This technique was further applied to the spatio-temporal characterization of scour at the base of a cylinder in [66]. Optical access to the scoured region might be an issue in supercritical flows, due to noise from the unsteady wavy water surface. The authors in [67] developed a bed level tracking system with micro cameras placed inside a pier to record the maximum scour depth under clear-water conditions. The study in [47] used an endoscopic camera placed inside the pier to record images of a graduated pier, registering the maximum scour depth in scour experiments conducted under live-bed conditions (Figure 6f). Similarly, the authors in [68] used a snake camera of 0.5 cm in diameter that slightly penetrated the flow surface (less than 1 cm) to look at a graduated strip on the upstream face of the pier. In these studies, algorithms for the automatic recognition of the scour hole bottom through digital image processing were developed. The ability of cameras placed inside the pier to record the maximum scour depth has emerged as a promising alternative for measuring the temporal evolution of the maximum scour depth at piers in supercritical flows. Indeed, surface waves and oscillations, as well as air entrainment occurring in supercritical flows, constrain the applicability of stereoscopic systems using cameras placed above the water surface. Recognition of the scour-hole bottom, and thus the detection of the maximum scour depth in the images, might be possible even in presence of suspended sediment particles because the color of the turbid water is different to the color of the bottom embedded foundation.

4.2. Field Techniques for Flow and Scour Dynamics

4.2.1. Approaching Flow Measurements

The main restrictions for typical flow measurements at large velocities are related to instrument intrusion in the flow. For supercritical flows, instruments can only be deployed from a fixed point (a bridge or cable way), and intrusive measurements generally become impossible above a flow speed of 5 m/s (for a current meter on a torpedo). Acoustic Doppler Current Profilers (ADCP), which are currently used in the field to measure flow fields, are limited to a relative velocity of a few meters per second between the apparatus and the river, making their use generally impossible for supercritical flows. Indeed, these instruments are highly sensitive to air entrainment around the apparatus. From our own experience, the result quality using ADCP starts to decrease above a flow speed of 3.5 m/s. It should also be noted that such high flows are often associated with high suspended load concentrations and woody debris [69], which make such a measurement very dangerous.

The non-intrusive counterpart instruments used to measure large flow velocities during flood events are radar velocimeters ([70]; Figure 7a) and video analysis systems, such as Large Scale Particle Image Velocimetry (LSPIV; [71]). The main limitation of these systems is that they can only provide a surface velocity, so complex 3D flows cannot be described. Moreover, radar velocimeters are not able to provide a flow direction. LSPIV has been applied successfully to measure the velocity field in rivers with high discharges [72,73]. However, a difficulty that appears close to bridge foundations in supercritical flows is that the water surface may become 3D due to standing waves or antidunes. Since surface velocity estimations are made on a plane, this type of surface could lead to significant errors. On the other hand, stereovision-based techniques could be of interest to describe the water surface of such rapid flows.

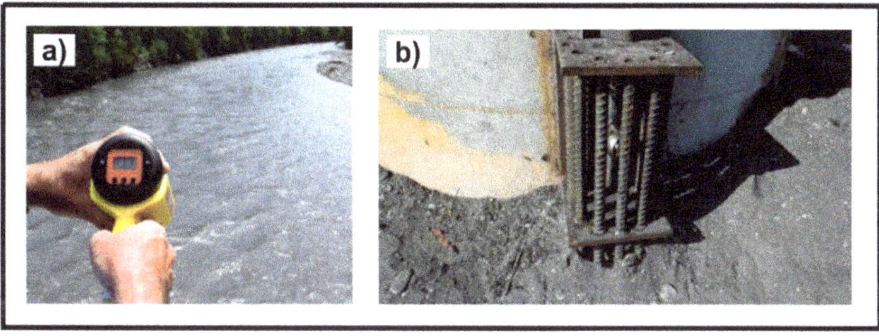

Figure 7. (**a**) Radar measurement from a bridge (Photo: P. Belleudy), and (**b**) a system for pier erosion monitoring used at NARLabs (National Applied Research Laboratories), Taiwan.

4.2.2. Bed Level Measurements

At present, only a few successful field measurements of scour during floods have been reported using the numbered-brick technique [5,74], which requires numbered bricks to be placed beneath the river bed by using excavators prior to the flood event. The location of the numbered-brick column can be accurately identified by using a total-station transit. After the flood has receded, excavators are again deployed to dredge the location of the preinstalled bricks to determine the number of bricks that have been washed away during the flood. To this end, one can then evaluate the maximum scour depth at the measuring location based on the remaining number of bricks [75]. Monitoring techniques for scour were reviewed in [76]. Such techniques include poles or sounding weights (with intrusion into the flow), float-out devices and tethered buried switches that can detect scour, and rods buried into the streambed (coupled with various sensors to measure the scour depth). These instrumentations may be robust, but they are limited to a single location and require expensive installation and/or maintenance.

Other techniques use sonars, fathometers (acoustic waves), or radars (electromagnetic waves) that are able to locally monitor the scour depth over time but are often limited to low flow velocities, water depths, and/or Suspended Sediment Matter (SSM) concentrations. Ground Penetrating Radars (GPR) allow a spatial view of the erosion but require manual operation and thus cannot be used during heavy-flood flows. Lastly, accelerometers provide a response for the bridge structure to scour. This structural dynamic response is able to provide a global view of the scour, which is of interest for early warning systems but is sensitive to the shape of scour holes and soil type and thus needs a site-specific calibration before monitoring [77].

As a consequence, a combination system based on structural dynamic response together with float-out or buried-typed devices would be the best actual monitoring approach to provide a quantitative description, over time, of the erosion depth at the foot of a bridge foundation in a supercritical flow.

Recent developments on active tracers [78] may allow their use around structures during a flood. Indeed, such systems could provide interesting results on gravel path and velocity but would be difficult to deploy during extreme events, such as those responsible for supercritical flows.

4.3. Physical Scale Models of Scour at Bridge Foundations

The authors in [79,80] analyzed the scale effects on scour, concluding that the use of laboratory flumes in developing accurate predictors of scour depth at full-scale piers is limited due to scale effects that may produce greater scour depths in the laboratory than at actual piers in rivers. The authors in [81] discussed the effects of sediment size scaling on the physical modeling of a bridge pier scour. Moreover, for model sediments smaller than d = 0.05 mm cohesion will reduce the scour [30]. Typical Froude scale models do not necessarily simulate the tractive forces and sediment erosion accurately because Froude scaling does not simulate viscous forces. Recently, researchers [82] showed that clear-water and live-bed scour under steady and unsteady flows in a subcritical regime is similar in the prototype and the model if the dimensionless flow work (as proposed in [83]) and the dimensionless grain diameter D^* are equal in both (the prototype and the model). Still, verifications of these scaling laws for supercritical cases are needed before such a scaling approach can be employed with confidence in physical modeling.

Additionally, important difficulties in the physical modeling of live-bed scour caused by a supercritical flow are the high water and sediment discharges to be supplied as upstream boundary conditions. While the water discharge may be overcome with large pumps, the sediment supply is a challenge in itself. The researchers in [6] supplied up to 400 l/s of water in order to reach their 2 m/s target velocity and 3 kg/s (10.8 t/h) of sediment during their experiments, which lasted up to 2 hours (Figure 5d–i). Moreover, the sediment needed to be regularly removed from the downstream basin in order not to generate backwater curves from downstream and keep the supercritical flow condition in the flume. Flow velocities were measured with propellers and ADV (Figure 5e), ensuring that the system did not produce important disturbances of the flow conditions. Because very loaded flows are prevented from using sonar or optical measurement systems (Figure 6f), the most adapted scour measurement methods appeared to be classical thin rods systems deployed from above the free surface (Figure 5f).

4.4. Advanced Numerical Tools for Simulation of Scour in Supercritical Flows

Numerical simulations play a significant role in the understanding of the flow hydrodynamics and scour around hydraulic structures, as they can explore conditions that cannot be reproduced experimentally or are inaccessible for measurement devices, complementing the observations, and providing additional insights into sediment-flow interactions and feedback [84]. In the following, numerical simulation techniques to be applied to supercritical scour processes are discussed, including the coupled hydrodynamics and sediment transport approach and the multi-phase flow approach.

4.4.1. Coupled Hydrodynamics and Sediment Transport Approach

First numerical simulations of scour around a bridge pier under a subcritical flow regime were performed about 25 years ago in [85], using the Reynolds-Averaged Navier–Stokes (RANS) equations for the hydrodynamics, coupled with the Exner equation for the bed's morphological evolution in steady flow conditions. Another important milestone was achieved in [86], which provided 3D morphodynamic simulations using a k-ω SST model, which was able to reproduce vortex shedding at the lee side of a surface mounted cylinder. Sumer [87], in his review on the mathematical modeling of scour, pointed out the free surface effects, the influence of small scale turbulence on sediment transport, and the potential effects of pore pressure on scour as the main avenues for future research. The study in [88] simulated scour around cylindrical and square piers, using a URANS (Unsteady Reynolds Averaged Navier Stokes) model coupled with an adapted version of the van Rijn model to estimate the bedload transport flux and the Exner equation. Better agreement was observed in the case where the shear drives the scour, as the HSV dynamics cannot be resolved with a URANS approach. Similar results were observed by [89], with the URANS and scour simulations past a surface mounted cylinder. This study showed that by considering the transport of suspended sediment, there is an improvement of the deposition patterns downstream of the pier. Recently, researchers in [90] proposed the use of a relaxation parameter to adjust the locally amplified bed shear stress due to the action of a horseshoe vortex to properly match the observed scour depth, solving the URANS–Exner equations system.

The work in [12,17,18] studied the dynamics of the HSV system at high-Reynolds numbers, using a hybrid URANS-LES turbulence model. These studies captured, for the first time, the intense velocity fluctuations and low-frequency bimodal oscillations, including the quasi-periodic vortex shedding and merging, and the formation of hairpin vortices, generated by a centrifugal instability that controls HSV dynamics. All these processes increase the instantaneous shear stress near the obstacle, allowing the development of models to study sediment dynamics from a Lagrangian perspective [23,91], as well as a model of bed evolution and bedform development, coupling the coherent-structure resolving model to the Exner equation [24], as shown in Figure 8a. These recent models have shown the potential of high-resolution 3D numerical simulations to predict different aspects of the scour process in subcritical conditions with good agreement. From these investigations, three key features have been identified: (1) Isotropic RANS turbulence models are inherently limited to predict the complex dynamics of an HSV system, as they yield a large turbulent viscosity, increasing the energy dissipation and suppressing the turbulence at the junction; (2) coherent structure resolving turbulence models can capture HSV dynamics, which can resolve the details of bedform formation and initial development of the scour hole in sand beds by turbulent structures; and (3) the main limiting assumption of these models stems from the prediction of instantaneous sediment fluxes from empirical formulas developed for uniform flows in steady and equilibrium conditions, adding ad-hoc formulations to represent the avalanches when the bed slope locally exceeds the angle of repose.

Using a Volume-Of-Fluid (VOF) approach, researchers [92] recently reported the first free surface resolving scour simulations using RANS–Exner type models. According to the author, accounting for the free surface effect improves the position of the HSV and the prediction of the mean velocity field, even at a Froude number as low as 0.2. The VOF technique emerges as a promising tool for the advanced numerical simulation of supercritical flows, including the scour at bridge foundations, following the coupled hydrodynamics and sediment transport approach.

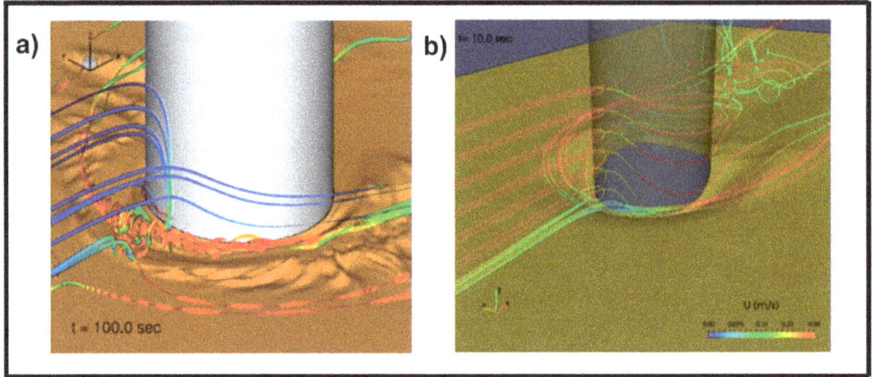

Figure 8. Instantaneous bed surface and streamlines around a cylindrical pier: (**a**) a subcritical flow with clear water conditions and the scour computed with the LES–Exner model, as in [24], and (**b**) a subcritical flow with live bed conditions computed with the turbulence averaged two-phase flow model, as in [93]. Streamlines are colored by the velocity magnitude.

4.4.2. Multi-Phase Flow Approach

Over the last decades, a new generation of sediment transport models has emerged: the Eulerian–Eulerian two-phase flow approach [94]. Unlike classical sediment transport models, the two-phase flow approach is based on the resolution of momentum balance for both the fluid and the sediment phases, the latter being seen as a continuum with a peculiar rheology. Very recently, using an open-source multi-dimensional two-phase flow model for sediment transport applications [95], the study in [93] presented the first two-phase flow RANS simulation for scour around a vertical cylinder. In Figure 8b, the scour mark induced by the HSV is clearly visible, while the downstream erosion induced by the lee–wake vortices was also observed. The two-phase flow approach is able to reproduce the scour dynamics induced by the HSV without empirical parametrization for the sediment fluxes and the avalanching process.

Further, the authors in [96] reported the first two-phase flow turbulence-resolving simulations. This opens new possibilities for simulating the complex interactions between sediment transport and HSV dynamics. Another important research possibility has been recently developed by [97] and concerns the development of a two-phase flow sediment transport model, including a free surface resolving capability. This model will allow one to reproduce the free surface features observed in the supercritical flow regime by [20,21] while solving sediment dynamics based on mechanical principles, as in [93].

Advances on scour modeling also consider computational techniques that efficiently calculate the dynamic coupling between the flow and the bed. RANS–Exner models coded in GPU (Graphics Processing Unit) have already been developed to improve the computation of these problems [98]. Future models that incorporate LES and Exner in the vicinity of bridge piers using these strategies show great promise in tackling large computational domains or fine resolutions in scour problems.

4.5. *Moving from the Local Scale Phenomena Up to Long Term Dynamics*

As detailed above, the scour process has been investigated mainly with laboratory and numerical approaches under several simplifying assumptions, neglecting the stochastic nature of floods. Indeed, the scour process should be considered a stochastic process controlled by the dynamics of floods, sediment transport, and riverbed evolution over time. All these processes act over a bridge foundation throughout the bridge's lifespan, and it is likely that the entire history of these events is responsible of the failure risk of a bridge rather than a specific flood event. An example of the potential time evolution of flood characteristics (here: discharge and Froude number) and scour depth over time is given in Figure 9.

The limitation of the current knowledge of this phenomenon is highlighted in [99], which documented the magnitude of flood events causing bridge collapses in the US. Most of these bridges were designed for a flood event with a return period of about 100 years, but they collapsed under a variety of events with a return period ranging from 1 to 1000 years. This is clear evidence of the limitations of the current methodologies and understanding of bridge design. Lately, a number of authors have explored alternative strategies for bridge failure predictions in order to account for the non-stationarity of the flow and the stochastic nature of floods. These studies highlighted that scour evolution over time is strongly controlled by factors like the shape of flood hydrographs [83], and also the time-sequence of floods occurring at a given location [100]. Moreover, hydrological variability has been amplified even more significantly by the effects of global climate change [101].

Now, the interactions between infrastructure and river morphodynamics are characterized by many factors that span a wide range of temporal and spatial scales. Besides the hydrographs, additional factors at larger scales include variations in sediment availability, vegetation, land use in the watershed, channel management and the effects of anthropic interventions (e.g., gravel mining), or other infrastructures built in the channel, such as reservoirs or intakes. The sum of these phenomena should be considered when designing a bridge foundation, but actual modeling schemes, as well as available information, are not able to fully describe such a complex dynamic. Therefore, the challenge in the coming years is to identify unifying principles able to interpret scour phenomena, both under subcritical and supercritical conditions, with the right compromise between model complexity and data availability [102–104]. Moreover, it is highly desired that forthcoming modeling approaches explicitly account for the stochastic nature of the flood process.

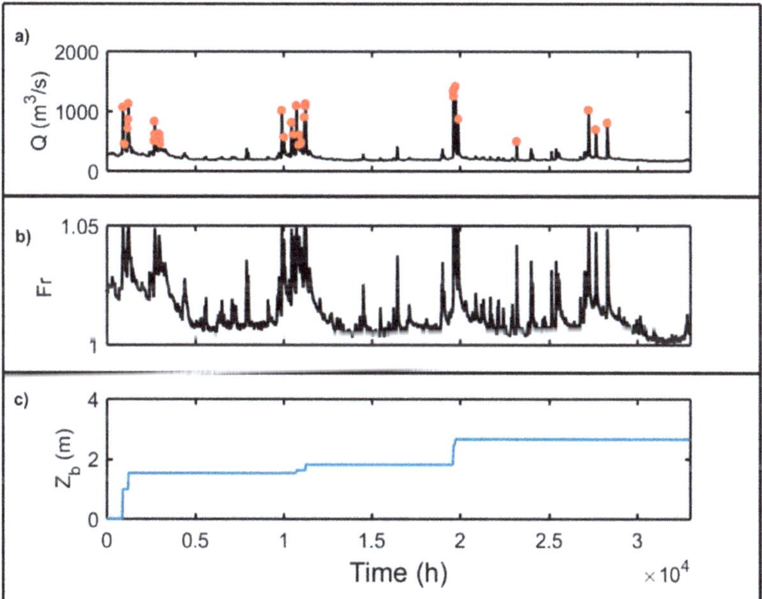

Figure 9. The synthetic time-series over a temporal window of approximately 4 years: (**a**) supercritical flow discharge, (**b**) the corresponding Froude number Fr, and (**c**) the evolution of the scour depth at the foot of a bridge pile, assuming it as an additive process. The graph displays the random nature of the flood events within a river system and its consequences in terms of the random evolution of scour depth over time, computed using the BRISENT model, as in [105].

5. Discussion and Conclusions

Bridges with foundations in river beds with supercritical flow regimes are frequent and can be found worldwide. Handling the scouring process that takes place in this type of flow (e.g., with bed sills [106]) is crucial for the structural integrity of bridges, transportation systems, and the safety of users. Scour around piers overlaps with several processes occurring at different spatial and temporal scales, such as bed degradation and aggradation, dynamic braiding or meandering, and the formation/destruction of antidunes, large woody debris [69], barbs [107–109], or spur dikes for preserving the desired water depth [110]. Moreover, extreme flood events produce rapid changes of the local riverbed morphology, affecting the scouring process at bridge foundations. The temporal dynamics of scour, then, are intimately linked with sediment transport. This paper focused on the knowledge, gaps, and open questions related to local scour at bridge foundations in supercritical flows. From our review, the controlling mechanisms of scour in supercritical flows differ from those controlling scour in subcritical flows, due to the appearance of a detached hydraulic jump or vertical wall-jet patterns [20]. Interactions with the horseshoe and wake vortex systems (and also with the sediment motion) remain nearly unexplored (except for the work in [21]). The unexpected scour depth in the available experiments [6,27] suggests that the maximum scour depth in supercritical flows might be of comparable magnitude to that occurring in subcritical flows, even though the flow intensity is much higher. Further research is needed to understand the controlling parameters of the scour in supercritical flows.

5.1. Measuring Techniques for Flow in the Laboratory

Laboratory scour experiments with supercritical flows should avoid intrusive measuring techniques due to their important effect on the flow field (especially creating surface waves). Efforts should be made to transfer some widely used optical techniques from fluid mechanics to the hydraulics of supercritical flows, such as high speed image based velocimetry, including two dimensional and stereo PIV/PTV, or even the use of tomographic velocimetry, taking advantage of techniques like index matching to access the flow field in the scour hole.

5.2. Measuring Techniques for Scour in the Laboratory

Scour hole geometry might be best measured with precision ultrasonic or optic distance sensors. However, important constraints appear for intrusive instruments, such as sonars, especially at the laboratory scale because the size of the instrument can alter the flow field and create surface waves. Thus, the use of video cameras placed inside transparent bridge elements is recommended to record the maximum scour depth at the front of the bridge foundation over time.

5.3. Measuring Techniques for Flow in the Field

Field measurements of the supercritical flow at bridges might provide a good idea of the flow velocity upstream of the impacted foundations. Accessibility is, by far, identified as the main constraint in the use of intrusive equipment, so radar velocimeters [70] and video analysis, such as LSPIV [71], are recommended. As these methods provide only the surface velocity, further research is needed to correlate this data with the corresponding flow field and features.

5.4. Measuring Techniques for Scour in the Field

A major issue for river engineers dealing with bridge foundation design is to evaluate all the processes that could affect erosion for a better assessment of the scouring intensity and, consequently, a better optimization of the bridge foundation cost. Three main processes affecting the bed level generally include (i) the overall long-term bed evolution linked to the river equilibrium (indeed, many rivers in the world suffers from erosion due to gravel mining or damming [111]); (ii) the natural river breath during a flood, leading to global erosion during the rising part of the flood and global deposition during the ebb part of the flood; and (iii) local erosion at the bridge foundation. Eventually, long term evolution should be tackled by being able to estimate the maximum erosion during an event but also to estimate the refilling of the scour after this event. Monitoring of very dynamic rivers remains quite difficult but obviously needs some investment to be able to verify the upscaling limits from laboratory experiments. The measurement of the local scour depth at real bridges in supercritical flows or during floods has succeeded until now by only using the numbered brick method [5,75]. Further research is needed to develop methods using the structural behavior of the bridge for scour measurement. These kinds of indirect methods would allow for scour monitoring in complex hydro-sedimentary conditions.

5.5. Physical Modeling of Scour in Supercritical Flows

Given the constraints of knowledge transfer from sub to supercritical scour, and the important difficulties for laboratory and field measurements of flow and scour at bridge foundations, the use of physical scale models appears to be a reasonable alternative for the analysis and design of particular cases. Important requirements include a wide set-up to prevent lateral confinement effects, high water and sediment discharges, and innovative measuring techniques in line with the involved rapid velocities. Moreover, important scale effects are expected in the case of model sediments with mean diameters smaller than 0.2 mm.

5.6. Numerical Modeling of Scour in Supercritical Flows

Classical coupled Navier-Stokes-Exner models have had some success in predicting the scour around bridge piles under subcritical flow conditions, even though some open questions remain to be answered. However, to the best of the authors' knowledge, no simulation of scour under supercritical flow conditions has ever been reported in the literature. Such simulations would be very challenging, particularly because of the very energetic flow conditions encountered in supercritical flows, the strong free surface dynamics, the multiple feedback mechanisms between the free surface dynamics, the flow hydrodynamics (e.g., HSV), and the sediment dynamics or the air entrainment. Some important breakthroughs can be expected in the near future by using multi-phase flow approaches in conjunction with high-resolution experimental data. From a numerical modeling standpoint, capturing the dynamics of a free surface is a critical prerequisite for numerical simulations of supercritical flows. To resolve the coupled flow field and interactions between the large-scale coherent structures of the HSV system and the free-surface, LES, or hybrid URANS–LES turbulence models should be tested to understand the flow physics and unsteady scour mechanisms. Eulerian–Lagrangian approaches seems unrealistic due to their very high computational cost. However, turbulence-resolving Eulerian–Eulerian simulations coupled with a free surface resolving capability will probably be possible in the near future.

Author Contributions: All eight authors (O.L., E.M., S.R., B.C., C.E., J.C., W.B. and S.M.) contributed to all aspects of present work in the framework of a one-week workshop at the University of Concepcion, Chile.

Funding: Financial support by the Chilean research council CONICYT through grants Fondecyt 1150997 and REDES 170021 is greatly acknowledged. C.E. acknowledges the support of Conicyt/Fondap grant 15110017.

Acknowledgments: The authors would like to thank Nicolas Rivière (University of Lyon) for his scientific advice.

Conflicts of Interest: The authors declare no conflict of interest.

References

1. Arneson, L.A.; Zevenbergen, L.W.; Lagasse, P.F.; Clopper, P.E. *Evaluating Scour at Bridges*, 5th ed.; FHWA-HIF-12-003, HEC-18; Books Express Publishing: Berkshire, UK, 2012.
2. Melville, B.; Coleman, S. *Bridge Scour*; Water Resources: Littleton, CO, USA, 2000.
3. Ettmer, B.; Bleck, M.; Broich, K.; Link, O.; Meyering, H.; Pfleger, F.; Stahlmann, A.; Unger, J.; Weichert, R.; Werth, K.; et al. *Scour Around Piers*; Merkblatt DWA-M529; Deutsche Vereinigung für Wasserwirtschaft, Abwasser und Abfall e.V. DWA: Hennef (Sieg), Germany, 2016. (In German)
4. MOP. *Highways Design Manual*; Chilean Ministry of Public Works: Santiago, Chile, 2000. (In Spanish)

5. Lu, J.Y.; Hong, J.H.; Su, C.C.; Wang, C.Y.; Lai, J.S. Field measurements and simulation of bridge scour depth variations during floods. *J. Hydraul. Eng.* **2008**, *134*, 810–821. [CrossRef]
6. Roux, S.; Misset, C.; Romieu, P. *Construction of a New Bridge over the Rivière des Galets*; Study on physical model; phase 1—tests on abutment scour; Technical Report CNR: Lyon, France, 2015.
7. Yang, H.C.; Su, C.C. Real-time river bed scour monitoring and synchronous maximum depth data collected during Typhoon Soulik in 2013. *Hydrol. Proc.* **2015**, *29*, 1056–1068. [CrossRef]
8. Camenen, B.; Dugué, V.; Proust, S.; Le Coz, J.; Paquier, A. Formation of standing waves in a mountain river and its consequences on gravel bar morphodynamics. In Proceedings of the 6th IAHR Symposium on River, Coastal and Estuarine Morphodynamics, Santa Fe, Argentina, 21–25 September 2009; Volume I, pp. 65–72.
9. Fruchart, F.; Delphin, G.; Reynaud, S.; Séraphine, C.; Cadudal, J.; Bonnet, P.; Lemahieu, H. Rivière des Galets Reunion Island—20 year experience of risk management. In Proceedings of the 12th Congress Interpraevent 2012, Grenoble, France, 23–26 April 2012.
10. Dargahi, B. The turbulent flow field around a circular cylinder. *Exp. Fluids* **1989**, *8*, 1–12. [CrossRef]
11. Devenport, W.J.; Simpson, R.L. Time-dependent and time-averaged turbulence structure near the nose of a wing-body junction. *J. Fluid Mech.* **1990**, *210*, 23–55. [CrossRef]
12. Paik, J.; Escauriaza, C.; Sotiropoulos, F. On the bimodal dynamics of the turbulent horseshoe vortex system in a wing-body junction. *Phys. Fluids* **2007**, *19*, 045107. [CrossRef]
13. Apsilidis, N.; Diplas, P.; Dancey, C.L.; Bouratsis, P. Time-resolved flow dynamics and Reynolds number effects at a wall–cylinder junction. *J. Fluid Mech.* **2015**, *776*, 475–511. [CrossRef]
14. Schanderl, W.; Manhart, M. Reliability of wall shear stress estimations of the flow around a wall-mounted cylinder. *Comput. Fluids* **2016**, *128*, 16–29. [CrossRef]
15. Schanderl, W.; Jenssen, U.; Manhart, M. Near-wall stress balance in front of a wall-mounted cylinder. *Flow Turbul. Combust.* **2017**, *99*, 665–684. [CrossRef]
16. Schanderl, W.; Jenssen, U.; Strobl, C.; Manhart, M. The structure and budget of turbulent kinetic energy in front of a wall-mounted cylinder. *J. Fluid Mech.* **2017**, *827*, 285–321. [CrossRef]
17. Paik, J.; Escauriaza, C.; Sotiropoulos, F. Coherent structure dynamics in turbulent flows past in-stream structures: Some insights gained via numerical simulation. *J. Hydraul. Eng.* **2010**, *136*, 981–993. [CrossRef]
18. Escauriaza, C.; Sotiropoulos, F. Reynolds number effects on the coherent dynamics of the turbulent horseshoe vortex. *Flow Turbul. Combust.* **2011**, *86*, 231–262. [CrossRef]
19. Dey, S. Sediment pick-up for evolving scour near circular cylinders. *Appl. Math. Model.* **1996**, *20*, 534–539. [CrossRef]
20. Rivière, N.; Vouillat, G.; Launay, G.; Mignot, E. Emerging obstacles in supercritical open-channel flows: Detached hydraulic jump versus wall-jet-like bow wave. *J. Hydraul. Eng.* **2017**, *143*, 04017011. [CrossRef]
21. Mignot, E.; Rivière, N. Bow-wave-like hydraulic jump and horseshoe vortex around an obstacle in a supercritical open channel flow. *Phys. Fluids* **2010**, *22*, 117105. [CrossRef]
22. Gond, L.; Perret, G.; Mignot, E.; Riviere, N. Analytical prediction of the hydraulic jump detachment length in front of mounted obstacles in supercritical open-channel flows. *Phys. Fluids* **2019**, *31*, 045101. [CrossRef]
23. Escauriaza, C.; Sotiropoulos, F. Lagrangian model of bed-load transport in turbulent junction flows. *J. Fluid Mech.* **2011**, *666*, 36–76. [CrossRef]
24. Escauriaza, C.; Sotiropoulos, F. Initial stages of erosion and bed form development in a turbulent flow around a cylindrical pier. *J. Geophys. Res. Earth Surf.* **2011**, *116*, 1–24. [CrossRef]
25. Dey, S.; Raikar, R.V. Characteristics of horseshoe vortex in developing scour holes at piers. *J. Hydraul. Eng.* **2007**, *133*, 399–413. [CrossRef]
26. Kumar, A.; Kothyari, U.C.; Raju, K.G.R. Flow structure and scour around circular compound bridge piers–A review. *J. Hydro Environ. Res.* **2012**, *6*, 251–265. [CrossRef]
27. Mignot, E.; Moyne, T.; Doppler, D.; Rivière, N. Clear-water scouring process in a flow in supercritical regime. *J. Hydraul. Eng.* **2016**, *142*, 04015063. [CrossRef]
28. Breusers, H.N.; Nicollet, G.; Shen, H.W. Local scour around cylindrical piers. *J. Hydraul. Res.* **1977**, *15*, 211–252. [CrossRef]
29. Debnath, K.; Chaudhuri, S. Bridge pier scour in clay-sand mixed sediments at near-threshold velocity for sand. *J. Hydraul. Eng.* **2010**, *136*, 597–609. [CrossRef]
30. Link, O.; Klischies, K.; Montalva, G.; Dey, S. Effects of bed compaction on scour at a bridge pier in sandy clay mixtures. *J. Hydraul. Eng.* **2013**, *139*, 1013–1019. [CrossRef]

31. Ng, K.W.; Chakradhar, R.; Ettema, R.; Kempema, E.W. Geotechnical considerations in hydraulic modeling of bridge abutment scour. *J. Appl. Water Eng. Res.* **2015**, *3*, 132–142. [CrossRef]
32. Harris, J.M.; Whitehouse, R.J. Scour development around large-diameter monopiles in cohesive soils: Evidence from the field. *J. Waterw. Port Coast. Ocean Eng.* **2017**, *143*, 04017022. [CrossRef]
33. Dey, S.; Barbhuiya, A.K. Clear-water scour at abutments in thinly armored beds. *J. Hydraul. Eng.* **2004**, *130*, 622–634. [CrossRef]
34. Dey, S.; Barbhuiya, A.K. Time variation of scour at abutments. *J. Hydraul. Eng.* **2005**, *131*, 11–23. [CrossRef]
35. Raikar, R.V.; Dey, S. Clear-water scour at bridge piers in fine and medium gravel beds. *Can. J. Civ. Eng.* **2005**, *32*, 775–781. [CrossRef]
36. Dey, S.; Raikar, R.V. Clear-water scour at piers in sand beds with an armor layer of gravels. *J. Hydraul. Eng.* **2007**, *133*, 703–711. [CrossRef]
37. Diab, R.; Link, O.; Zanke, U. Geometry of developing and equilibrium scour holes at bridge piers in Gravel. *Can. J. Civ. Eng.* **2010**, *37*, 544–552. [CrossRef]
38. Pandey, M.; Sharma, P.K.; Ahmad, Z.; Singh, U.K. Experimental investigation of clear-water temporal scour variation around bridge pier in gravel. *Environ. Fluid Mech.* **2018**, *18*, 871–890. [CrossRef]
39. Kothyari, U.C.; Ranga Raju, K.G. Scour around spur dikes and bridge abutments. *J. Hydraul. Res.* **2001**, *39*, 367–374. [CrossRef]
40. Coleman, S.E.; Lauchlan, C.S.; Melville, B.W. Clear-water scour development at bridge abutments. *J. Hydraul. Res.* **2003**, *41*, 521–531. [CrossRef]
41. Dey, S.; Barbhuiya, A.K. Flow Field at a Vertical-Wall Abutment. *J. Hydraul. Eng.* **2005**, *131*, 1126–1135. [CrossRef]
42. Kothyari, U.C.; Hager, W.H.; Oliveto, G. Generalized approach for clear-water scour at bridge foundation elements. *J. Hydraul. Eng.* **2007**, *133*, 1229–1240. [CrossRef]
43. Ettema, R.E. Scour at Bridge Piers. Ph.D. Thesis, The University of Auckland, Auckland, New Zealand, 1980.
44. Raudkivi, A.J. Functional Trends of Scour at Bridge Piers. *J. Hydraul. Eng.* **1986**, *112*, 1–13. [CrossRef]
45. Dargahi, B. Controlling mechanism of local scouring. *J. Hydraul. Eng.* **1990**, *116*, 1197–1214. [CrossRef]
46. Gobert, C.; Link, O.; Manhart, M.; Zanke, U. Discussion of coherent structures in the flow field around a circular cylinder with scour hole by G. Kirkil, S.G. Constantinescu and R. Ettema. *J. Hydraul. Eng.* **2010**, *119*, 82–84. [CrossRef]
47. Ettmer, B.; Orth, F.; Link, O. Live-bed scour at bridge piers in a lightweight polystyrene bed. *J. Hydraul. Eng.* **2015**, 1491–1495. [CrossRef]
48. Oliveto, G.; Hager, W.H. Morphological evolution of dune-like bed forms generated by bridge scour. *J. Hydraul. Eng.* **2014**, *140*, 06014009. [CrossRef]
49. Lachaussée, F.; Bertho, Y.; Morize, C.; Sauret, A.; Gondret, P. Competitive dynamics of two erosion patterns around a cylinder. *Phys. Rev. Fluids* **2018**, *3*, 012302. [CrossRef]
50. Jain, S.C.; Fischer, E.E. Scour around bridge piers at high velocities. *J. Hydraul. Div.* **1980**, *106*, 1827–1842.
51. Oliveto, G.; Hager, W. Temporal evolution of clear-water pier and abutment scour. *J. Hydraul. Eng.* **2002**, *128*, 811–820. [CrossRef]
52. Revil-Baudard, T.; Chauchat, J.; Hurther, D.; Barraud, P.-A. Investigation of sheet-flow processes based on novel acoustic high-resolution velocity and concentration measurements. *J. Fluid Mech.* **2015**, *767*, 1–30. [CrossRef]
53. Brevis, W.; Niño, Y.; Jirka, G.H. Integrating cross-correlation and relaxation algorithms for particle tracking velocimetry. *Exp. Fluids* **2011**, *50*, 135–147. [CrossRef]
54. Wise, D.J.; Brevis, W. Experimental and numerical assessment of turbulent flows through arrays of multi-scale, surface mounted cubes. In Proceedings of the Inaugural UK Fluid Conference, Imperial College London, London, UK, 7–9 September 2016.
55. Auel, C.; Albayrak, I.; Boes, R.M. Turbulence characteristics in supercritical open channel flows: Effects of Froude number and aspect ratio. *J. Hydraul. Eng.* **2014**, *140*, 04014004. [CrossRef]
56. Wiederseiner, S.; Andreini, N.; Epely-Chauvin, G.; Ancey, C. Refractive-index and density matching in concentrated particle suspensions: A review. *Exp. Fluids* **2011**, *50*, 1183–1206. [CrossRef]
57. Bai, K.; Katz, J.; Meneveau, C. Turbulent flow structure inside a canopy with complex multi-scale elements. *Bound. Layer Meteorol.* **2015**, *155*, 435–457. [CrossRef]

58. Fromant, G.; Mieras, R.S.; Revil-Baudard, T.; Puleo, J.A.; Hurther, D.; Chauchat, J. On bed load and suspended load measurement performances in sheet flows using acoustic and conductivity profilers. *JGR* **2018**, *123*, 2546–2562.
59. Aubourg, Q.; Mordant, N. Investigation of resonances in gravity-capillary wave turbulence. *Phys. Rev. Fluids* **2016**, *1*, 023701. [CrossRef]
60. Nichols, A.; Rubinato, M. Remote sensing of environmental processes via low-cost 3d free-surface mapping. In Proceedings of the 4th IAHR Europe Congress, Liege, Belgium, 27–29 July 2016.
61. Cobelli, P.J.; Maurel, A.; Pagneux, V.; Petitjeans, P. Global measurement of water waves by Fourier transform profilometry. *Exp. Fluids* **2009**, *46*, 1037. [CrossRef]
62. Link, O. Untersuchung der Kolkung an Einem Schlanken Zylindrischen pfeiler in sandigem boden. Ph.D. Thesis, Institut für Wasserbau und Wasserwirtschaft, Technische Universität, Darmstadt, German, 2006.
63. Link, O.; Pfleger, F.; Zanke, U. Characteristics of developing scour-holes at a sand-embedded cylinder. *International, J. Sediment Res.* **2008**, *23*, 258–266. [CrossRef]
64. Lanckriet, T.; Puleo, J.A.; Waite, N. A conductivity concentration profiler for sheet flow sediment transport. *IEEE J. Ocean. Eng.* **2013**, *38*, 55–70. [CrossRef]
65. Bouratsis, P.P.; Diplas, P.; Dancey, C.L.; Apsilidis, N. High-resolution 3-D monitoring of evolving sediment beds. *Water Resour. Res.* **2013**, *49*, 977–992. [CrossRef]
66. Bouratsis, P.; Diplas, P.; Dancey, C.L.; Apsilidis, N. Quantitative spatio-temporal characterization of scour at the base of a cylinder. *Water* **2017**, *9*, 227. [CrossRef]
67. Chang, W.Y.; Lai, J.S.; Yen, C.L. Evolution of scour depth at circular bridge piers. *J. Hydraul. Eng.* **2004**, *130*, 905–913. [CrossRef]
68. Link, O.; Castillo, C.; Pizarro, A.; Rojas, A.; Ettmer, B.; Escauriaza, C.; Manfreda, S. A model of bridge pier scour during flood waves. *J. Hydraul. Res.* **2017**, *55*, 310–323. [CrossRef]
69. Wallerstein, N.P. Dynamic model for constriction scour caused by large woody debris. *Earth Surf. Process. Landf.* **2003**, *28*, 49–68. [CrossRef]
70. Welber, M.; Le Coz, J.; Laronne, J.; Zolezzi, G.; Zamler, D.; Dramais, G.; Hauet, A.; Salvaro, M. Field assessment of noncontact stream gauging using portable surface velocity radars (SVR). *Water Resour. Res.* **2016**, *52*, 1108–1126. [CrossRef]
71. Fujita, I.; Muste, M.; Kruger, A. Large-scale particle image velocimetry for flow analysis in Hydraul. *Eng. Appl. J. Hydraul. Res.* **1998**, *36*, 397–414. [CrossRef]
72. Le Coz, J.; Hauet, A.; Pierrefeu, G.; Dramais, G.; Camenen, B. Performance of image-based velocimetry (LSPIV) applied to flash-flood discharge measurements in Mediterranean rivers. *J. Hydrol.* **2010**, *394*, 42–52. [CrossRef]
73. Dramais, G.; Le Coz, J.; Camenen, B.; Hauet, A. Advantages of a mobile LSPIV method for measuring flood discharges and improving stage–discharge curves. *J. Hydro Environ. Res.* **2011**, *5*, 301–312. [CrossRef]
74. Su, C.C.; Lu, J.Y. Measurements and prediction of typhoon-induced short-term general scours in intermittent rivers. *Nat. Hazards* **2013**, *66*, 671–687. [CrossRef]
75. Hong, J.H.; Guo, W.D.; Chiew, Y.M.; Chen, C.H. A new practical method to simulate flood-induced bridge pier scour—A case study of Mingchu bridge piers on the Cho-Shui River. *Water* **2016**, *8*, 238. [CrossRef]
76. Prendergast, L.J.; Gavin, K. A review of bridge scour monitoring techniques. *J. Rock Mech. Geotech. Eng.* **2014**, *6*, 138–149. [CrossRef]
77. Bao, T.; Liu, Z. Vibration-based bridge scour detection: A review. *Struct. Control Health Monit.* **2017**, *24*, e1937. [CrossRef]
78. McNamara, J.P.; Borden, C. Observations on the movement of coarse gravel using implanted motion-sensing radio transmitters. *Hydrol. Process.* **2004**, *18*, 1871–1884. [CrossRef]
79. Ettema, R.; Melville, B.; Barkdoll, B. Scale effect in pier-scour experiments. *J. Hydraul. Eng.* **1998**, *124*, 639–642. [CrossRef]
80. Ettema, R.; Kirkil, G.; Muste, M. Similitude of large-scale turbulence in experiments on local scour at cylinders. *J. Hydraul. Eng.* **2006**, *132*, 33–40. [CrossRef]
81. Lee, S.; Sturm, T. Effect of sediment size scaling on physical modeling of bridge pier scour. *J. Hydraul. Eng.* **2009**, *135*, 793–802. [CrossRef]
82. Link, O.; Henríquez, S.; Ettmer, B. Physical scale modeling of scour around bridge piers. *J. Hydraul. Res.* **2019**, *57*, 227–237. [CrossRef]

83. Pizarro, A.; Ettmer, B.; Manfreda, S.; Rojas, A.; Link, O. Dimensionless effective flow work for estimation of pier scour caused by flood waves. *J. Hydraul. Eng.* **2017**, *143*, 06017006. [CrossRef]
84. Escauriaza, C.; Paola, C.; Voller, V.R. Computational models of flow, sediment transport and morphodynamics in rivers. In *Gravel-Bed Rivers: Processes and Disasters*, 1st ed.; Tsutsumi, D., Laronne, J.B., Eds.; John Wiley and Sons Ltd.: Hoboken, NJ, USA, 2017.
85. Olsen, N.R.; Melaaen, M.C. Three-dimensional calculation of scour around cylinders. *J. Hydraul. Eng.* **1993**, *119*, 1048–1054. [CrossRef]
86. Roulund, A.; Sumer, B.M.; Fredsøe, J.; Michelsen, J. Numerical and experimental investigation of flow and scour around a circular pile. *J. Fluid Mech.* **2005**, *534*, 351–401. [CrossRef]
87. Sumer, M.B. Mathematical modeling of scour: A review. *J. Hydraul. Res.* **2007**, *45*, 723–735. [CrossRef]
88. Khosronejad, A.; Kang, S.; Sotiropoulos, F. Experimental and computational investigation of local scour around bridge piers. *Adv. Water Resour.* **2012**, *37*, 73–85. [CrossRef]
89. Baykal, C.; Sumer, B.M.; Fuhrman, D.R.; Jacobsen, N.G.; Fredsøe, J. Numerical investigation of flow and scour around a vertical circular cylinder. *Philos. Trans. R. Soc. A* **2015**, *373*, 20140104. [CrossRef]
90. Quezada, M.; Tamburrino, A.; Niño, Y. Numerical simulation of scour around circular piles due to unsteady currents and oscillatory flows. *Eng. Appl. Comput. Fluid Mech.* **2018**, *12*, 354–374. [CrossRef]
91. Link, O.; González, C.; Maldonado, M.; Escauriaza, C. Coherent structure dynamics and sediment motion around a cylindrical pier in developing scour holes. *Acta Geophys.* **2012**, *60*, 1689–1719. [CrossRef]
92. Zhou, L. Numerical Modeling of Scour in Steady Flows. Ph.D. Thesis, École centrale de Lyon, Lyon, France, 2017. Available online: https://tel.archives-ouvertes.fr/tel-01598600/document (accessed on 9 August 2018).
93. Nagel, T. Numerical Study of Multi-Scale Flow Sediment, Structure Interactions Using a Multiphase Approach. Ph.D. Thesis, Université de Grenoble-Alpes, Grenoble, France, 2018.
94. Jenkins, J.T.; Hanes, D.M. Collisional sheet flows of sediment driven by a turbulent fluid. *J. Fluid Mech.* **1998**, *370*, 29–52. [CrossRef]
95. Chauchat, J.; Cheng, Z.; Nagel, T.; Bonamy, C.; Hsu, T.-J. Sedfoam- 2.0: A 3-D two-phase flow numerical model for sediment transport. *Geosci. Model Dev.* **2017**, *10*, 4367–4392. [CrossRef]
96. Cheng, Z.; Hsu, T.-J.; Chauchat, J. An eulerian two-phase model for steady sheet flow using large-eddy simulation methodology. *Adv. Water Resour.* **2018**, *111* (Suppl. C), 205–223. [CrossRef]
97. Kim, Y.; Cheng, Z.; Hsu, T.-J.; Chauchat, J. A numerical study of sheet flow under monochromatic non-breaking waves using a free surface resolving Eulerian two-phase flow model. *J. Geophys. Res. Oceans* **2018**, *123*, 4693–4719. [CrossRef]
98. Juez, C.; Lacasta, A.; Murillo, J.; García-Navarro, P. An efficient GPU implementation for a faster simulation of unsteady bed-load transport. *J. Hydraul. Res.* **2016**, *54*, 275–288. [CrossRef]
99. Flint, M.M.; Fringer, O.; Billington, S.L.; Freyberg, D.; Diffenbaugh, N.S. Historical analysis of hydraulic bridge collapses in the continental United States. *J. Infrastruct. Syst.* **2017**, *23*, 04017005. [CrossRef]
100. Tubaldi, E.; Macorini, L.; Izzuddin, B.A.; Manes, C.; Laio, F. A framework for probabilistic assessment of clear-water scour around bridge piers. *Struct. Saf.* **2017**, *69*, 11–22. [CrossRef]
101. Fischer, E.M.; Knutti, R. Observed heavy precipitation increase confirms theory and early models. *Nat. Clim. Chang.* **2016**, *6*, 986. [CrossRef]
102. Grayson, R.; Blöschl, G. *Spatial Patterns in Catchment Hydrology: Observations and Modeling*; CUP Archive: Cambridge, UK, 2001.
103. Kirchner, J.W. Getting the right answers for the right reasons: Linking measurements, analyses; models to advance the science of hydrology. *Water Resour. Res.* **2006**, *42*. [CrossRef]
104. Manfreda, S.; Mita, L.; Dal Sasso, S.F.; Samela, C.; Mancusi, L. Exploiting the use of physical information for the calibration of a lumped hydrological model. *Hydrol. Process.* **2017**, *32*, 1420–1433. [CrossRef]
105. Pizarro, A.; Samela, C.; Fiorentino, M.; Link, O.; Manfreda, S. BRISENT: An entropy-based model for bridge-pier scour estimation under complex hydraulic scenarios. *Water* **2017**, *9*, 889. [CrossRef]
106. Guan, D.; Melville, B.W. and Friedrich, H. Live-bed scour at submerged weirs. *J. Hydraul. Eng.* **2014**, *141*, 04014071. [CrossRef]
107. Fox, J.F.; Papanicolaou, A.N.; Hobbs, B.; Kramer, C.; Kjos, L. Fluid-sediment dynamics around a barb: An experimental case study of a hydraulic structure for the Pacific Northwest. *Can. J. Civil Eng.* **2005**, *32*, 853–867. [CrossRef]

108. Fox, J.F.; Papanicolaou, A.N.; Kjos, L. Eddy taxonomy methodology around a submerged barb obstacle within a fixed rough bed. *J. Eng. Mech.* **2005**, *131*, 1082–1101. [CrossRef]
109. Papanicolaou, A.N.; Bressan, F.; Fox, J.; Kramer, C.K.; Kjos, L. Role of structure submergence on scour evolution in gravel bed rivers: Application to slope-crested structures. *J. Hydraul. Eng.* **2018**, *144*, 03117008. [CrossRef]
110. Duan, J.G.; He, L.; Fu, X.; Wang, Q. Mean flow and turbulence around experimental spur dike. *Adv. Water Resour.* **2009**, *32*, 1717–1725. [CrossRef]
111. Kondolf, G.M. Hungry water: Effects of dams and gravel mining on river channels. *Environ. Manag.* **1997**, *21*, 533–551. [CrossRef]

Article

Experimental Investigation of Local Scour Protection for Cylindrical Bridge Piers Using Anti-Scour Collars

Shunyi Wang, Kai Wei *, Zhonghui Shen and Qiqi Xiang

Department of Bridge Engineering, Southwest Jiaotong University, Chengdu 610031, China
* Correspondence: kaiwei@home.swjtu.edu.cn; Tel.: +86-182-0286-7260

Received: 18 June 2019; Accepted: 18 July 2019; Published: 21 July 2019

Abstract: Local scour of bridge piers is one of the main threats responsible for bridge damage. Adopting scour countermeasures to protect bridge foundations from scour has become an important issue for the design and maintenance of bridges located in erodible sediment beds. This paper focuses on the protective effect of one active countermeasure named an "anti-scour collar" on local scour around the commonly used cylindrical bridge pier. A cylindrical pier model was set up in a current flume. River sand with a median particle size of 0.324 mm was selected and used as the sediment in the basin. A live-bed scour experimental program was carried out to study the protective effect of an anti-scour collar by comparing the local scour at a cylindrical bridge pier model with and without collar. The effects of three design parameters including collar installation height, collar external diameter and collar protection range, on the scour depth and scour development were investigated parametrically. According to the experimental results, it can be concluded that: the application of an anti-scour collar alleviates the local scour at the pier effectively; and the protection effect decreases with an increase in the collar installation height, but increases with an increase in the collar external diameter and the protection range. Design suggestions for improving the scour protective effect of the anti-scour collar are summarized and of great practical guiding significance to the development of anti-scour collars for bridge piers.

Keywords: cylindrical pier; local scour; anti-scour collar; bridge; experiment; scour protection; scour development; optimal design

1. Introduction

The local scour of bridge piers is one of the main natural causes to bridge structure damage [1,2]. A survey conducted by the US Federal Highway Administration (FHWA) in 1973 showed that more than 20% of pier damage and 70% of abutment damage in a total number of 383 bridge accidents were caused by scour [3]. Wardhana and Hadipriono [4] studied over 500 instances of bridge damage that occurred between the year 1989 and 2000 in the United States and found that more than 50% of the bridge failures were caused by scour due to flood. In a Transportation Research Board's (TRB's) National Cooperative Highway Research Program (NCHRP) report, scour was implicated in up to 60% of bridge failures [5]. Yi et al. [6] reviewed 160 bridge collapse accidents that occurred from the year 2000 to 2014 in China and found that more than 30% of the bridge accidents were caused by scour. Scour is a natural phenomenon caused by the erosive action of flowing streams on erodible beds. Scour can remove the sediment around the bridge foundation, change the structural natural frequencies, reduce the lateral bearing capacity of the foundation and even reduce the seismic capacity of the bridge [7–11].

Scour is usually classified as general scour, contraction scour or local scour. Compared with the first two types of scour, the local scour is generally taken as the main hazard responsible for bridge damage [2,12]. Local scour is the flow erosion due to the blockage of the bridge pier or foundation, where sediment is carried away from the bridge piers or foundations and the scour hole is formed near the piers or foundations [1]. Many researchers have carried out theoretical analysis and experimental research on the local scour of piers. Ataie-Ashtiani and Beheshti [13] studied the local scour around pile groups under steady clear-water scour conditions experimentally. Zhang et al. [14] explored the local scour mechanism of cylindrical piers. Khosronejad et al. [15] carried out experiments and numerical simulations to investigate the clear-water scour around three bridge piers with different cross sections. Sumer et al. [16] summarized some recent research about scour around coastal structures, including flow and scour processes with the subheadings, and sediment behavior close to the structure with the subheadings and further examined the local scour around a pile subject to combined waves and current through testing [17]. Bouratsis et al. [18] carried out a quantitative spatio-temporal characterization study on scour at the base of a cylinder.

Local scour of a bridge foundation is almost inevitable in the actual river or ocean environment. Therefore, adopting some appropriate scour countermeasures to reduce the local scour hole around the bridge pier and protect the foundation of the bridge structure has become an important issue for the design and maintenance of bridges located in the erodible sediment beds. Traditional countermeasures used in bridge engineering can be divided into two categories: (a) passive countermeasures and (b) active countermeasures [2]. Passive countermeasures mainly include: the enlarged foundation [19], the concrete protection [20,21], the partial grouted riprap [5] and the riprap [22,23]. These measures aim to increase the resistance capacity of the bridge foundation to the inflow shear stress by laying a protective layer on the riverbed around the bridge foundation, so as to protect the erosion-prone sediment in the lower layer. Passive countermeasures have been widely used in current engineering practice, but they cannot eliminate the fundamental causes of scour. On the contrary, active countermeasures focus on the disturbing the flow field and thereby reducing the effect of inflow, and thus attract research interest. Some active countermeasures reported in the previous research consist of: the anti-scour collar [24,25], the ring-wing pier [26], the slot [27] and the sacrificial piles [28], etc. The active measures aim to improve the original hydraulic characteristics, such as changing the direction and reducing the flow speed, to weaken the effect of downflow and horseshoe vortexes and hence protect the bridge foundation from local scour.

Tafarojnoruz et al. [29] reviewed various flow-altering scour countermeasures and concluded that the difficulties in the field applications of active countermeasures limited their practical use. Among the active methods, the anti-scour collar can be a useful and simple countermeasure for local scour around cylindrical bridge piers, which can reduce the horseshoe vortex strength by diverting and acting as an obstacle in the downflow passage. Previous literature shows that placing a collar around the bridge pier can reduce the depth of scour [30]. The anti-scour collar external diameter and installation height proved to be the most important factors influencing its protection efficiency [2]. Chen et al. [31] proposed a new type of collar named hooked-collar, and found that that maximal downflow is highly reduced along with a corresponding decrease in horseshoe vortex strength for the experiments with the hooked-collar, compared to cases without the collar. Chiew [24] investigated the local scour of a pier using a collar, a slot, and their combination and found that placing either a collar or a slot around the bridge pier can reduce the scour depth, and their combination can further reduce the scour depth. Gaudio et al. [32] carried out clear-water tests on the bridge pier scour with various combinations of collars with some other countermeasures and concluded that an improper combination of two countermeasures may be less effective than each individual countermeasure. Similar studies about collars in reducing the scour depth can be found in several studies [25,27,29,33–35]. However, these studies seldom focused on the distribution of the scour depth around piers and how to design the anti-scour collar and optimize its protective effect on the local scour of piers.

In order to address the issues related to the optimal design of anti-scour collars and the local scour depth distribution around cylindrical bridge piers with anti-scour collars, this study is organized as follows: (1) A scour experimental program is firstly presented to study the protective effect of anti-scour collars at a local scour of a cylindrical bridge pier model; (2) local scour of the pier model with and without anti-scour collar is studied experimentally; (3) the effect of collar installation height, collar external diameter and collar protection range on its scour protective effect were investigated; and (4) the design of anti-scour collars on the characteristics and development of local scour hole around the pier model are discussed parametrically.

2. Experimental Setup

2.1. Testing Facilities

The experiments were carried out in the large-sized current flume at Southwest Jiaotong University. As shown in Figure 1, the flume is 60 m long, 2 m wide and 1.8 m high with side walls comprised of steel frames and toughened glass, which provides a clear and wide visual range during the testing. There is a scour section in the large-sized current flume, which is 7.5 m long, 2 m wide, 0.35 m deep. A test rack is installed above the sand pool and used to fix the testing physical models. A flow pump is installed at the inflow point to generate steady current in the flume. An outlet is set in the tail part. The water flows into the outlet through an annular channel under the flume and backs to the pump to realize the flow circulation.

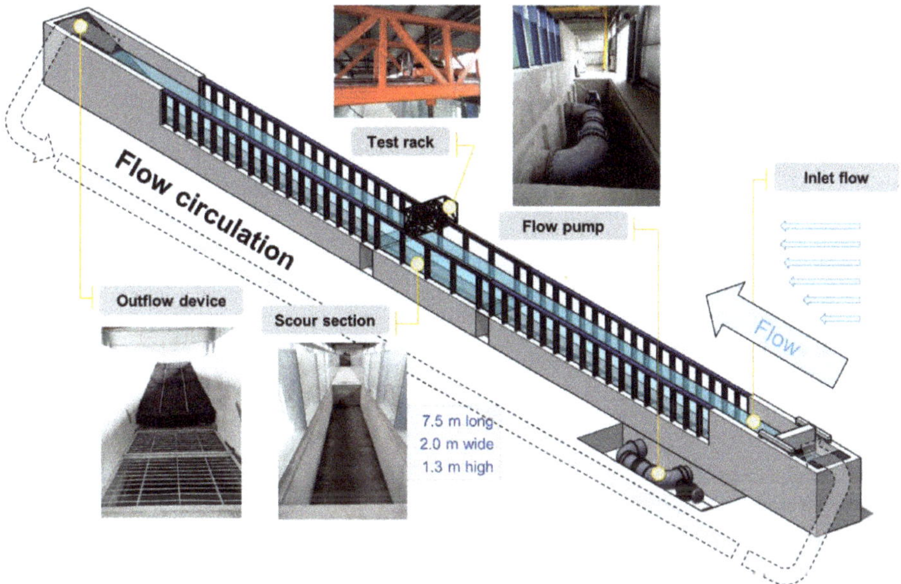

Figure 1. The testing facility: Photo and detail of the current flume.

The layout of the scour testing facilities is shown in Figure 2. As shown in Figure 2a, the bridge pier model is inserted 20 cm into the riverbed. The top of the pier model is fixed with the test rack by the bolts to ensure the structural rigidity in the scour testing process. An Acoustic Doppler Velocimeter (ADV) was used to measure the flow velocity, and the maximum allowable flow velocity was 4 m/s. As shown in Figure 2b,c, ADV can be moved up and down to measure the velocity distribution along the water depth in this test. The ADV device in this study can also be used to measure the scour depth and obtain the shape of the scour pit around the foundation. Figure 2d,e shows the setup of ADV for the measurement of the range and depth of scour hole around piers.

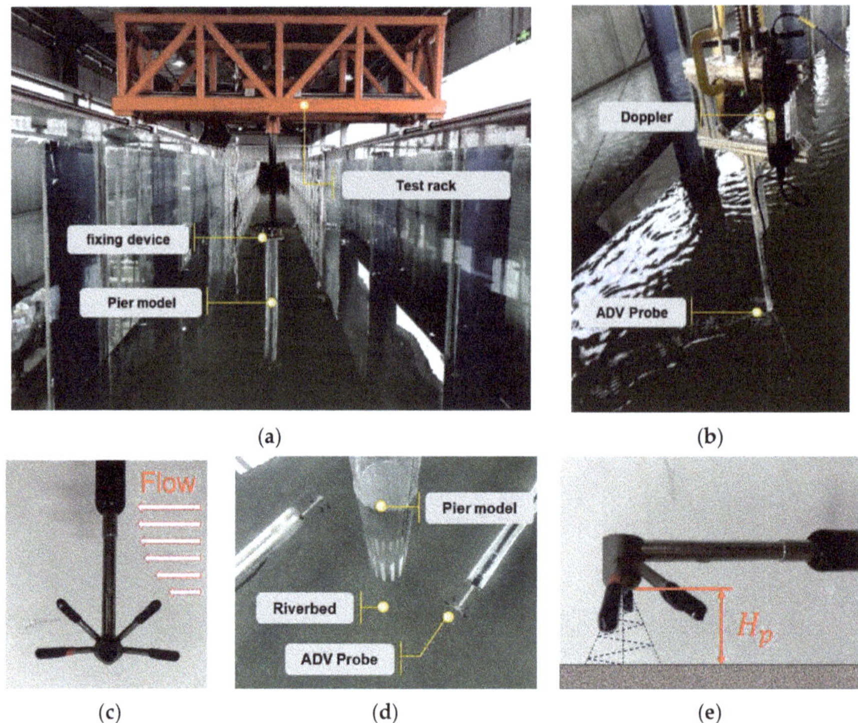

Figure 2. Layout of the testing facilities: (**a**) installation of test model; (**b**) Acoustic Doppler Velocimeter (ADV) flow velocity measurement; (**c**) velocity measurement probe details; (**d**) scour depth measurement by ADV; (**e**) depth measurement probe details.

2.2. Description of Anti-Scour Collar

The commonly used highway bridges in the valley river areas always adopt cylindrical piers with a diameter of 2 m, as illustrated in Figure 3. A cylindrical pile with a diameter of 2 m was also used for scour analysis in some literature [36]. In this study, a length scale of 1:20 is adopted to design the model and flow conditions. The pier model was made of cylindrical acrylic tube with an outer diameter of 10 cm and a thickness of 1 cm.

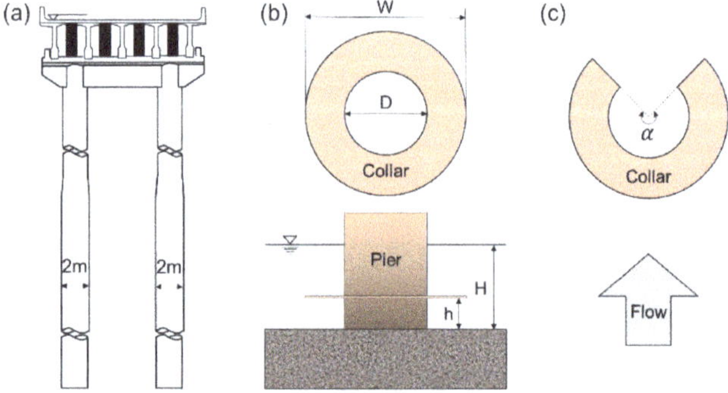

Figure 3. Schematic diagram of anti-scour collar: (**a**) the pier with 2 m diameter in highway bridges; (**b**) the installation of 360° collar in pier; and (**c**) indication of collar with a protection angle a (°).

The downward flow and horseshoe vortex are two main hydraulic effects that affect the local scour. The anti-scour collar installed at the bottom of the pier can protect the sediment particles far from erosion by reducing the downward flow particle velocity and the horseshoe vortex.

In this study, the anti-scour collar used in the physical model tests is an acrylic circle plate with a thickness of 0.5 cm. As shown in Figure 3, three key design parameters of the collar need to be addressed, including collar installation height h (a distance from the collar to the riverbed), collar external diameter W and collar protection range α. D is the outer diameter of pier, H is the water depth.

2.3. Flow Velocity Determination

Figure 4a shows the flow velocity curve measured by the ADV probe at the position of 10 cm below water surface with a water depth of 50 cm and a flow velocity of 0.35 m/s. The sampling frequency of the ADV is 100 Hz. At each height, the mean value of three 20 s time histories of the flow velocity was calculated and used to remove the effect of the noise of the ADV velocity data.

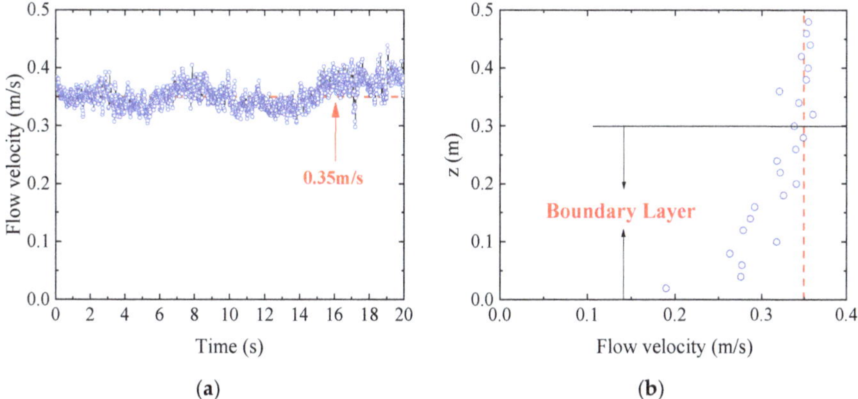

Figure 4. The flow velocity measurement by ADV: (**a**) determination of underwater 10 cm flow velocity; and (**b**) velocity distribution at different heights.

The flow velocity at different heights was measured by the above method with height increment of 2 cm. The velocity distribution along the height z is shown in Figure 4b. The height of the boundary layer is about 30 cm. When the distance between the ADV probe and water surface is less than 20 cm, the measured velocity is basically stable near the maximum value. This phenomenon has also been found in previous similar studies [37].

3. Experimental Program

3.1. Threshold Velocity Calculation of Sediment

Natural uniform river sand was selected as the riverbed material for the tests. The median particle size of the sand is 0.324 mm, and the density of the river sediment is 2.65 g/cm^3. Before the start of each test, the riverbed sediment in the scour section was smoothed and kept still for a period of time. Only one type of sand was used in all the tests carried out in this study. The water depth and flow velocity were kept constant at 0.5 m and 0.35 m/s, respectively.

The critical Shields parameter was calculated by the improved empirical formula by Soulsby and Whitehouse [38,39]

$$\theta_{cr} = \frac{0.30}{1+1.2D_*} + 0.055[1 - \exp(-0.020D_*)] \quad (1)$$

where the non-dimensional particle size D_* is defined as $D_* = [g(s-1)/v^2]^{1/3}d_{50}$ with ν (=10^{-6} m^2/s) being the kinematic viscosity of water. The Shields parameter due to total friction was obtained according to the measured velocity profile. The Shields parameter θ_s due to skin friction is defined as

$$\theta_s = \frac{\tau_s}{\rho g(s-1)d_{50}} = \frac{U_{fs}^2}{g(s-1)d_{50}} \tag{2}$$

where τ_s is the shear stress due to skin friction experienced by the sea bed from the flow, ρ is the water density, g is the acceleration due to gravity, s is the specific gravity of sand, d_{50} is the median particle size of sand, $U_{fs} = (\tau_s/\rho)^{1/2}$ is friction velocity associated with skin friction. The shear stress due to skin friction is calculated by the empirical formula for flat sand surface [39]

$$\tau_s = \rho C_D \overline{U}^2 \tag{3}$$

where the logarithmic relationship for C_D is $C_D = \{\kappa/[ln(z_{0s}/h) + 1]\}^2$, κ (=0.4) is the Karman constant, $\bar{U}$ is the depth averaged flow velocity, z_{0s} (=d_{50}/12) is the roughness height due to skin friction and h is the water depth.

Table 1 summarizes the parameters used in the tests based on the above calculation and the observed sediment start-up phenomena in the experiment. The threshold velocity of sediment at different water depths calculated by the above equations is shown in Figure 5. The threshold velocity of sediment at the water depth of 0.5 m is about 0.303 m/s. Considering that the ripple development effect may affect the scour tests, the test flow velocity is set to be 0.35 m/s, which is only slightly larger than the actual threshold velocity of sediment. The sediment bed was carefully leveled and flattened before the test to ensure that the initial riverbed surface was the same under every working condition. Under such conditions, we observed no ripples in front of the pier model, only ripples behind the pier. In the test, the authors increased the flow velocity incrementally and observed the start-up of sediments. The observed threshold velocity in the test was very close to the threshold velocity calculated by above formula. It can be concluded that this formula is applicable to the sediments used in this paper. According to the skin friction Shields parameter, the tests are all under live-bed conditions. Live-bed scour was also confirmed by the sand ripples observed in the latter tests.

Table 1. Parameters of sediment used in the tests.

Median particle size of sand d_{50} (mm)	0.324
d_{10} (mm)	0.114
d_{30} (mm)	0.228
d_{60} (mm)	0.369
Uniformity coefficient $C_u = d_{60}/d_{10}$	3.237
Coefficient of curvature $C_c = {d_{30}}^2/(d_{60*}\ d_{10})$	1.236
Specific gravity s	2.65
Critical Shields parameter θ_{cr}	0.036
Water depth h (m)	0.5
Approximate depth averaged velocity $\bar{U}$ (m/s)	0.325 (Maximum is 0.35)
Shields parameter due to skin friction θ_s	0.041
Logarithmic relationship C_D	0.002
Roughness length for skin friction z_{0s} (m)	2.7×10^{-5}
Angle of repose ϕ (°)	32.9
Threshold velocity U (m/s)	0.303

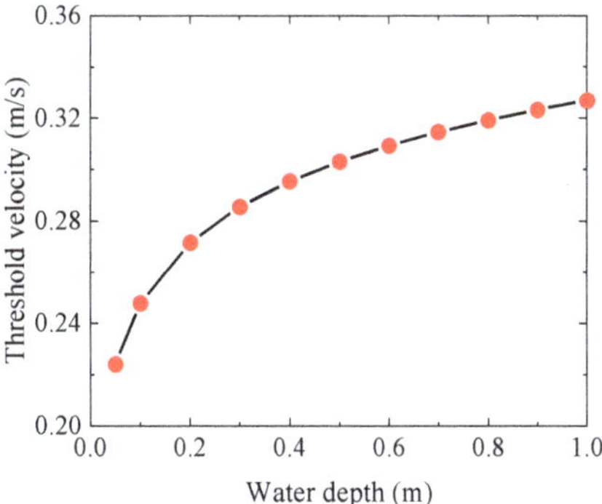

Figure 5. The threshold velocity of selected sediment at different water depths.

3.2. Calculation of Equilibrium Scour Time

Melville and Chiew [40] considered the temporal development of clear-water local scour depth at cylindrical bridge piers in uniform sand beds and put forward a formula to estimate the time taken for equilibrium scour depth development. Bateni et al. [41] presented a Bayesian neural network technique to predict the equilibrium scour depth around a bridge pier and the time variation of scour depth. Melville et al. [42] has divided the local scour process into three stages: initial stage, main scour stage and equilibrium condition. In order to understand the duration required for the scour test, time scale of scour around a vertical pile of a height the same as the water depth is calculated based on the following empirical formula proposed by Sumer et al. [43]

$$S(t) = S_0(1 - \exp^{-t/T}) \tag{4}$$

where S is the scour depth, S_0 is the equilibrium scour depth, t is time and T is defined as the time scale of scour. Sumer et al. [43] found that time scale of scour around a vertical slender pile follows

$$T = \frac{D^2}{\left[g(s-1)d_{50}^3\right]^{1/2}} T^* \tag{5}$$

$$T^* = \frac{\delta}{2000D} \theta_s^{-2.2} \tag{6}$$

where D is the representative dimension of the vertical pile (diameter for a circular pile or dimension perpendicular to flow for a rectangular cylinder), δ is the boundary layer thickness and T^* is the non dimensional time scale of scour. By using δ = 0.30 m, Equations (5) and (6), the time scale for D = 0.1 m is T = 720. By substituting $S(t)/S_0$ = 0.95 and the above calculated time scales into Equation (4), the time taken for scour reaching 95% of equilibrium scour depth for D = 0.1 m is estimated as t= 36 min.

Yang et al. [44] carried out a study on the evolution of hydrodynamic characteristics with scour holes developing around pile groups, which demonstrates a high scouring rate for the early stage and the scouring phenomenon almost reached equilibrium after 240 min. Zhang et al. [45] established a set of three-dimensional numerical models to investigate the mechanisms of local scour around three adjacent piles with different arrangements under steady currents by as short as 500 s scour time. The total scour testing time of each test is thus set to be 120 min, which is much larger than the calculated approximate equilibrium scour time.

3.3. List of Testing Conditions

For the conditions given in Table 1, a series of experiments was conducted to investigate the effect of anti-scour collar parameters. Table 2 lists all the working conditions carried out in this study. Based on the scour test of a single cylindrical pier without collar, the optimal design of the collar is carried out to achieve the maximum protection efficiency. Three collar design parameters (collar installation height, external diameter and protection range) are considered to design the experimental scheme. According to the results of threshold velocity and approximate equilibrium scour time, a live-bed scour testing scheme was adopted in this test. Under various tests of cylindrical piers with different water depth and flow velocity, the working conditions of 0.35 m/s flow velocity and 0.5 m water depth were selected.

Table 2. Parameters for the test conditions.

Case	Anti-Scour Collar	Installation Height h/H	External Diameter W/D	Protection Angle α (°)
1	w/o.	—	—	—
2	w.	0.10	2.5	360
3	w.	0.04		
4	w.	0		
5	w.	−0.04		
6	w.	0.04	3.0	360
7	w.		2.0	
8	w.		1.5	
9	w.	0.04	2.5	270
10	w.			180

As shown in Figure 6, 12 measuring points were evenly arranged around the pier by 30° to monitor the scour depth and measure the development of the scour hole for different testing cases. In the experiment, a GoPro camera was used for data acquisition of scour depth around the pier model in water, and an ADV velocimeter is used to judge the evenness of the initial riverbed and to measure the length and width of the scour hole. The measured scour width is the distance between the edges of the scour hole on both sides, while the scour length is the distance between the edges of the scour hole upstream and the highest point of the riverbed behind the pier. In the 60 min, the measuring point reading was recorded every 10 min, and then the data was recorded every 30 min.

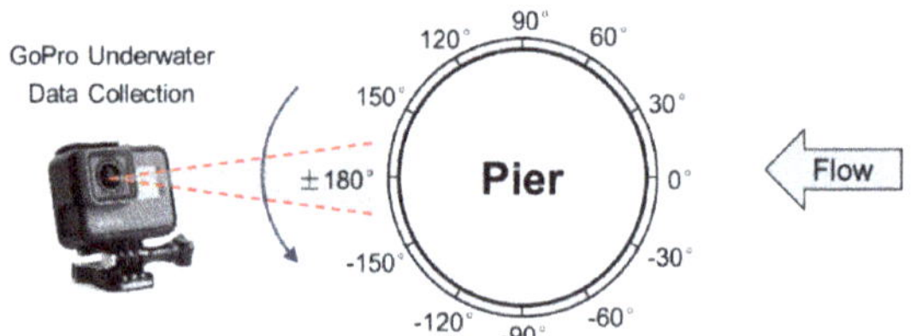

Figure 6. The layout of measuring points around the cylindrical pier.

4. Results and Discussion

The test results show that the maximum scouring depth can be replaced by the maximum value of the measuring points reading. A local scour test of the cylindrical pier model without an anti-scour collar (the unprotected case) was conducted first to obtain the reference scour characteristics for the pier without any protection. The rest of the testing cases of the pier model with different anti-scour collar designs were then carried out and compared with the unprotected case to show the scour protection efficiency (factor β in percent).

4.1. Characteristics of Local Scour for a Cylindrical Pier without Protection

In order to eliminate the effect of geometry of the pier on the results [46], the dimensionless scour depth *S/D* measured at twelve measuring points under different scouring times are plotted in Figure 7a, respectively. The shape of the scour hole around the pier model is symmetric around the current flow direction. Because the flow velocity and bed shear stress behind the pier were smaller than that in front of the pier due to the shadowing effect of the pier, the scour depth is decreasing from the upstream (0°) to the downstream pier side (180°). The scour hole developed continuously with the testing time. At 10 min of scour, the maximum scour depth occurred at the 60° side of the pier, while the minimum scour depth appeared at the 180° position, which is just in the back of the pier. After 10 min, the position of the measured maximum scour depth gradually moved from the side to the front of the pier, and the final maximum scour depth at 120 min occurred at the front of the pier.

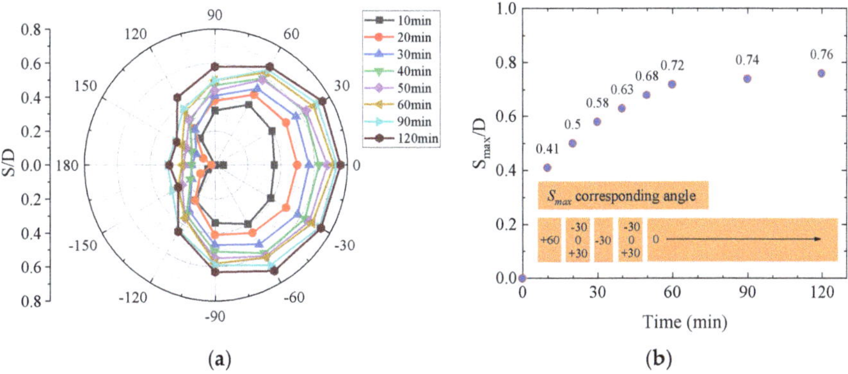

Figure 7. Depth scour data of a single cylindrical pier: (**a**) the time-varied scour development at twelve measuring points; and (**b**) the maximum scour depth of a single pier.

The maximum dimensionless scour depth *S/D* around the pier as a function of scour testing time is given in Figure 7b. In the first 30 min, scour occurred quickly; the maximum scour depth and the length and width of the scour hole increased rapidly. The growth rate of the maximum scour depth slowed down with the increase of testing time and reached close to zero in the approximate equilibrium condition. Similar phenomena have also been reported in the references [31,43]. The maximum dimensionless scour depth *S* of the sediment hole around the pier is 0.58 at 30 min and increases to 0.72 at 60 min, which is more than 90% of the scour depth at 120 min. Although scour still occurred in the later period, the growth of both scour depth and hole's range are very small. The minimum scour depth occurred at 150° at 120 min. As the photos of the scour hole at 120 min shown in Figure 8, the minimum scour depth is approximately 30% of the maximum scour depth of the final scour hole. According to the development of scour depth as a function of the experimental time for all working conditions, although the scour depth will still increase after 120 min in the tests, the increase of depth is very slight and the increasing rate is significantly slower than that of the first 60 min, which can be treated as the time to approximate equilibrium condition.

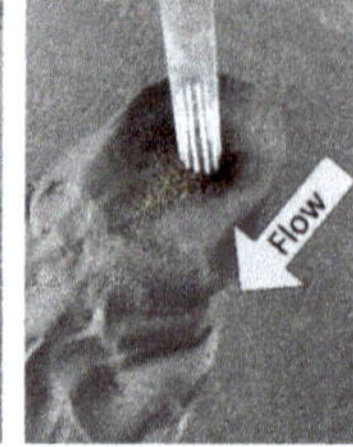

Figure 8. The scour hole in the approximate equilibrium condition.

The maximum bed shear stress and the maximum scour depth appeared in front of the upstream side of the pier [40,43]. Due to the blocking effect of the pier, turbulent flow is generated around the pier and accelerated the development of scour in front of the pier. They all found that the maximum scour depth occurred in the upstream front of the cylindrical pier.

4.2. Effect of Collar Installation Height

Testing of Cases 2 to 5 was conducted to investigate the influence of collar installation height on the scour protection effect. In these cases, the external diameter W/D and protection angle a of the anti-scour collar equaled 2.5 and 360°, respectively. The installation height of the anti-scour collar h/H was set to be 0.1, 0.04, 0 and −0.04 for Case 2 to 5, respectively. The negative height indicates that the collar was located under the riverbed. The development of maximum scour depth S_{max}/D at different collar installation heights is shown in Figure 9a. The anti-scour collar reduced the local scour depth and had a protective effect on the cylindrical pier regardless of the installation height. The results of scour depth as a function of collar installation height agree with the experimental results of Chiew [24] and Ettema [30].

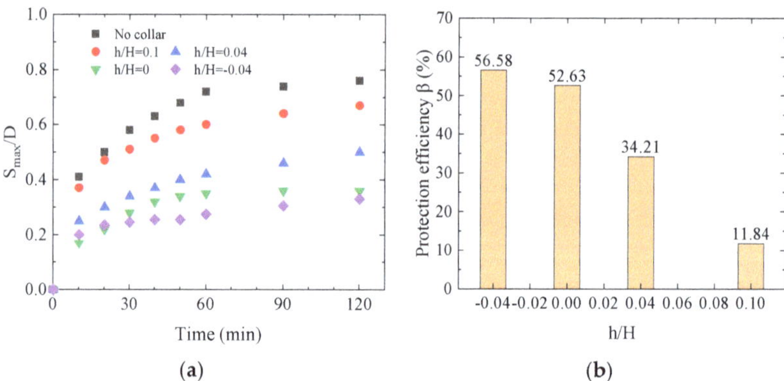

Figure 9. Scour protection effect at different collar installation height h/H: (**a**) the maximum scour depth development; and (**b**) the protection efficiency.

In order to assess the efficiency of the collar, we define a protection efficiency factor $\beta = (S_{t,max} - S_i)/S_{t,max}$, where $S_{t,max}$ is the maximum scour depth of the unprotected case in 120 min, S_i is the maximum scour depth of Case i in 120 min. The larger the factor β is, the better the protective effect is. The protection efficiency factor β versus the installation height is given in Figure 9b. The photos of scour holes for Case 2 to 5 at t = 120 min are shown in Figure 10. The development of scour depth around the pier model as a function of collar installation height is illustrated in Figure 11.

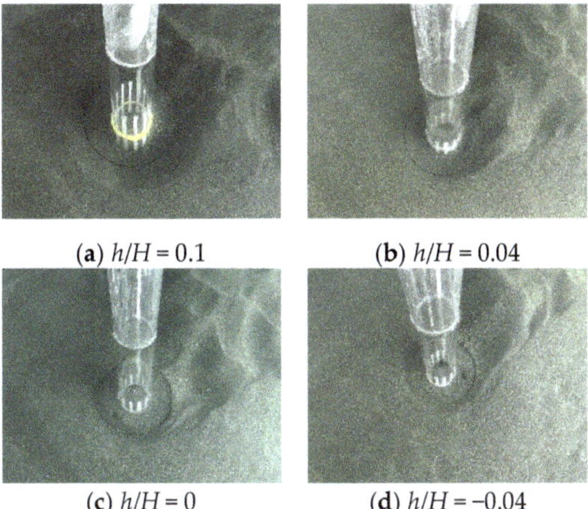

Figure 10. The scour hole form of different collar installation height: (**a**–**d**).

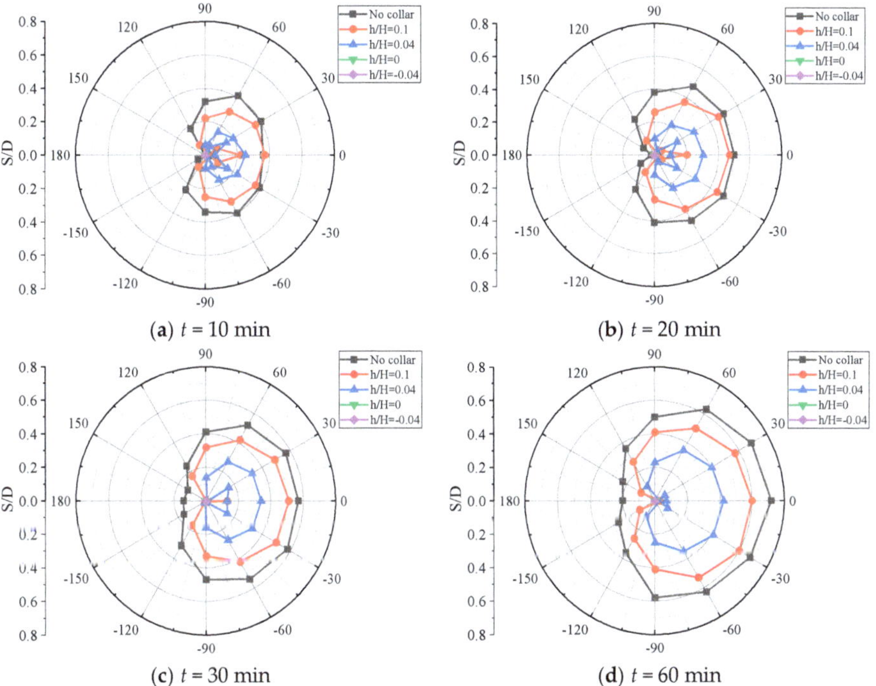

Figure 11. *Cont.*

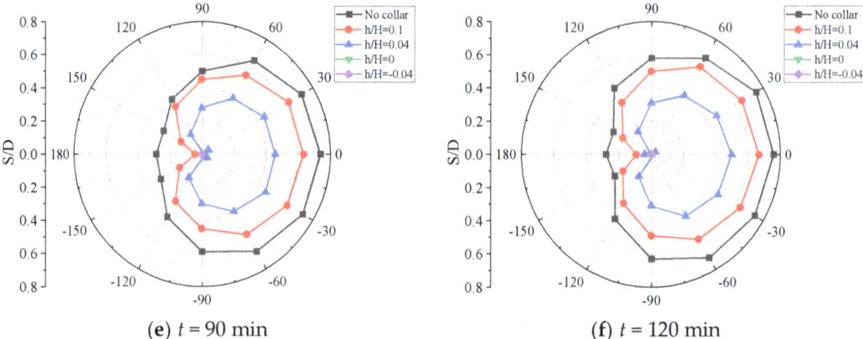

Figure 11. The dimensionless scour depth development of different collar installation height h/H at each measuring point.

When the collar was located above riverbed, the closer the collar was to the riverbed, the better the protective effect. The protective effect is increased with the decrease of the installation height for all the measuring points around the pier. The maximum scour depth occurred at the 0° point, which located in front of the pier. The location of the maximum scour is the same to the model without an anti-scour collar.

When the anti-scour collar was located at the riverbed surface ($h/H = 0$), there was not any scour of sediment found at all twelve measuring points in the first 120 min testing. The anti-scour collar protected the sediment under the collar from being washed away, but the sediment at the downstream edge of the collar was washed away. Although the scour hole was developed to the sediment under the collar with the increase of scour testing time after 120 min, the maximum scour depth always occurred at two side points outside the collar behind the pier.

When the anti-scour collar was embedded into the riverbed ($h/H = -0.04$), the sediment above the collar was quickly removed away in the beginning of the test. The maximum scour depth equaled to the value of h. With the increase of testing time, the maximum scour depth occurred at two side points outside the collar behind the pier, which was similar to that of the case $h/H = 0$. Therefore, the maximum scour depth of $h/H = -0.04$ condition at the beginning stage was larger, and then was smaller than that of $h/H = 0$. According to the comparison of final scour depth, the case with $h/H = -0.04$ has the best protective effect. Moreover, there was almost no scour around the collar edge except that the sediment on the upper part of collar was washed out at the beginning, when the anti-scour collar was embedded into the riverbed.

It should be noted that the anti-scour collar cannot be embedded too deep. Otherwise all the sediment above the collar will be removed. Because of the existence of general scour and natural evolution scour, it is suggested to install the anti-scour collar at the general scour line to prevent local scour as much as possible.

4.3. *Effect of Collar External Diameter*

The effect of the collar external diameter on the scour protective effect was investigated experimentally at a certain collar installation height $h/H = 0.04$. Four anti-scour collars with different external diameters ($W/D = 3.0$, 2.5, 2.0 and 1.5) were considered. The development of maximum scour depth S_{max}/D as a function of the collar external diameter is plotted in Figure 12a. The protection efficiency factor β (defined as $(S_{t,max} - S_i)/S_{t,max}$ in above section) versus the external diameter is shown in Figure 12b. It is clear that the external diameter of anti-scour collar affects the scour protective effect significantly. The protective effect is increased with the increase of the collar external diameter for all twelve measuring points. The scour development related to the increase of collar external diameter has also been reported in the previous studies [33–35].

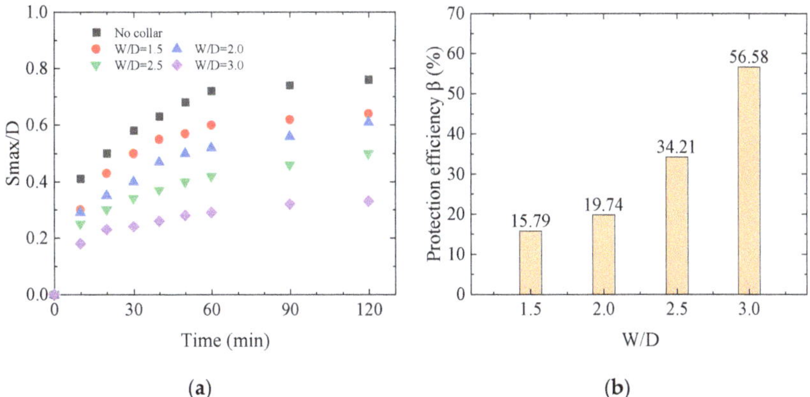

Figure 12. Scour protection effect at different collar external diameter *W/D*: (**a**) the maximum scour depth development; and (**b**) the protection efficiency.

The photos of the scour hole at 120 min for four cases are given in Figure 13, respectively. According to Figure 14, the increase of the collar external diameter could reduce not only the maximum scour depth, but also the scour hole range. When the external diameter *W/D* was set to be 3.0, the edge of the scour hole was located inside the range of the collar. The sediment deposited behind the pier. With the decrease of the collar external diameter, its protective effect was decreased. When the collar external diameter *W/D* was 2.5, the scour hole range was similar with the collar size, which was larger than the case of 3.0 value diameter. With the collar external diameter decreasing, the scour hole was becoming larger than the collar size, and its range was larger.

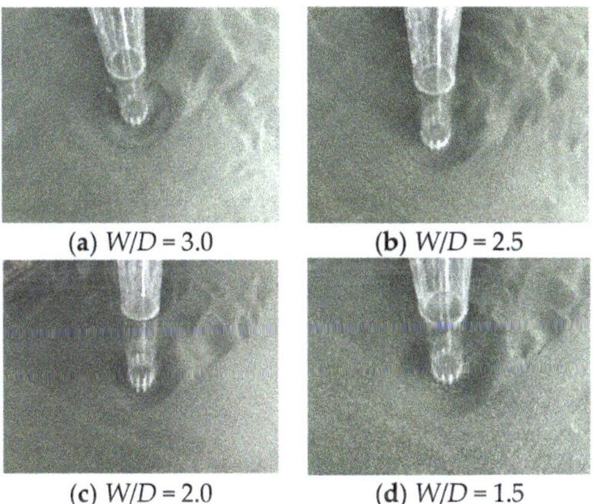

Figure 13. The scour hole form of different collar external diameters at 120 min.

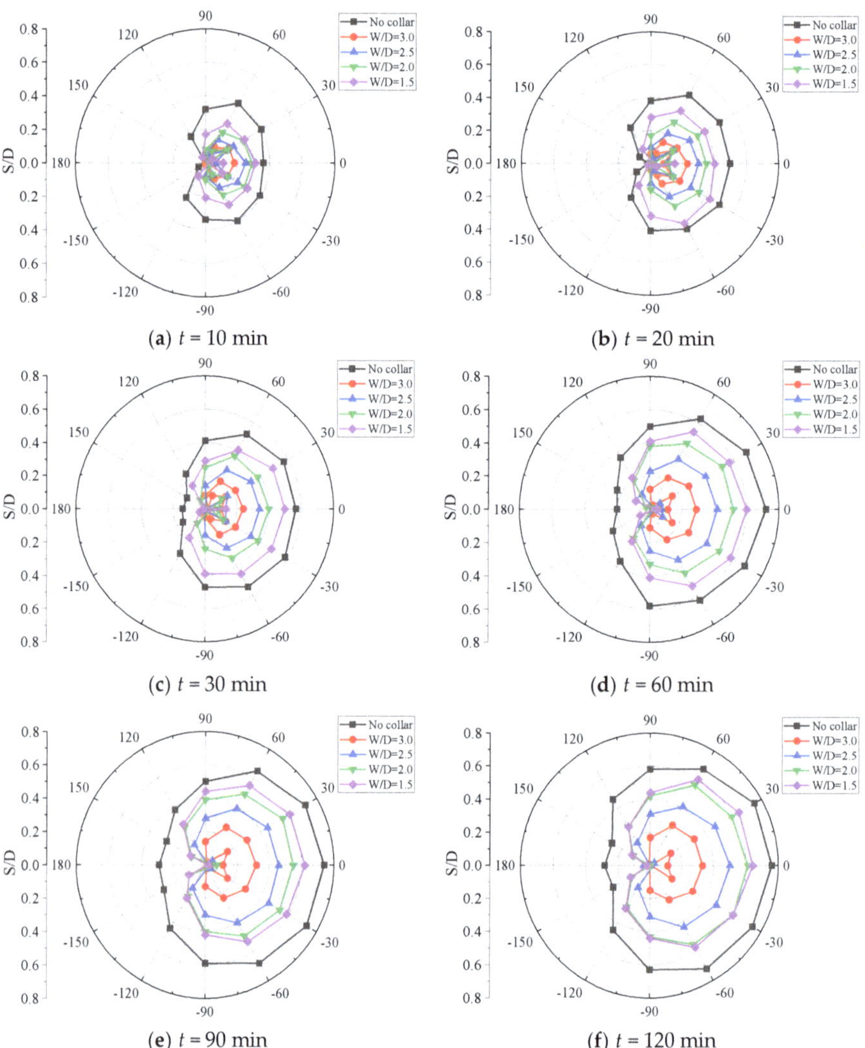

(**a**) t = 10 min (**b**) t = 20 min

(**c**) t = 30 min (**d**) t = 60 min

(**e**) t = 90 min (**f**) t = 120 min

Figure 14. The dimensionless scour depth development of different collar external diameters W/D at each measuring point.

The development of scour depth as a function of scouring time at twelve measuring points is shown in Figure 14 for different anti-scour collar external diameters, respectively. The maximum scour depth occurred in front of the pier regardless of the external diameter. Although anti-scour collars with larger diameters resulted in better protective effects, it is not possible to install a collar with infinite external diameter due to hydraulic and economic reasons. According to this study, an anti-scour collar with an external diameter that equals three times the diameter of the pier can reduce more than 50% of the scour for the pier without any protection. However, it should be noted that there should be a limit between the scour reduction and the cost due to the increase of collar diameter, especially when it is applied in the actual long-span bridge pier with a large diameter [47].

4.4. Effect of Collar Protection Range

Considering downward flow in front of the pier is one of the reasons that causes local scour of sediment. In the existing literature, there are many studies on the installation height and external diameter of anti-scour collars [24,25,27,29–33], but there are few studies focused on the protection range of anti-scour collars. It is interesting to determine if we can protect the sediment by reducing the protective range of the collar. In order to discuss the effect of the protective range of the collar, three protective ranges: 180 (half circle), 270 (3/4 circle) and 360 (full circle) are considered on the basis of anti-scour collars with installation height $h/H = 0.1$ and external diameter $W/D = 2.5$.

The development of maximum scour depth S_{max}/D for different collar protection ranges are given in Figure 15a. The protection efficiency factor β $(=(S_{t,max} - S_i)/S_{t,max})$ versus the protection range of collars is shown in Figure 15b. It can be concluded from the results that the protection range affects the protective effect as well. The scour protective effect of the collar with 360° protection range was the best. With the decrease of protection range, the protection effect was weakened.

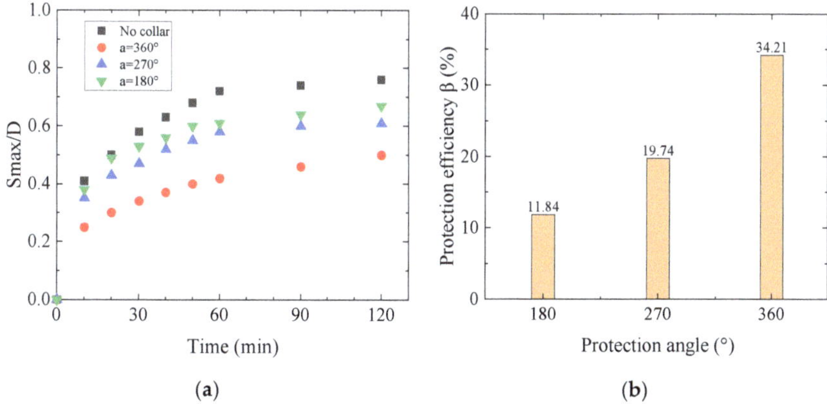

Figure 15. Scour protection effect at different collar protection range: (**a**) the maximum scour depth development; and (**b**) the protection efficiency.

The scour holes formed at 120 min of different protection angles are shown in Figure 16. The photos of scour holes at 120 min for four collar protection ranges are given in Figure 15. The development of scour depth as a function of scouring time at twelve measuring points is shown in Figure 17. When the protection range of the collars decreased from 360° to 270°, the protective efficiency decreased distinctly. When it continued reducing from 270° to 180°, the decrease in protective effect did not greatly change. Therefore, it is recommended to use anti-scour collars with a full circle protection range.

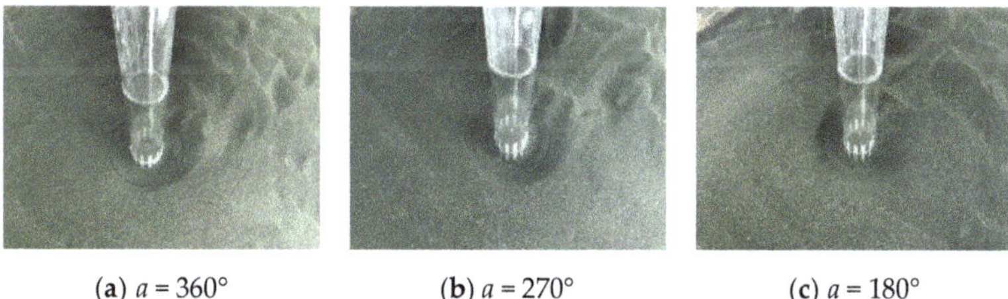

(**a**) $a = 360°$ (**b**) $a = 270°$ (**c**) $a = 180°$

Figure 16. The scour hole form of different collar protection angle at 120 min.

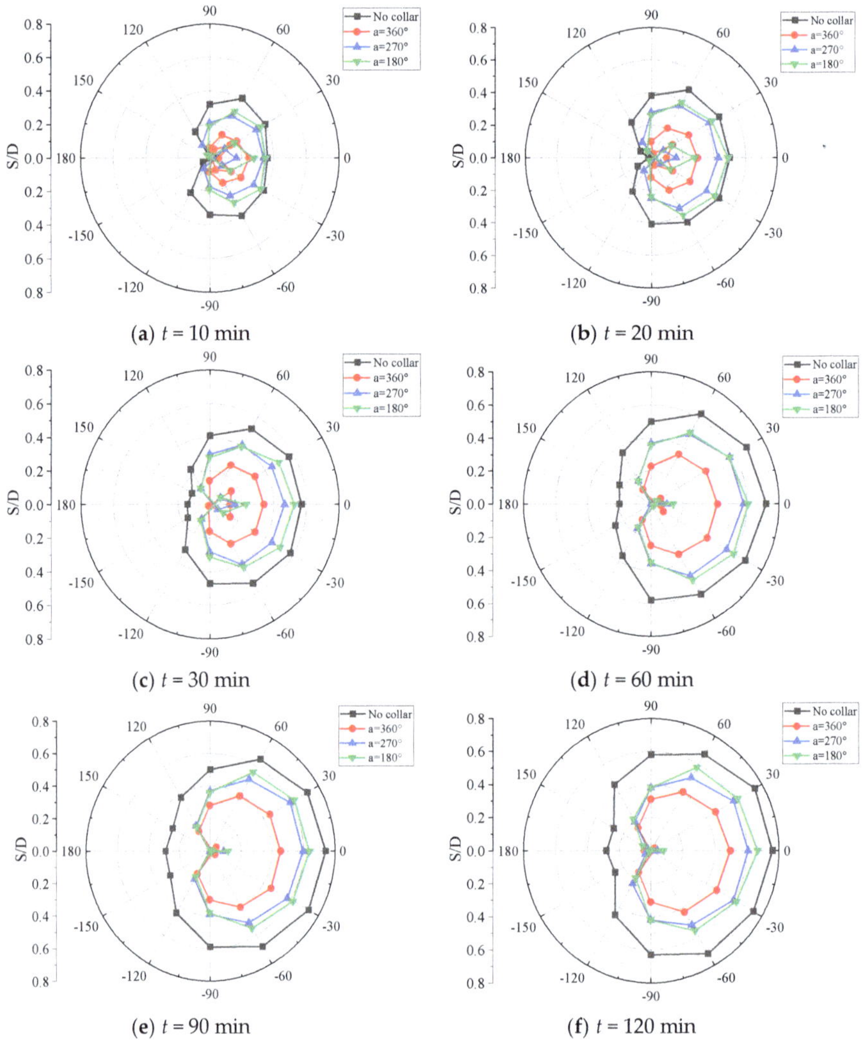

(**a**) t = 10 min (**b**) t = 20 min

(**c**) t = 30 min (**d**) t = 60 min

(**e**) t = 90 min (**f**) t = 120 min

Figure 17. The scour depth development of different collar protection angles at each measuring point.

5. Conclusions

Scour around single cylindrical piers and the effects of anti-scour collars were investigated by carrying out laboratory tests. The scour mechanism and the time-varied development of scour depth were studied. Experimental investigations regarding local scour using anti-scour collars on cylindrical piers were carried out. The protective efficiency of anti-scour collars with different design parameters, including collar installation height, collar external diameter and collar protection range, were investigated and discussed respectively. The main findings of this study can be summarized as follows:

(1) In the scour test of a cylindrical pier without protection, the maximum scour depth first appears in the front side of the pier, then moves to the front of the pier, while the minimum scour depth appears in the downstream side of the pier. The shape of the local scour hole around the pier is approximately symmetric with the flow direction.

(2) The anti-scour collar has clear protective effects on pier scour. The decrease of the installation height and the increase of both collar external diameter and the protection range all contribute to the increase the protection effect of the anti-scour collar.
(3) Three design suggestions for improving the scour protective effect of anti-scour collars on piers are given based on the experimental results: (1) the collar is suggested to be embedded into the riverbed and installed under the general scour level; (2) the collar with an external diameter three times the pier diameter can reduce more than half of the scour depth of a pier without protection; (3) the protection range of the anti-scour collar is suggested to be a full circle of 360° for a better protective effect.

It should be noted that there are still many limitations in this study. The parametric study of anti-scour collars here is carried out only under one thickness of anti-scour collars and one constant current condition. The change of both flow velocity and water depth and the type of sand are not considered. The application of two or more collars is not considered as well. These limitations can be further addressed by experimental or numerical simulation study. Notwithstanding the shortcomings listed above, the experimental results of anti-scour collars are of great practical guiding significance to the optimal design of anti-scour collars and can provide a useful testing reference for further improvement of active scour countermeasures for bridges.

Author Contributions: S.W., K.W., Z.S., Q.X. designed and supervised the experiments; S.W. conducted the analysis on the experimental data; K.W. provided motivation and contributed interpretation of results; S.W. and K.W. wrote the paper; and all authors participated in final review and editing of the paper.

Funding: This research was funded by National Natural Science Foundation of China (Grants No. 51708455), the Fundamental Research Fund for Central Universities (A1920502051907-2-001) and Key Research and Development Project of Sichuan Province (2019YFG0460).

Acknowledgments: The authors sincerely thank the anonymous reviewers for their valuable suggestions in improving the manuscript.

Conflicts of Interest: The authors declare no conflict of interest.

References

1. Barbhuiya, A.K.; Dey, S. Local scour at abutments: A review. *Sadhana* **2004**, *29*, 449–476. [CrossRef]
2. Wang, C.; Yu, X.; Liang, F. A review of bridge scour: Mechanism, estimation, monitoring and countermeasures. *Nat. Hazards* **2017**, *87*, 1881–1906. [CrossRef]
3. Richardson, E.V.; Harrison, L.J.; Richardson, J.R.; Davies, S.R. *Evaluating Scour at Bridges*; Publ. FHWA-IP-90-017, Federal Highway Administration, US Department of Transportation: Washington, DC, USA, 1993.
4. Wardhana, K.; Hadipriono, F.C. Analysis of recent bridge failures in the United States. *J. Perform. Constr. Facil.* **2003**, *17*, 144–150. [CrossRef]
5. Lagasse, P.F. *1998 Scanning Review of European Practice for Bridge Scour and Stream Instability Countermeasures*; NCHRP Research Results Digest-241; Transportation Research Board: Washington, DC, USA, 1999.
6. Yi, R.; Zhou, R. Reason and risk of bridge collapse in recent 15 years. *Transp. Sci. Technol.* **2015**, *5*, 61–64. (In Chinese)
7. Federico, F.; Silvagni, G.; Volpi, F. Scour vulnerability of river bridge piers. *J. Geotech. Geoenviron. Eng.* **2003**, *129*, 890–899. [CrossRef]
8. Ye, A.; Zhang, X.; Liu, W. Effects of riverbed scouring depth on the seismic response of bridges on pile foundations. *China Civ. Eng. J.* **2007**, *40*, 58–62. (In Chinese)
9. Wang, Z.; Dueñas-Osorio, L.; Padgett, J.E. Influence of scour effects on the seismic response of reinforced concrete bridges. *Eng. Struct.* **2014**, *76*, 202–214. [CrossRef]
10. Liang, F.; Zhang, H.; Huang, M. Influence of flood-induced scour on dynamic impedances of pile groups considering the stress history of undrained soft clay. *Soil Dyn. Earthq. Eng.* **2017**, *96*, 76–88. [CrossRef]
11. Zhang, J.; Wei, K.; Qin, S. An efficient numerical model for hydrodynamic added mass of immersed column with arbitrary cross—Section. *Ocean Eng.* **2019**, *187*, 106192. [CrossRef]
12. Melville, B.W.; Coleman, S.E. *Bridge Scour*; Water Resources Publication: Highlands Ranch, CO, USA, 2000.

13. Ataie-Ashtiani, B.; Beheshti, A. Experimental investigation of clear-water local scour at pile groups. *J. Hydraul. Eng. ASCE* **2006**, *132*, 1100–1104. [CrossRef]
14. Zhang, X.; LV, H.; Shen, B. Experimental studies on local scour mechanism of cylinder bridge piers. *Hydro-Sci. Eng.* **2012**, *2*, 34–41. (In Chinese)
15. Khosronejad, A.; Kang, S.; Sotiropoulos, F. Experimental and computational investigation of local scour around bridge piers. *Adv. Water Resour.* **2012**, *37*, 73–85. [CrossRef]
16. Sumer, B.M.; Whitehouse, R.J.; Tørum, A. Scour around coastal structures: A summary of recent research. *Coast. Eng.* **2001**, *44*, 153–190. [CrossRef]
17. Sumer, B.M.; Fredsøe, J. Scour around pile in combined waves and current. *J. Hydraul. Eng.* **2001**, *127*, 403–411. [CrossRef]
18. Bouratsis, P.; Diplas, P.; Dancey, C.L.; Apsilidis, N. Quantitative spatio-temporal characterization of scour at the base of a cylinder. *Water* **2017**, *9*, 227. [CrossRef]
19. Parola, A.C.; Mahavadi, S.K.; Brown, B.M.; Khoury, A.E. Effects of rectangular foundation geometry on local pier scour. *J. Hydraul. Eng. ASCE* **1996**, *122*, 35–40. [CrossRef]
20. Parker, G.; Toro-Escobar, C.; Voigt, R.L. *Countermeasures to Protect Bridge Piers from Scour*; NCHRP project 24-7, Transportation Research Board by St. Anthony Falls Laboratory; University of Minnesota: Minneapolis, MN, USA, 1998.
21. Lagasse, P.F.; Zevenbergen, L.W.; Schall, J.D.; Clopper, P.E. *Bridge Scour and Stream Instability Countermeasures: Experience, Selection, and Design Guidance*, 2nd ed.; Technical Report of US. Department of Transportation; US Department of Transportation: Washington, DC, USA, 2001.
22. Lauchlan, C.S.; Melville, B.W. Riprap protection at bridge piers. *J. Hydraul. Eng. ASCE* **2001**, *127*, 412–418. [CrossRef]
23. Lagasse, P.F.; Clopper, P.E.; Zevenbergen, L.W.; Girard, L.W. *Countermeasures to Protect Bridge Pier from Scour*; NCHRP Report-593; Transportation Research Board: Washington, DC, USA, 2007.
24. Chiew, Y.M. Scour protection at bridge piers. *J. Hydraul. Eng. ASCE* **1992**, *118*, 1260–1269. [CrossRef]
25. Zarrati, A.R.; Gholami, H.; Mashahir, M.B. Application of collar to control scouring around rectangular bridge piers. *J. Hydraul. Res.* **2004**, *42*, 97–103. [CrossRef]
26. Cheng, L.Y.; Mou, X.Y.; Wen, H.; Hao, L.Z. Experimental research on protection of ring-wing pier against local scour. *Adv. Sci. Technol. Water Resour.* **2012**, *32*, 14–19. (In Chinese)
27. Kumar, V.; Raju, K.G.R.; Vittal, N. Reduction of local scour around bridge piers using slots and collars. *J. Hydraul. Eng. ASCE* **1999**, *125*, 1302–1305. [CrossRef]
28. Melville, B.W.; Hadfield, A.C. Use of sacrificial piles as pier scour countermeasures. *J. Hydraul. Eng. ASCE* **1999**, *125*, 1221–1224. [CrossRef]
29. Tafarojnoruz, A.; Gaudio, R.; Dey, S. Flow-altering countermeasures against scour at bridge piers: A review. *J. Hydraul. Res.* **2010**, *48*, 441–452. [CrossRef]
30. Ettema, R. *Scour at Bridge Piers*; Report No. 216; School of Engineering, University of Auckland: Auckland, New Zealand, 1980.
31. Chen, S.C.; Tfwala, S.; Wu, T.Y.; Chan, H.C.; Chou, H.T. A hooked-collar for bridge piers protection: Flow fields and scour. *Water* **2018**, *10*, 1251. [CrossRef]
32. Gaudio, R.; Tafarojnoruz, A.; Calomino, F. Combined flow-altering countermeasures against bridge pier scour. *J. Hydraul. Res.* **2012**, *50*, 35–43. [CrossRef]
33. Karami, H.; Hosseinjanzadeh, H.; Hosseini, K.; Ardeshir, A. Scour and three-dimensional flow field measurement around short vertical-wall abutment protected by collar. *KSCE J. Civ. Eng.* **2017**, *22*, 141–152. [CrossRef]
34. Moncada-M, A.T.; Aguirre-Pe, J.; Bolivar, J.C.; Flores, E.J. Scour protection of circular bridge piers with collars and slots. *J. Hydraul. Res.* **2009**, *47*, 119–126. [CrossRef]
35. Heidarpour, M.; Afzalimehr, H.; Izadinia, E. Reduction of local scour around bridge pier groups using collars. *Int. J. Sediment Res.* **2010**, *25*, 411–422. [CrossRef]
36. Vasquez, J.A.; Walsh, B.W. CFD simulation of local scour in complex piers under tidal flow. In Proceedings of the 33rd IAHR Congress: Water Engineering for a Sustainable Environment, Vancouver, BC, Canada, 9–14 August 2009; pp. 913–920.
37. Zhao, M.; Zhu, X.; Cheng, L.; Teng, B. Experimental study of local scour around subsea caissons in steady currents. *Coast. Eng.* **2012**, *60*, 30–40. [CrossRef]

38. Soulsby, R.L.; Whitehouse, R.J.S. Threshold of sediment motion in coastal environments. In Proceedings of the 13th Australasian Coastal and Engineering Conference and 6th Australasian Port and Harbour Conference, Cristchurch, New Zealand, 7–11 September 1997.
39. Soulsby, R. *Dynamics of Marine Sands*; Tomas Telford Ltd.: London, UK, 1997; 249p.
40. Melville, B.W.; Chiew, Y.M. Time scale for local scour at bridge piers. *J. Hydraul. Eng.* **1999**, *125*, 59–65. [CrossRef]
41. Bateni, S.M.; Jeng, D.S.; Melville, B.W. Bayesian neural networks for prediction of equilibrium and time-dependent scour depth around bridge piers. *Adv. Eng. Softw.* **2007**, *38*, 102–111. [CrossRef]
42. Melville, B.M. *Local Scour at Bridge Sites*; University of Auckland, School of Engineering: Auckland, New Zealand, 1975.
43. Sumer, B.M.; Christiansen, N.; Fredsoe, J. Time scale of scour around a vertical pile. In Proceedings of the 2nd International Offshore and Polar Engineering Conference, ISOPE, San Francisco, CA, USA, 14–19 June 1992; Volume 3, pp. 308–315.
44. Yang, Y.; Qi, M.; Li, J.; Ma, X. Evolution of hydrodynamic characteristics with scour hole developing around a pile group. *Water* **2018**, *10*, 1632. [CrossRef]
45. Zhang, Q.; Zhou, X.L.; Wang, J.H. Numerical investigation of local scour around three adjacent piles with different arrangements under current. *Ocean Eng.* **2017**, *142*, 625–638. [CrossRef]
46. Melville, B.W.; Raudkivi, A.J. Effects of foundation geometry on bridge pier scour. *J. Hydraul. Eng.* **1996**, *122*, 203–209. [CrossRef]
47. Ti, Z.; Zhang, M.; Li, Y.; Wei, K. Numerical study on the stochastic response of a long-span sea-crossing bridge subjected to extreme nonlinear wave loads. *Eng. Struct.* **2019**, *196*, 109287. [CrossRef]

Article

Examination of Blockage Effects on the Progression of Local Scour around a Circular Cylinder

Priscilla Williams, Ram Balachandar * and Tirupati Bolisetti

Department of Civil and Environmental Engineering, University of Windsor, Windsor, ON N9B 3P4, Canada; williamq@uwindsor.ca (P.W.); tirupati@uwindsor.ca (T.B.)
* Correspondence: rambala@uwindsor.ca

Received: 30 October 2019; Accepted: 11 December 2019; Published: 13 December 2019

Abstract: An evaluation of scour estimation methods has indicated that the effects of blockage ratio are neglected in both scour modelling and development of new predictive methods. The role of channel blockage on the mechanism and progression of local scour is not well understood, and further analysis is required in order to incorporate this effect into scour estimation. In the present investigation, local scour experiments were carried out under varying blockage ratio. The results were compared with data from literature in order to explore the effects of blockage ratio (D/b, where D is the pier diameter, and b is the channel width) on equilibrium scour depth (d_{se}/D, where d_{se} is the depth of scour at equilibrium). It was determined that D/b had a small influence on both d_{se}/D and the progression of scour depth (d_s/D) when relative coarseness $D/d_{50} < 100$ (where d_{50} is the median diameter of sediment), and that the influence appeared to be amplified when $D/d_{50} > 100$. The efficacy of scour estimation methods used to predict the progression of local scour was also dependent on D/d_{50}. A method of scour estimation used to predict d_{se}/D was evaluated, and it was similarly found to be particularly effective when $D/d_{50} < 100$. In future work, further experiments and analysis in the range of $D/d_{50} > 100$ are required in order to establish the role of D/b under prototype conditions and to refine existing scour estimation methods.

Keywords: scour and erosion; time scale of local scour; blockage ratio

---

1. Introduction

1.1. The Cost of Scour

Scour and erosion have been well established as leading causes of bridge failures. Over half of bridge failures in the United States alone have been attributed to scour [1,2]. Damage to roadway infrastructure due to scour can consist of minor erosion or complete failure of a bridge. Restoration of an overwater bridge of any size can require significant expenditure, cause disruption of local traffic and pose appreciable risk to surrounding ecosystems. In addition to capital for reconstruction, costs include rerouting of traffic and potential erection of temporary service bridges which can exceed the cost of replacement itself by 50 percent [3]. Furthermore, it has been estimated that indirect losses incurred by the general public, local business, and industry are five times greater than reconstruction costs alone [4]. Most crucially, due to the sudden nature of collapses caused by scour, failure of this type can result in loss of life.

In 1987, riprap protection around one pier of the Schoharie Creek Bridge on Interstate 90 over Schoharie Creek in New York failed due to spring flooding. The unprotected pier footing failed in tension, causing collapse of the pier and two spans of the bridge, resulting in the deaths of 10 motorists [5]. In 1995, a road bridge on Interstate 5 over Arroyo Pasajero in California collapsed, similarly under flood conditions. The collapse resulted in the deaths of seven motorists. After investigation by the Federal Highway Administration (FHWA), the cause of failure was found to be bed degradation due to the presence of flooding [4].

In 2013, a single pier of the Bonnybrook Bridge over the Bow River in Alberta, Canada, was undermined due to scour during an unprecedented rainfall event, causing derailment of six cars of a passing Canadian Pacific Railway freight train. Fortunately, the collapse did not result in any fatalities. The derailed train cars were transporting industrial chemicals (including petroleum products) and flammable liquids which were contained during the incident [6,7]. A subsequent investigation by the Government of Canada [7] stated: "If measures are not taken to inhibit local scour, especially at bridges with spread footing foundations, there is an increased risk that high water events will lead to bridge failures."

1.2. The Mechanism of Local Scour

When a structure such as a circular cylinder is introduced into a fluvial environment, there are several features which are induced in the surrounding flow field. One such feature is the downflow, which is formed when the approach flow decelerates leading up to the upstream face of the cylinder. Due to the nature of the velocity profile of the approach flow in an open channel, the pressure changes along the stagnation line, driving the flow downwards. This feature is known as the downflow, which impinges upon the bed at the base of the cylinder [8,9].

A horseshoe vortex (HSV) is formed when the downflow rolls up to form a vortex tube at the junction between the base of the cylinder and the bed. The legs of the HSV wrap around the cylinder extending in the downstream direction and are occasionally broken up and shed. A necklace or collar vortex is formed when the adverse pressure gradient (APG) associated with the stagnation line causes flow separation in the near-bed region. The boundary layer on the bed surface around the pier separates and the vorticity from the approach flow causes formation of a necklace vortex in the region of maximum shear stress in the vicinity of the cylinder. The necklace vortex is broken up in the wake region by the high bed shear stress and interaction with the wake vortices. The necklace vortex contributes to scour [10], and the HSV is one of the primary mechanisms by which sediment is removed from around the base of the cylinder. The vortical motion of the HSV entrains sediment from the bed into the flow around the cylinder. The size and strength of the HSV are related to the size of the scour hole around the cylinder, and both will continue to increase until equilibrium is reached. This is the point at which the strength of the HSV is no longer sufficiently high to continue to remove sediment from the bottom of the scour hole or when the critical shear stress of sediment at the bottom of the scour hole is no longer exceeded [8,9].

The vortices in the von Kármán vortex street are formed due to the shear layers which are detached from either side of the cylinder. Flow velocity is maximized along the separating streamline, and scour is thus initiated along the sides of the cylinder where the bed shear stress is also highest. The wake vortices shed alternately from the cylinder and carry the entrained sediment from the HSV region past the cylinder into the wake region. Downstream of the cylinder, the size of the wake vortices increases, causing them to weaken and deposit the sediment in dune-like formations [9]. From this discussion, it can be inferred that the strength and structure of the downflow, the horseshoe vortex, and the wake vortices are unsteady and highly influential on local scour around a cylinder. It is important to make note of the significant variation in structure, strength, and scale of each of the aforementioned turbulence structures.

1.3. Scale Effects in Hydraulic Modelling

Experimental modelling of local scour around a cylinder has been comprehensively explored for an appreciable range of flow, structure, and bed material characteristics. Typical scour experiments involve installation of a cylinder in a sediment recess filled with bed material of a prescribed size within a recirculating flume, from which point scour is allowed to progress until equilibrium is reached. The geometry of the scour formation is then measured to varying levels of detail, where the maximum depth of scour d_{se} (typically located near the upstream face of the cylinder) is the primary quantity of interest. Under prototype conditions, this would theoretically be taken as the minimum required foundation head, or the depth below which pier foundations should be placed in order to avoid the possibility of structural failure due to a loss of lateral support from the bed material. In practice, foundation head is determined on the basis of empirical equations, which have been developed by curve-fitting large quantities of laboratory data acquired through an experimental methodology similar to what is described above. Dimensional analysis has indicated that the maximum depth of scour normalized with pier diameter d_{se}/D can be evaluated from a set of dimensionless variables which can be further reduced under certain conditions [9,11]. For fully turbulent subcritical flow aligned with a circular cylinder in well-graded erodible sediment, relative scour depth d_{se}/D can be evaluated as described in Equation (1):

$$d_{se}/D = f\ (U/U_c,\ h/D,\ D/d_{50}) \tag{1}$$

In Equation (1), U is the average velocity of approach flow, U_c is the critical velocity for incipient motion of sediment, h is the flow depth, D is the cylinder diameter, and d_{50} is the median sediment diameter. The relationship between each dimensionless parameter and d_{se}/D has been well established in the literature [3,9]. However, analysis has indicated that commonly used empirical equations in the form of Equation (1) have a tendency to overestimate d_{se}/D values acquired from laboratory measurements [11]. Scale effects, which arise due to the imbalance in force ratios between a model and prototype, are certainly partially responsible for this discrepancy. This is particularly obvious in scaling of relative coarseness, D/d_{50}, which cannot be equated between the laboratory and the field. If sediment size were to be scaled similarly to cylinder diameter, the bed material would be in the size range for cohesive sediment, and flow-sediment interactions would not be accurately replicated in the model. Therefore, bed material size in the approximate range of d_{50} in the field is used for modelling, and the experimental value of D/d_{50} is significantly reduced [11].

There are other model effects to which the poor performance of scour-predicting equations can be attributed. In a laboratory flume experiment, bed sediment is typically well-graded, and the approach flow is well regulated and usually two-dimensional in the central region of flow at the position of the cylinder. These are controlled conditions under which natural river flow rarely, if ever, occurs. Therefore, the differences between a value of d_{se}/D estimated using an equation derived from laboratory results and an actual maximum depth of scour in the field can be understood. It has also been shown that prediction of scour in experiments with similar values of each governing parameter described in Equation (1) yields different values of d_{se}/D, which implies that there are additional significant influences in scour modelling which have not been incorporated into scour estimation [12,13].

1.4. The Influence of Blockage Ratio D/b on Local Scour

One such influence which has been previously explored in physical scour modelling is blockage ratio, D/b, where b is the channel width. While the effect of D/b on flow around bluff bodies has been widely investigated for a fixed bed condition (e.g., [14,15]), the effect of D/b on local scour has not been clearly established. This is partially due to the generally significant relative width in naturally occurring rivers which mostly eliminates channel blockage as a concern for scour in the field. Nonetheless, in a comprehensive review of pier scour processes, Ettema et al. [9] stated that estimation of d_{se}/D at a pier can be "complicated" by close channel bank proximity.

The effect of channel blockage in scour experiments has been erroneously defined as negligible when D/b is less than ten percent [8]. Laboratory flumes are usually constrained in width by facility size, and pier diameter D is generally chosen such that relative coarseness D/d_{50} is high enough to induce a measurable scour formation. Blockage ratio in experiments is therefore of greater concern than in the field. The influence of blockage ratio on scour around circular cylinders has been investigated by Hodi [16], D'Alessandro [17], and Tejada [18]. A review of these investigations, in which scour experiments were conducted for varying D/b, can be found in Williams et al. [19] (see Figure 1). The results of this investigation as well as those of both D'Alessandro [17] and Tejada [18] reported changes in d_{se}/D when D/b was ten percent or less, which does not agree with the assumed threshold stated above. Examination of the literature indicates that there are many experiments for which the effects of blockage have been ignored, despite having $D/b > 0.10$. Since code-specified design equations were derived from such experiments, a correction factor for the effect of D/b in prediction of d_{se}/D was presented by Williams et al. [19].

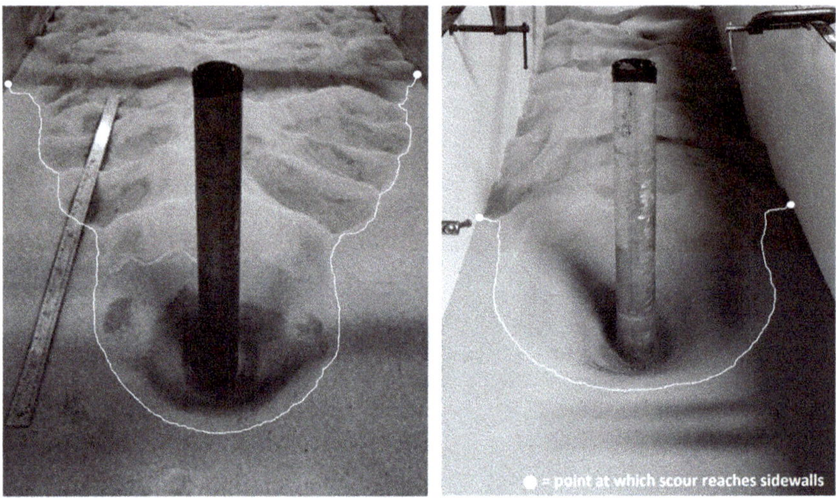

Figure 1. Equilibrium scour formation for $D/b = 0.05$ (**left**) and 0.10 (**right**) [19].

However, the results reported by Hodi [16], D'Alessandro [17], Tejada [18], and Williams et al. [19] indicate that while changes to the scour formation (i.e., the extent of the scour hole and the shape of the dune) can be considerable due to the changes in D/b, changes to d_{se}/D are minimal in the context of foundation head design.

1.5. Time Scale of Local Scour

The definition of the aforementioned equilibrium state of local scour varies slightly among investigations. As previously described, the most commonly used definition is the point at which the bed shear stress within the scour hole reaches the critical shear stress of the bed material [19,20]. In the field, a point of true equilibrium according to this description is rarely reached, and so the progression of local scour under peak flood discharge is of particular interest for the design of foundation head in prototypes. For instance, the duration of peak discharge for which a foundation head is established may not reach the time to equilibrium, resulting in an overdesigned pier [20]. As such, the time scale of local scour around a circular cylinder is of particular interest to the hydraulic engineer.

There have been several experimental investigations into the progression of local scour under particular hydraulic conditions [20–24]. It has generally been established that scour depth increases rapidly in the early stages of the scour process, and the relationship between elapsed time and scour depth becomes asymptotic as the point of equilibrium is approached. The main objective of some studies has been to develop empirical relations for estimation of the time scale of local scour [21,23].

Guan et al. [24] acquired particle image velocimetry (PIV) measurements in the flow field surrounding a cylinder during the process of local scour. Bed measurements and PIV measurements were taken at various times up to a point of equilibrium, from which the evolution of the horseshoe vortex system was investigated. At 0.5 h, one clockwise vortex was observed near the leading face of the scour hole; when 48 h had elapsed, the system had evolved to generate two clockwise vortices and one smaller counterclockwise vortex [24].

1.6. Description of the Current Investigation

The above review of the literature has indicated the importance of understanding both blockage effects on local scour and the time scale of scour. Further experimentation and analysis will allow for clarification and quantification of these effects for future incorporation into design methods used in practice.

In the present work, the effect of channel blockage D/b on the progression of local scour was further investigated. Local scour tests were carried out for varying D/b and D/d_{50}, and experimental results were compared with data from literature in order to isolate the influence of blockage ratio on scour depth. Predictive methods from literature were evaluated for the tests under consideration.

2. Materials and Methods

Experiments were carried out at the University of Windsor in Windsor, Canada. The laboratory facility contains a horizontal flume that is 10.5 m in length, 0.84 m in depth, and 1.22 m in width. A schematic of the flume is shown in Figure 2. The flume was fitted with two flow conditioners upstream of the test section. As shown in Figure 2, a PVC ramp led to the test section, which was a sediment recess of 3.68 m in length and 0.23 m in depth, encompassing the width of the flume. The sediment recess was filled with granular material with d_{50} = 0.74 mm, standard deviation of particle size $\sigma_g = \sqrt{d_{84}/d_{16}} = 1.34$, coefficient of uniformity $C_u = d_{60}/d_{10} = 1.56$, and coefficient of gradation $C_c = d_{30}^2/(d_{10} \times d_{60}) = 0.96$. (Here, d_{XX} is the sediment size for which XX percent is finer). The material was classified as poorly graded sand according to ASTM standards. The critical velocity for incipient motion of sediment (U_c) for the bed material was evaluated using standard methods which have been described in previous works [13,19].

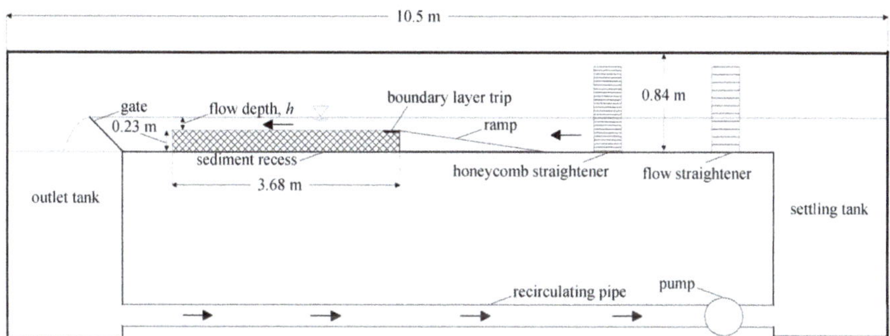

Figure 2. Schematic of the laboratory flume used for experimentation.

A boundary layer trip was located at the beginning of the sediment recess. Flow measurements in the absence of the cylinder indicated that the streamwise velocity was self-similar in the streamwise direction in the test section at the location of the cylinder. Further details are provided in Williams [25]. The depth of flow at the cylinder was adjusted by a gate located at the downstream end of the flume, preceding the outlet tank. The flow was serviced by a 60 HP centrifugal pump. The flow was calibrated with 30°, 60°, and 90° v-notch weirs, using methods described in the U.S. Department of the Interior Bureau of Reclamation Water Measurement Manual [26]. The Kindsvater–Shen relationship and 8/15 triangular weir equation were used to calculate flow rate and develop the performance curve for the flume pump in the absence of the installed test section [26]. The orientation of the flume and experimental measurements corresponded to X in the streamwise direction, Y in the vertical direction, and Z in the spanwise or transverse direction. The bed level was taken as zero in the vertical direction for all experiments, and the geometric centre of the cylinder was taken as the origin in the XZ plane for local scour tests.

In the current investigation, three local scour tests were carried out under varying conditions (see Table 1). In Table 1, t_e is the time to equilibrium. Prior to testing, movable PVC walls were installed in the flume and adjusted to the desired width, b (see Figure 3). This was done in order to alter the blockage ratio D/b while maintaining cylinder diameter D thus isolating the effects of D/b. The sediment in the test section was levelled using a trowel, and the necessary pier was installed in the centre of the channel. The flume was then filled with water to the desired depth (for which the resolution of measurement was accurate within ±0.5 mm) [25], and the pump was started and brought up to the required flow rate corresponding to an approximate flow intensity (U/U_c) of 0.85, used in order to maintain clear-water conditions for local scour.

Table 1. Experimental conditions for tests of the current investigation.

ID	b (m)	U (m/s)	U/U_c	h/D	D/d_{50}	D/b	t_e (min)	d_{se}/D
E1	1.22	0.254	0.83	2.0	81	0.05	2880	1.42
E2	0.4	0.262	0.86	2.0	81	0.15	1440	0.91
E3	1.22	0.254	0.83	2.1	76	0.05	1440	1.48

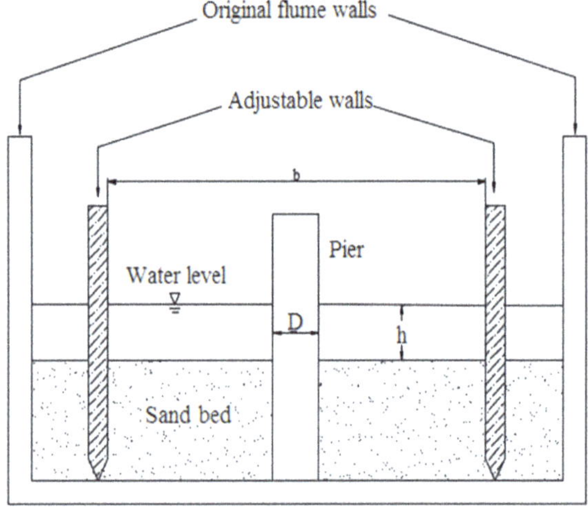

Figure 3. Schematic of the experimental setup.

The tests were conducted for a period of 24 h, after which the pump was slowed gradually in order to avoid disturbance of the bed material and then shut off. The flume was then drained slowly in order to avoid disturbance of the scour formation and a Leica laser distance meter was used to measure bed profiles in the streamwise direction along the centreline in the XY plane ($Z/D = 0$) and around the contour of the bed formation in the XZ plane ($Y/D = 0$). The uncertainty of the acquired bed measurements due to the accuracy of the laser distance meter was determined to be ±0.05 mm from the resolution of the measurements.

3. Results

3.1. Time Series for Test E3

In Figure 4, a time series for test E3 is given. The tests were conducted under the conditions for test E3 (see Table 1 for details) for 1, 2, 4, 8, 16, and 24 h, after which bed profile measurements along the symmetry plane at $Z/D = 0$ were acquired. It can be seen that the depth of the scour hole upstream of the cylinder increased incrementally with time throughout the first 24 h of the experiment. However, between 16 and 24 h, the profiles in the upstream region were very similar, showing minimal differences.

In terms of the relative scour depth, d_{se}/D, and its relation to the foundation head, this indicates that 24 h can be seen as an acceptable time to equilibrium, t_e, under the described experimental conditions.

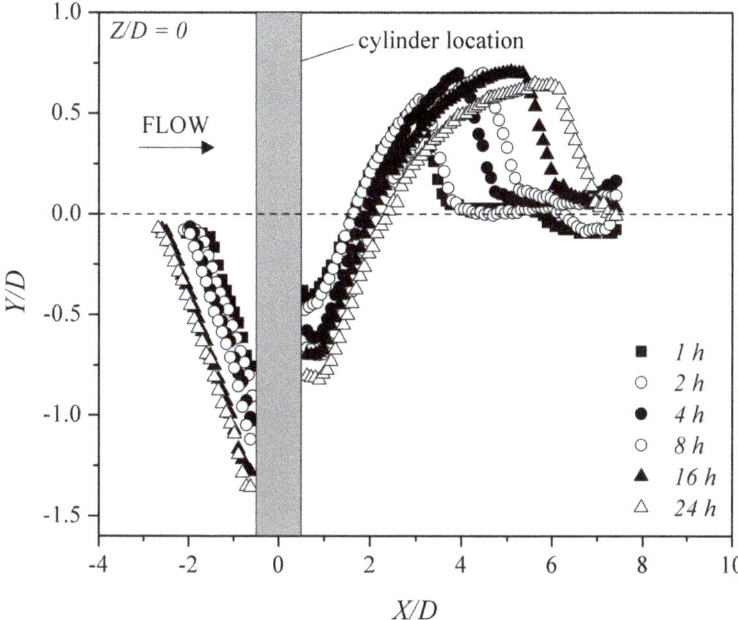

Figure 4. Bed profile measurements in the *XY* (symmetry) plane at $Z/D = 0$ for $t = \{1, 2, 4, 8, 16, 24\}$ h.

Downstream of the cylinder, it can be seen that the dune-like primary deposit increased in height and length with time until 16 h elapsed. At 24 h, the profile further increased in length, but the height of the dune decreased slightly when compared with the dune formed after 16 h. It can be seen that changes in the scour formation in the downstream region were more significant than changes in the upstream region with time; however, since the primary quantity for the foundation head design is d_{se}/D located near the upstream face of the cylinder, the best indication of an equilibrium condition with respect to design can be taken from this region.

This is further indicated by comparison of the scour profiles in the *XY* plane between tests E1 and E3 in Figures 5 and 6. For this pair of tests, the relative coarseness D/d_{50}, flow shallowness h_0/D, and blockage ratio D/b were very similar (refer to Table 1); only the duration of the tests (48 h for test E1 compared with 24 h for test E3) differed among the two tests.

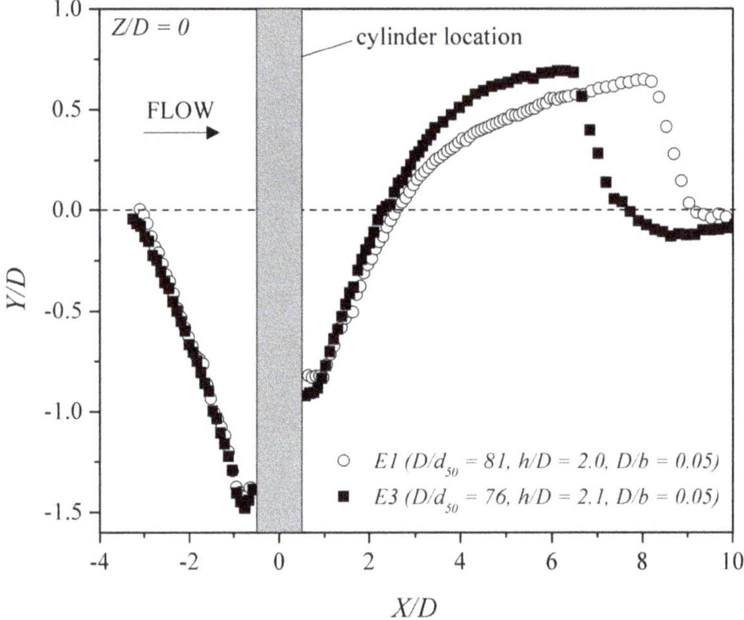

Figure 5. Bed profile measurements in the *XY* (symmetry) plane at *Z/D* = 0 for tests E1 and E3.

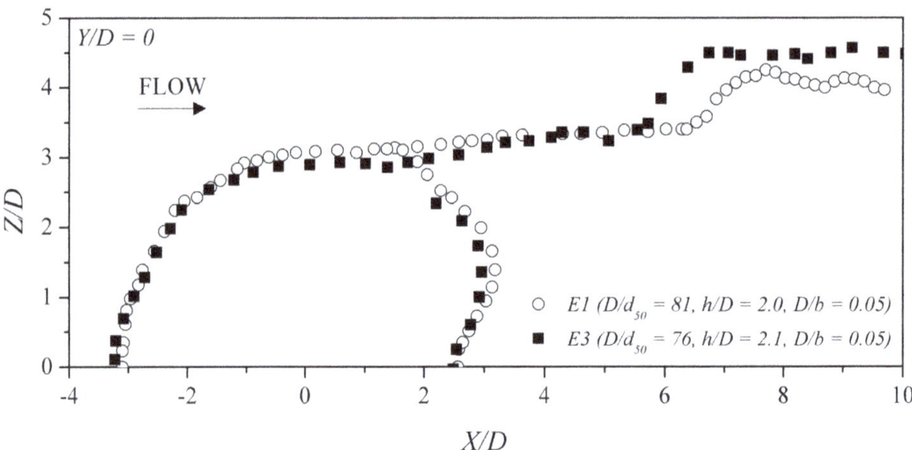

Figure 6. Bed profile measurements in the *XZ* plane at *Y/D* = 0 for tests E1 and E3.

As was indicated by the time series shown in Figures 4–6, when all other scour-governing parameters are held constant, the scour formation upstream of the cylinder does not change significantly as time progresses beyond 24 h. From the bed profiles in the *XY* plane, the scour hole profiles in the upstream region were very similar between tests E1 and E3. Downstream of the cylinder, the length of the dune increased, and the height of the dune decreased with time. This was also observed in the time series profiles in Figure 4.

The contour profiles of the scour formation in Figure 6 similarly show that the geometry of the scour hole did not change significantly with time upstream of the cylinder. Downstream of the cylinder, the width of the dune changed slightly. However, as previously mentioned, since changes in the downstream region do not affect design of the foundation head, the equilibrium state is evaluated primarily based on the region in which the depth of scour is at a maximum (i.e., upstream).

3.2. Effect of Blockage Ratio D/b on Equilibrium Scour Formation

Figures 7 and 8 show profiles of the equilibrium scour formations for tests E1 and E2. For this pair of tests, all scour-governing parameters (U/U_c, D/d_{50}, and h/D) were held constant. Movable sidewalls were installed in the flume in order to alter flume width b and, therefore, blockage ratio D/b as well. Test E1 was conducted for 48 h, and test E2 was conducted for 24 h. However, the results of the previous section indicated that the scour formation in the vicinity of the cylinder was virtually unchanged beyond 24 h, and any significant changes were only observed in the downstream region. Therefore, the following discussion will pertain mainly to the scour formation upstream and close to the cylinder.

In Figure 7, the depth of the scour hole is shown to be higher for the test with the lower blockage ratio. This is indicative of the strong effect of sidewall proximity on the mechanism of local scour. In test E1, the channel was significantly wide (b = 1.22 m) such that the spanwise pressure gradient imposed by the sidewalls did not significantly affect the flow field mechanisms surrounding the cylinder (i.e., the HSV and the wake vortices). Therefore, scour was allowed to progress unimpeded.

In contrast, the width of the channel for test E2 was approximately one-third of that of test E1 (b = 0.4 m). It is then very likely that the secondary flows (boundary-induced currents at right angles to the main flow) were capable of interacting with the flow field mechanisms surrounding the cylinder, particularly the wake vortices. Furthermore, the aforementioned spanwise pressure gradient that is a noted feature of horizontally confined flows was more likely to influence the cylinder as the confinement increased. All of these effects of sidewall proximity served to disrupt and weaken the removal and deposition of sediment, reducing the size of the scour hole and the primary deposit.

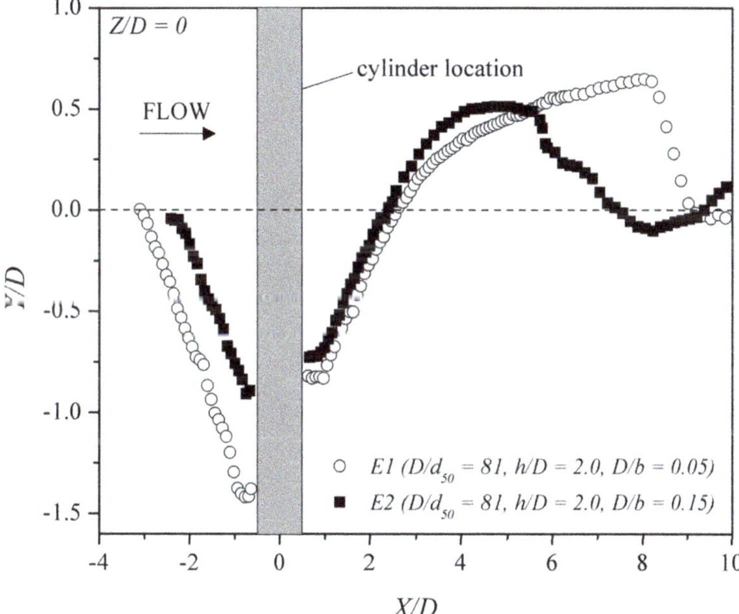

Figure 7. Bed profile measurements in the XY (symmetry) plane at Z/D = 0 for tests E1 and E2.

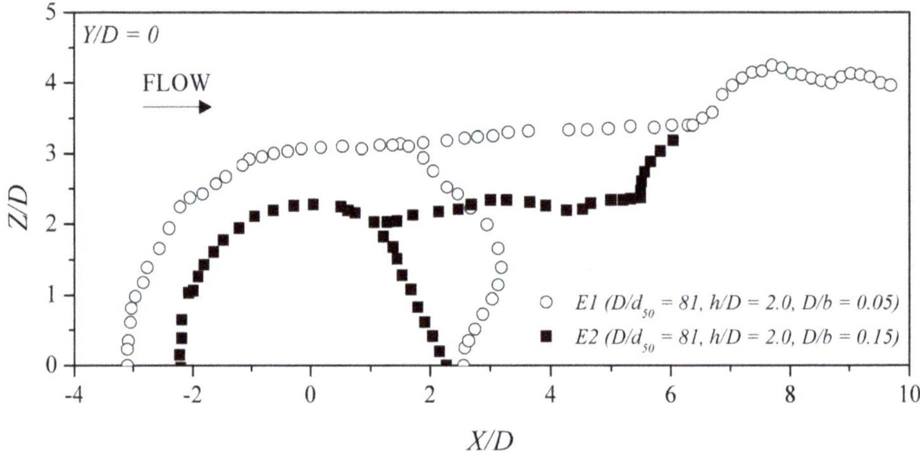

Figure 8. Bed profile measurements in the *XZ* plane at *Y/D* = 0 for tests E1 and E2.

Although the above discussion is restricted to the upstream region due to the previously mentioned time differences between tests E1 and E2, the alteration in the dune formation for test E2 can also be attributed to the effect of channel blockage. Beyond the crest of the dune for test E2, the form of the primary deposit differs from that of test E1. Since the wake region is in closer proximity to the sidewalls, it is reasonable that the effect of horizontal confinement would affect the deposition of sediment downstream of the cylinder in addition to the removal of sediment in the vicinity of the cylinder as previously discussed.

This is further illustrated in Figure 8 which shows the effect of *D/b* on the extents of the scour formation in the horizontal plane at *Y/D* = 0. It can be seen that the size of the scour hole was much smaller in plain view for test E2 with *D/b* = 0.15 than for test E1 with *D/b* = 0.05. Again, the increase in sidewall proximity appears to have suppressed the lateral progression of scour. This is also observed downstream of the scour hole, where the primary deposit became narrow as horizontal confinement increased.

3.3. *Effect of Blockage Ratio D/b on the Progression of Local Scour*

In Figure 9, a description of the progression of relative scour depth d_s/D with dimensionless time t/t_e (where t is time and t_e is the time to equilibrium) is given. Test E3 of the present investigation was included as well as two tests from the investigation of Yanmaz and Altinbilek [22]. All tests had similar values of D/d_{50} and h/D; only D/b differed within the presented data set. For all tests, it can be seen that d_s/D increased rapidly, attaining most of the maximum relative scour depth within half of the time to equilibrium. The values of the measured equilibrium scour depth (at $t/t_e \approx 1$) are shown to be very similar among all tests as well. The figures also include curves calculated using predictive methods for the progression of d_s/D with time from Melville and Chiew [21] (Equation (2)) and Aksoy et al. [23] (Equation (3)).

$$d_s/d_{se} = \exp\{-0.03|(U_c/U)\ln(t/t_e)|^{1.6}\} \tag{2}$$

$$d_s/D = 0.8 \times (U/U_c)^{3/2} \times (h/D)^{0.15} \times (\log T_s)^{0.6} \tag{3}$$

$$T_s = td_{50} \times (\Delta g d_{50})^{0.5}/D^2 \tag{4}$$

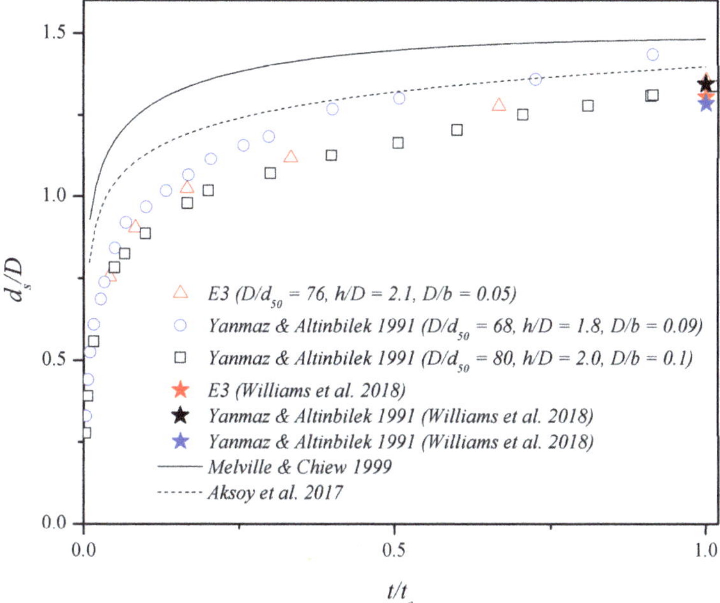

Figure 9. Comparison of d_{se}/D with t/t_e for test E3 and data from [22] with $D/d_{50} \approx 76$ as well as estimation curves [21,23] and points [19] from literature.

In Equation (4), $\Delta = (\rho_s - \rho)/\rho$, where ρ_s is the density of the bed material, and ρ is the density of water. In Figure 9, the experimental parameters for test E3 were used to calculate the predictive curves. In Figure 10, the parameters for the test by D'Alessandro [17] were used for the prediction. The value of d_{se}/D calculated using the predictive method presented in Williams et al. [19] (Equation (5)) was also included for each test as shown by the starred data points.

$$d_{se}/D = 0.76{k_c}^{1.69} \times (h/D)^{0.32} \tag{5}$$

In Equation (5), k_c is the ratio between the velocity along the separating streamline U_s and U_c. In Figure 9, it can be seen that there were small differences in d_{se}/D observed in the middle of the scour process (i.e., approximately between $0.1 < t/t_e < 0.7$). However, there was no specific trend which can be noted between the development of scour and blockage ration D/b. In the described middle section of the scour process, the scour depth was slightly lower for $D/b = 0.1$ when compared with $D/b = 0.05$; however, d_{se}/D was slightly higher for $D/b = 0.09$ than either of the other tests. (Although unlikely, this could possibly be attributed to the small changes in D/d_{50} or h/D among the tests.). In general, for $D/d_{50} < 100$, it can be concluded that changes in D/b have a very small influence on both the progression of local scour as well as the equilibrium depth of scour.

The predictive method described by Equation (2) is represented by the solid curve in Figures 9 and 10, and this method slightly over-predicted the depth of scour (d_s) throughout the time to equilibrium. The method described by Equation (3) over-predicted the depth of scour for $t/t_e < 0.5$, beyond which the curve approached the results of Yanmaz and Altinbilek [22]. Equation (2) showed good prediction of d_{se}/D for all tests.

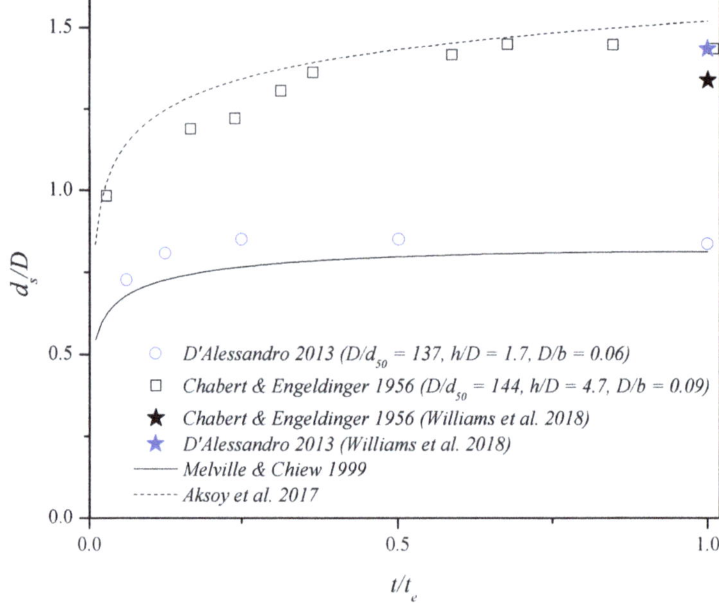

Figure 10. Comparison of d_{se}/D with t/t_e and data from [17,27] with $D/d_{50} \approx 137$ as well as estimation curves [21,23] and points [19] from literature.

However, the results presented in Figure 10 indicate that the effect of D/b was amplified as D/d_{50} increased. The results presented in Figure 10 had similar values of U/U_c and D/d_{50}. The value of h/D was also altered among the tests; however, the literature indicated that the effect of flow depth on d_{se}/D is minimal when $h/D > 1.4$ [3]. Therefore, in terms of influencing parameters, only D/b differed among the tests. In contrast to the data set presented in Figure 9, there were significant changes noted with changing D/b. For $D/b = 0.09$, the depth of scour was significantly higher throughout the progression to equilibrium when compared with $D/b = 0.06$. This was also in contrast to the results of the previous section which indicated that the relative scour depth increased as D/b decreased. This indicates that the effect of blockage ratio on local scour is also dependent on the value of D/d_{50}, and the effect is amplified when $D/d_{50} > 100$.

This is in agreement with the results of Tejada [18] who conducted a series of experiments with varying sizes of bed material (d_{50}) and similarly reported that the effect of D/b was minimal when $D/d_{50} < 100$ when compared with $D/d_{50} > 100$. This is reasonable, since larger values of relative coarseness are usually indicative of smaller sediment particles which would be more susceptible to changes in the flow field surrounding a cylinder.

Interestingly, Equation (2) provides a good estimation of the progression of scour for the results of Chabert and Engeldinger [27], while Equation (3) provides a better estimation for the test of D'Alessandro [17]. Furthermore, Equation (5) shows a much closer prediction of the test of Chabert and Engeldinger [27] than D'Alessandro [17]. This also indicates that further investigation on higher values of D/d_{50} is required in order to establish the role of D/b on local scour.

4. Discussion and Conclusions

The present investigation explored the progression of local scour around a circular cylinder placed in an erodible bed and the effects of channel blockage ratio D/b therein. It was established that, under the experimental conditions of this work, equilibrium of sediment removal in the vicinity of the cylinder was reached at 24 h. Beyond this period, changes in the scour formation were observed primarily in the downstream region which did not contribute significantly to the design of the foundation head for local scour.

At $D/d_{50} < 100$, the experimental results of the present investigation indicated that the size of the scour hole decreased with increasing blockage ratio which can likely be attributed to a weakening of the flow field mechanisms due to the sidewall proximity. In order to fully understand this phenomenon, detailed flow measurements in the flow field surrounding the cylinder are required in future work.

When compared with data from the literature, the progression of local scour and the relative scour depth did not appear to be significantly affected by changes in the blockage ratio for small values of relative coarseness (i.e., $D/d_{50} < 100$). However, analysis of data with high values of relative coarseness ($D/d_{50} > 100$) showed that blockage ratio had a greater effect on scour depth. Furthermore, comparison of the results of the present investigation with the literature results was particularly limited, since such analysis requires that all scour-governing parameters are identical. Future experimental design corresponding to previously published results would assist in accomplishing similar analysis for a broader range of conditions in future work. The authors recommend that further experimentation on the effect of D/b is required, particularly for a wider range of D/d_{50} which approaches field conditions (i.e., $D/d_{50} > 100$). In order to achieve this, use of larger-sized flumes with finer bed material is required.

Various predictive methods from the literature were used to estimate the progression of local scour to varying degrees of efficacy. The scour-predicting equation presented in Williams et al. [19] was shown to provide a better estimation of d_{se}/D for $D/d_{50} < 100$ than for larger values of relative coarseness, indicating that a scaling factor may be required in order to implement such predictive methods in practical design.

Larger values of D/d_{50} are particularly desirable in this venture, since field values of relative coarseness are usually greater than 200. Once the effect of sidewall proximity on scour depth has been established at a prototype level, further evaluation can be carried out in order to incorporate this effect into scour estimation.

Author Contributions: The experimental work in the present investigation was carried out by P.W. The analysis and manuscript compilation were conducted jointly by the authors.

Funding: This research was funded by the Natural Sciences and Engineering Research Council of Canada through the Discovery Grant program to Ram Balachandar and Tirupati Bolisetti.

Conflicts of Interest: The authors declare no conflict of interest.

Notation

ASTM	American Society for Testing and Materials
b	channel width
C_c	coefficient of sediment gradation
C_u	coefficient of sediment uniformity
D	cylinder diameter
D/b	blockage ratio
D/d_{50}	relative coarseness
d_{se}	maximum equilibrium scour depth
d_{se}/D	relative scour depth
d_{10}	sediment size of which 10 percent is finer
d_{16}	sediment size of which 16 percent is finer
d_{30}	sediment size of which 30 percent is finer
d_{50}	median sediment diameter
d_{60}	sediment size of which 60 percent is finer

d_{84}	sediment size of which 84 percent is finer
g	gravitational acceleration
h	flow depth
h/D	flow shallowness
k_c	U_s/U_c
t	time
t_e	time to equilibrium of local scour
T_s	dimensionless time parameter
U	mean streamwise velocity of approach flow
U/U_c	flow intensity
U_c	critical velocity for incipient motion of sediment
U_s	streamwise velocity along the separating streamline
X	streamwise coordinate direction
Y	vertical coordinate direction
Z	spanwise coordinate direction
Δ	relative density, $(\rho_s - \rho)/\rho$
ρ	density of fluid
ρ_s	density of bed material
σ_g	standard deviation of sediment particle size

References

1. Shirhole, A.M.; Holt, R.C. Planning for a comprehensive bridge safety program. *Transp. Res. Rec.* **1991**, *1290*, 39–50.
2. Wardhana, K.; Hadipriono, F.C. Analysis of recent bridge failures in the United States. *J. Perform. Constr. Fac.* **2003**, *17*, 144–150. [CrossRef]
3. Melville, B.W.; Coleman, S.E. *Bridge Scour*; Water Resources Publication: Highlands Ranch, CO, USA, 2000.
4. Lagasse, P.F.; Thompson, P.L.; Sabol, S.A. *Guarding against Scour*; Civil Engineering: New York, NY, USA, 1995; p. 56.
5. LeBeau, K.H.; Wadia-Fascetti, S.J. Fault Tree analysis of schoharie creek bridge collapse. *J. Perform. Constr. Fac.* **2007**, *21*, 320–326. [CrossRef]
6. Graveland, B.; Krugel, L. Train Derails In Calgary On Bridge, Carries Petroleum Product. Available online: http://www.huffingtonpost.ca/2013/06/27/train-derails-calgary-bridge_n_3509067.html (accessed on 8 March 2018).
7. Government of Canada. Railway Investigation Report R13C0069–Transportation Safety Board of Canada. Available online: http://www.bst-tsb.gc.ca/eng/rapports-reports/rail/2013/r13c0069/r13c0069.html (accessed on 29 May 2019).
8. Chiew, Y.M. Local Scour at Bridge Piers. Ph.D. Thesis, University of Auckland, Auckland, New Zealand, 1984.
9. Ettema, R.; Melville, B.W.; Constantinescu, G. *Evaluation of Bridge Scour Research: Pier Scour Processes and Predictions*; Transportation Research Board of the National Academies: Washington, DC, USA, 2011.
10. Nasif, G.; Balachandar, R.; Barron, R.M. Characteristics of flow structures in the wake of a bed-mounted bluff body in shallow open channels. *J. Fluids Eng.* **2015**, *137*. [CrossRef]
11. Ettema, R.; Melville, B.W.; Barkdoll, B. Scale effect in pier-scour experiments. *J. Hydraul. Eng.* **1998**, *124*, 639–642. [CrossRef]
12. Williams, P.; Balachandar, R.; Bolisetti, T. Evaluation of Local Bridge Pier Scour Depth Estimation Methods. In Proceedings of the 24th Canadian Congress of Applied Mechanics, Saskatoon, SK, Canada, 2–6 June 2013; pp. 2–6.
13. Williams, P.; Bolisetti, T.; Balachandar, R. Evaluation of governing parameters on pier scour geometry. *Can. J. Civ. Eng.* **2016**, *44*, 48–58. [CrossRef]
14. Ramamurthy, A.S.; Lee, P.M. Wall effects on flow past bluff bodies. *J. Sound Vib.* **1973**, *31*, 443–451. [CrossRef]
15. Ramamurthy, A.S.; Balachandar, R.; Vo, D.N. Blockage correction for sharp-edged bluff bodies. *J. Eng. Mech-ASCE* **1989**, *115*, 1569–1576. [CrossRef]
16. Hodi, B. Effect of Blockage and Densimetric Froude Number on Circular Bridge Pier Local Scour. Master's Thesis, University of Windsor, Windsor, ON, Canada, 2009.

17. D'Alessandro, C. Effect of Blockage on Cylindrical Bridge Pier Local Scour. Master's Thesis, University of Windsor, Windsor, ON, Canada, 2013.
18. Tejada, S. Effects of Blockage and Relative Coarseness on Clear Water Bridge Pier Scour. Master's Thesis, University of Windsor, Windsor, ON, Canada, 2014.
19. Williams, P.; Bolisetti, T.; Balachandar, R. Blockage correction for pier scour experiments. *Can. J. Civ. Eng.* **2018**, *45*, 413–417. [CrossRef]
20. Mia, M.F.; Nago, H. Design method of time-dependent local scour at circular bridge pier. *J. Hydraul. Eng.* **2003**, *129*, 420–427. [CrossRef]
21. Melville, B.W.; Chiew, Y.M. Time scale for local scour at bridge piers. *J. Hydraul. Eng.* **1999**, *125*, 59–65. [CrossRef]
22. Yanmaz, A.M.; Altinbilek, H.D. Study of time-dependent local scour around bridge piers. *J. Hydraul. Eng.* **1991**, *117*, 1247–1268. [CrossRef]
23. Aksoy, A.O.; Bombar, G.; Arkis, T.; Guney, M.S. Study of the time-dependent clear water scour around circular bridge piers. *J. Hydrol. Hydromech.* **2017**, *65*, 26–34. [CrossRef]
24. Guan, D.; Chiew, Y.M.; Wei, M.; Hsieh, S.C. Characterization of horseshoe vortex in a developing scour hole at a cylindrical bridge pier. *Int. J. Sediment Res.* **2019**, *34*, 118–124. [CrossRef]
25. Williams, P. The Role of Approach Flow and Blockage Ratio on Local Scour around Circular Cylinders with and without Countermeasures. Ph.D. Thesis, University of Windsor, Windsor, ON, Canada, 2019.
26. U.S. Department of the Interior Bureau of Reclamation. *Water Measurement Manual*; US Government Printing Office: Washington, DC, USA, 2001.
27. Chabert, J.; Engeldinger, P. *Etude Des Affouillements Autour Des Piles de Points (Study of Scour at Bridge Piers)*; Bureau Central d'Etudes les Equipment d'Outre-Mer, Laboratoire National d'Hydraulique: Paris, France, 1956.

Article

Turbulent Flow Structures and Scour Hole Characteristics around Circular Bridge Piers over Non-Uniform Sand Bed Channels with Downward Seepage

Rutuja Chavan [1], Paola Gualtieri [2] and Bimlesh Kumar [3,*]

[1] Department of Civil Engineering, VJTI, Mumbai 400019, India
[2] Department of Civil, Archictural and Environmental Engineering (DICEA), University of Naples Federico II (UNINA), 80125 Napoli NA, Italy
[3] Department of Civil Engineering, Indian Institute of Technology, Guwahati 781039, India
* Correspondence: bimk@iitg.ac.in; Tel.: +91-361-2582420

Received: 6 June 2019; Accepted: 18 July 2019; Published: 30 July 2019

Abstract: In alluvial rivers bridge piers often cause local scour, a complex phenomenon as a result of the interaction between turbulent flow and bed material. In this paper, the results of an experimental study on the scour hole characteristics around single vertical pier sets on a non-uniform sand bed, under no seepage, and with downward seepage conditions, are described. In case of downward seepage, turbulent statistics, such as Reynolds stress, higher order moments, TKE-flux, and consequently sediment transport, decrease upstream of the pier, while increasing on both sides of it, where the enhanced erosive capacity of the flow results in an increase in the scour hole width. Moreover, the scour hole length shifts downstream. Empirical equations for the evaluation of scour hole characteristics, such as the length, width, area, and volume, including the downward seepage parameter, are proposed and experimentally tested. Model predictions give reasonably good agreement with the experimental data.

Keywords: downward seepage; pier; scour; turbulent statistics

1. Introduction

Bridges play an important role in the transportation of goods and people across rivers. In civil engineering, one of the most important issues is to protect bridge piers from collapse. In fact, their foundation may be threatened by localized scour, as a result of the flow constriction of the cross-sectional area, and the subsequent increase of the flow velocity [1]. Thus, in order to ensure the protection of such structures, it is of a high priority to predict the flow field and sediment transport around the bridge piers.

Many researchers have studied the vortice systems around piers, stating that they primarily affect the local scour [2–6]. Melville and Coleman [5] stated that the flow field around a bridge pier is characterized by down-flow, surface roller, and wake vortices, such as the horseshoe vortex, at the base of the pier, and wake vortices behind the pier. According to Melville [2] and Chiew [4], a horseshoe vortex increases the flow velocity near the bed, and the wake vortex carries the eroded bed material downstream. However, the effect of the pier on flow separation, and the consequent sediment transport, depends on turbulence.

Some studies have investigated the stochastics nature of turbulent flow around a pier [2,7–11], but there still remains a lack information concerning the trend of higher order moments for the fluctuating velocities.

As the safety of the pier depends on the depth and volume of the scoured region around it, most researchers focus on the estimation of the maximum scour depth around the pier [5,12–15], developing, from laboratory and field data, empirical relations different from each other, for model or equation parameters. Melville and Coleman [5] proposed a scour depth prediction model depending on several parameters, such as flow depth, pier geometry, and flow intensity. Chavan and Kumar [14], from the experimental data, developed an empirical equation for the estimation of the scour depth in alluvial channels with a downward seepage.

Alluvial channels have granular permeable boundaries; hence, the flow field is a complex interaction between the surface and subsurface flow, as water is either seeping in (upward seepage) or seeping out (downward seepage) from the channel [15]. Shukla and Misra [16] quantified the water loss as a result of seepage as nearly 45% of the total flow volume. Tanji and Kielen [17] observed that the seepage losses in the earthen channels in semi-arid areas range between 20% and 50%. Kinzli et al. [18], and Martin and Gates [19] estimated that water loss as a result of downward seepage was around 15% and 40%, respectively. Moreover, seepage affects the channel morphology [20–22] and increases the streamwise velocity near the bed [23–26].

The aim of this study is to deepen the description of the turbulent flow field around a single vertical bridge pier set in a seepage-affected alluvial channel, and to develop empirical relations for evaluating scour hole characteristics such as length, width, area, and volume, including the downward seepage parameters.

2. Experimental Set-Up and Procedure

2.1. Experimental Facility

Experimental measurements have been carried out in a recirculating plexi-glassed tilting flume, with dimensions of 20 m long, 1 m wide, and 0.72 m deep, supplied by an overhead tank, through a regulating valve (Figure 1). The slope of the channel was 0.5%. Detailed information on the experimental flume can be found in Chavan et al. [9]. Starting 2 m from the inlet, the bed of the flume was made porous, placing a fine mesh on a pressure chamber at a length 15.2 m, width of 1 m, and depth of 0.22 m. Two electromagnetic flow meters measured the seepage discharge. The test section was 5 m, that is, it is 5 m to 10 m from the tailgate of the flume.

For the fine mesh, two river sands of median diameters (d_{50}), 0.395 mm and 0.5 mm, and standard deviation (σ_g), 1.85 and 1.65, respectively, were selected. Both sands had standard deviation values higher than 1.4, so they were non-uniform [27]. The sand bed thickness was 17 cm. Two perspex circular piers with a diameter of (d) 75 mm and 90 mm, respectively, which were both 150 mm high, were alternatively set at the center of the test section, 7.5 m from the tail end of the flume (i.e., 7.5, 0.5, and 0). Both diameters of the piers were less than 10% of the channel width, so as to avoid a sidewall effect on the scour depth [14].

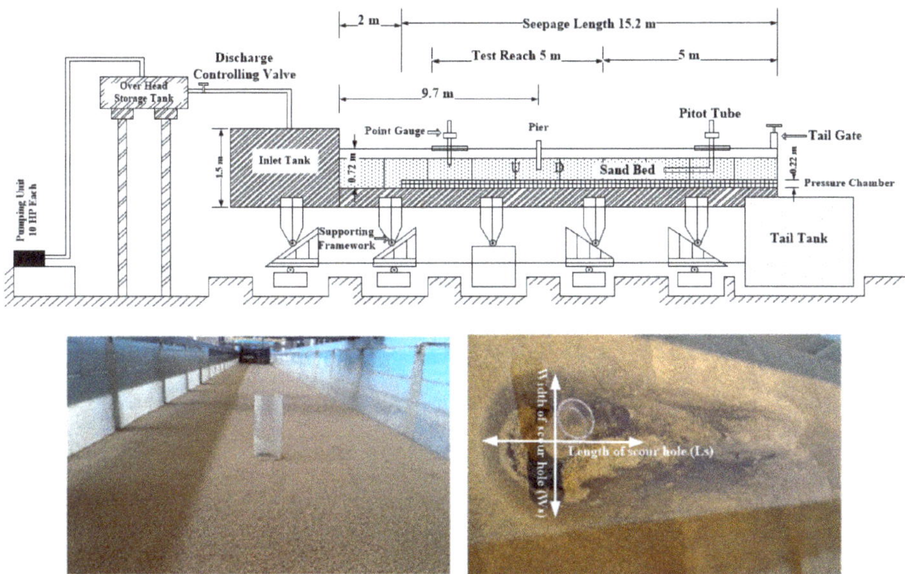

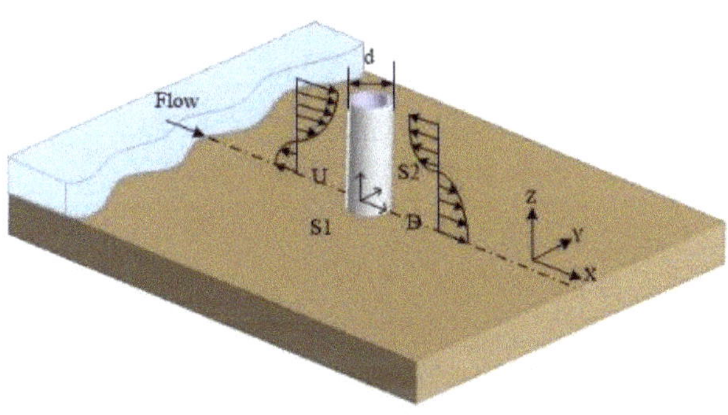

AA' Section View

Figure 1. Schematic diagram of the experimental set-up.

2.2. Experimental Measurements

For each river sand and pier, experimental measurements of the instantaneous velocity have been carried out along four vertical sections, set upstream (U), downstream (D), and laterally against (S1 and S2) the pier; each section was 6 cm away from the face of the pier, with five discharges, and, for each one, having three seepage percentages, that is, 0%, 10%, and 15%, respectively. Such percentages were selected so as to obtain a seepage velocity less or equal to 1% of the mean velocity in the channel [28]. The combinations resulted in 15 test runs, as specified in Table 1.

Table 1. Experimental conditions.

Expt. Run	Pier dia. D (mm)	Standard Deviation of River Sand σ_g	Flow Depth h (m)	Flow Rate Q (m^3/s)	Reynolds Number (Re)	Froude Number (Fr)	Temp. (°C)	Seepage Percentage
1	75 and 90 mm	1.65 and 1.85	0.118	0.032	31,860	0.25	28	0
2	75 and 90 mm	1.65 and 1.85	0.118	0.032	31,860	0.25	28	10
3	75 and 90 mm	1.65 and 1.85	0.118	0.032	31,860	0.25	28	15
4	75 and 90 mm	1.65 and 1.85	0.121	0.034	34,001	0.2579	28	0
5	75 and 90 mm	1.65 and 1.85	0.121	0.034	34,001	0.2579	28	10
6	75 and 90 mm	1.65 and 1.85	0.121	0.034	34,001	0.2579	28	15
7	75 and 90 mm	1.65 and 1.85	0.123	0.036	36,039	0.2667	28	0
8	75 and 90 mm	1.65 and 1.85	0.123	0.036	36,039	0.2667	28	10
9	75 and 90 mm	1.65 and 1.85	0.123	0.036	36,039	0.2667	28	15
10	75 and 90 mm	1.65 and 1.85	0.126	0.038	38,052	0.272	28	0
11	75 and 90 mm	1.65 and 1.85	0.126	0.038	38,052	0.272	28	10
12	75 and 90 mm	1.65 and 1.85	0.126	0.038	38,052	0.272	28	15
13	75 and 90 mm	1.65 and 1.85	0.129	0.04	39,990	0.28	28	0
14	75 and 90 mm	1.65 and 1.85	0.129	0.04	39,990	0.28	28	10
15	75 and 90 mm	1.65 and 1.85	0.129	0.04	39,990	0.28	28	15

The instantaneous velocity was measured using an Acoustic Doppler Velocimeter (ADV), with an acoustic frequency of 10 MHz and a sampling rate of 200 Hz for 5 min at each of the data points. Each test run duration was 24 h. The duration of the experimental runs was set depending upon the criteria of the equilibrium state for the scour depth, given by Kumar et. al. [29]. During the experimental measurements, the signal-to-noise ratio (SNR) was kept at 15 or above, and the average correlation coefficient between the transmitted and received signal was less than 70%. The ADV uncertainty was evaluated by collecting 10 pulses over 5 min near the bed ($z \approx 5$ mm). The results are shown in Table 2, where $\overline{U}$, $\overline{V}$, and $\overline{W}$ are the mean time velocities in an x (longitudinal), y (transverse), and z (vertical) direction, respectively, and u', v', and w' are the corresponding fluctuating velocities, respectively.

Table 2. Uncertainty associated with Acoustic Doppler Velocimeter (ADV) data.

	$\overline{U}$ (m/s)	$\overline{V}$ (m/s)	$\overline{W}$ (m/s)	$\left(\overline{u'u'}\right)^{0.5}$	$\left(\overline{v'v'}\right)^{0.5}$	$\left(\overline{w'w'}\right)^{0.5}$
Standard deviation	2.22×10^{-3}	9.9×10^{-4}	4.2×10^{-4}	1.0×10^{-3}	9.1×10^{-4}	3.21×10^{-4}
Uncertainty %	0.11	0.085	0.021	0.095	0.09	0.034

The ADV data were contaminated with spikes because of interference between the transmitted and received signals. Acceleration threshold methods [30] were applied in order to remove the spikes, by fitting velocity power spectra with Kolmogorov's −5/3 scaling law in the inertial subrange, with threshold values ranging between 1–1.5.

The measurement of bed deformations under water typically utilizes high-frequency sound wave (ultrasonic) technology. The bed profiles around the pier were measured using Sea Tech 5 MHz Ultrasonic Ranging System (URS), with four pairs of transducers fixed to a mechanical trolley moving at a constant speed. The sound waves transmitted from the sensors were reflected by a solid object, the time for the sound waves to travel from the sensor to the object and back to the sensor was recorded, and the distance from the sensor to the reflective object was obtained. The URS uncertainty was evaluated collecting 15 measurements near the bed ($z \approx 5$mm). The results are in Table 3.

Table 3. Uncertainty associated with Ultrasonic Ranging System (URS) data.

Statistical Parameters	**Standard Deviation**	**Standard Deviation about the Mean**	**Uncertainty %**
	0.00095	0.01834	0.07264

3. Results

3.1. Turbulence Characteristics

The distribution of the turbulent quantities, such as the sediment transport, is affected by the interaction between the bridge pier and flow. Chavan et al. [9,31] observed that the velocity and Reynolds stresses are negative upstream of the pier and close to bed, and downstream of the pier and close to the free surface, because of reversal flow, which causes a horseshoe vortex at the bed and a wake vortex at the free surface. In this paper, in order to deeply analyze the turbulent quantities and sediment transport around a bridge pier, upstream (U), downstream (D) and laterally against (S1 and S2) the pier, the distributions of the higher order moments (i.e., turbulence intensity and skewness), turbulent kinetic energy budget, and fluxes of turbulent kinetic energy, have been experimentally studied. In particular, as in the lateral sections, the turbulence quantities' distributions are similar [9]; hereafter, only the lateral section S1 will be considered, in order to avoid the repetition of results.

3.1.1. Turbulence Intensities

Around the pier, the distributions of the fluctuating components of the streamwise and vertical instantaneous velocity (σu and σw, respectively), expressed as root mean square (RMS), have been evaluated for the different discharges and percentages of seepage. In Figure 2, σu and σw, scaled with the shear velocity (u^*), are expressed as $u^* = \sqrt{}(\tau/\rho$; where $\tau = \gamma RS$ is the bed shear stress, ρ is the flow density, R is hydraulic radius, and S is the bed slope), versus the distance (z) from the bed, scaled with the flow depth (h), represented as $h+$, are shown.

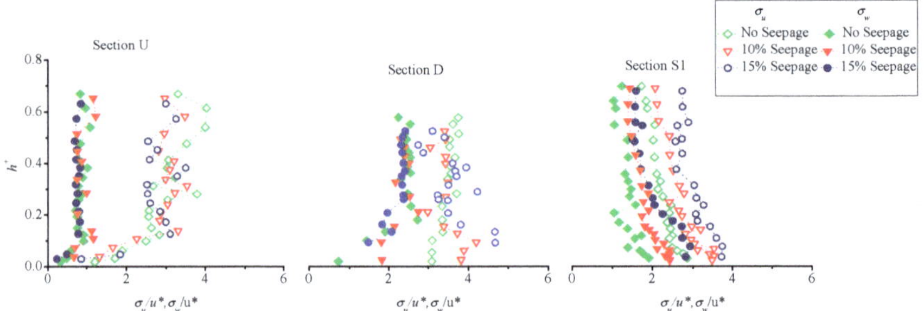

Figure 2. Non-dimensional distributions of turbulence intensities.

Upstream of the pier (U), σu and σw show similar trends. In particular, σu is lower near the bed, because of the reversal flow. With respect of the zero seepage, σu is reduced by 15% in case of 10% seepage, and by 22% averagely in case of 15% seepage. For the different seepage percentages, σu and σw attain higher values at the edge of the scour hole ($h+ \approx 0.1$), then, increasing in $h+$; while σw shows a nearly constant value, σu slightly decreases, oscillating around a constant value.

Downstream of the pier (D), where there is no obstruction to the flow, for the different seepage percentages, σw shows an increasing trend, attaining higher values than in the section U, because of the wake vortices behind the pier. Near the free surface, σw slightly decreases, because of the reduced strength of the wake vortices in the presence of the lateral flow through the channel boundaries. For the zero seepage percentage, σu shows a nearly constant value, while it slightly decreases, with scattering trends, for the seepage percentages of 10% and 15%, respectively.

Laterally to the pier (S1), in the case of 10% and 15% seepage percentages, near the bed ($h+ < 0.2$), σu and σw increase at an average about 20%–30% and 25%–35% with no seepage value, respectively, because of the enhanced turbulence near the bed.

3.1.2. Skewness

Despite many studies on the hydrodynamics of the flows around bridge piers and downward seepage, there is still a lack of knowledge about higher order moments of velocity fluctuations, which contain information related to the flux of the Reynolds normal stress, retaining the sign details [32].

The third order correlation, M_{lm}, is expressed as [33] follows:

$$M_{lm} = \overline{\hat{u}^l \hat{w}^m}$$

where, $\hat{u} = \frac{u'}{\left(\overline{u'^2}\right)^{0.5}}$, $\hat{w} = \frac{w'}{\left(\overline{w'^2}\right)^{0.5}}$ and $(l + m) = 3$.

In particular, M_{30} ($\hat{u}^3$)and M_{03} ($\hat{w}^3$) express the longitudinal and vertical flux of streamwise RNS ($\overline{u'u'}$) and vertical RNS ($\overline{w'w'}$), respectively, and M_{12} and M_{21} indicate the advection of ($w'w'$) and ($u'u'$), in x-direction and z-direction, respectively. They are shown in Figure 3, together with the vertical profiles of the skewness factors in the streamwise ($Su = \overline{u'^3}/u_*{}^3$) and vertical ($Sw = \overline{w'^3}/u_*{}^3$) direction, respectively.

Upstream of the pier (U), near the bed, positive $M30$ and $M12$ and negative $M03$ and $M21$ show the flux of streamwise RNS and the diffusion of vertical RNS in streamwise direction, and the flux of vertical RNS and the diffusion of streamwise RNS in a downward direction, respectively. By increasing

$h+$, $M30$ and $M12$ become negative, and $M03$ and $M21$ become positive. The positivity of $M30$ and $M12$ decreases in the seepage runs, showing that the flux of the streamwise RNS and the diffusion of the vertical RNS results in a lower mobility of the bed material.

Near the bed, the positive and negative values of the streamwise and vertical skewness factors confirm that the transport of the turbulent kinetic energy are in streamwise and downward directions, respectively.

In section D, near the bed, $M30$ is negative, while $M03$ is positive, suggesting that the streamwise and vertical RNS fluxes are in an opposite direction to the flow, and in an upward direction, respectively, showing ejections. By increasing $h+$, $M30$ and $M03$ attain an opposite sign, becoming positive and negative, respectively. Near the bed, $M12$, is positive, while $M21$, is negative, showing that the diffusion of vertical and streamwise RNS is in the flow and in downward directions, respectively. By increasing $h+$, $M12$ becomes negative, while $M21$ becomes positive, showing the diffusion of vertical and streamwise RNS in the opposite direction to the flow, and in the upward directions, respectively. In section D, near the bed, positive Sw and negative Su show that the transport of the turbulent kinetic energy is in the opposite direction to the flow, and in an upward direction, respectively, increasing the seepage percentage. The results confirm the dominance of the secondary currents downstream from the piers, because of the formation of wake vortices; however, downward seepage limits the strength of the wake vortices.

In the Section S1, near the bed, $M30$ and $M12$ increase with the seepage percentage, because of the increased particle mobility in the seepage runs. Similarly, the increased $M03$ and $M21$ in the seepage runs, correspond to the increasing vertical flux of RNS, and the advection of streamwise RNS in a downward direction, respectively. In the section of S1, in the seepage runs, the distributions of the skewness that factor their increase with $h+$, clearly show the increased movement of bed particles. These results show the prevalence of a high turbulent activity, resulting in enhanced sediment transport, laterally to the pier, in the case of seepage runs.

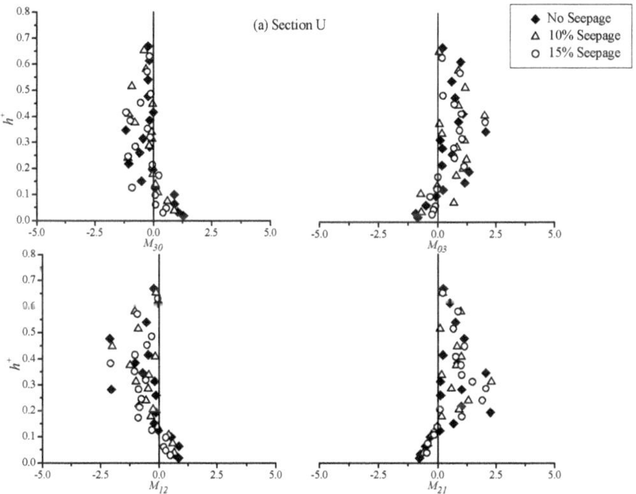

Figure 3. *Cont.*

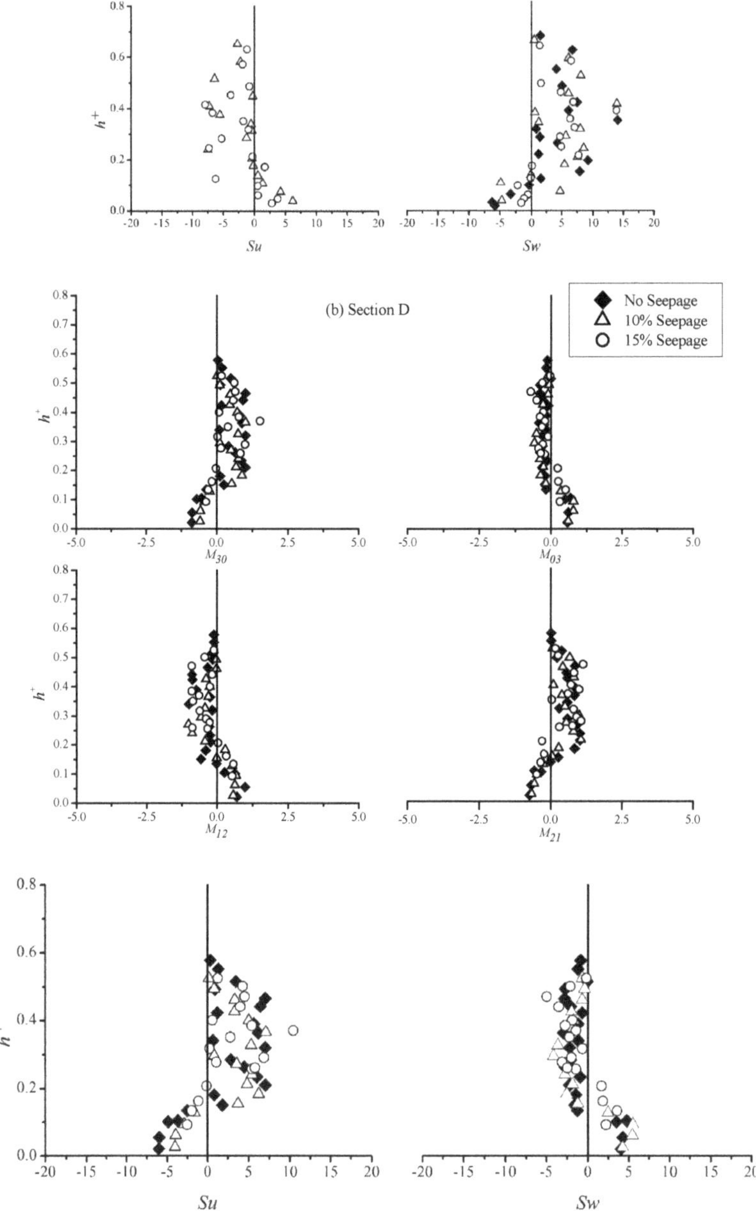

Figure 3. *Cont.*

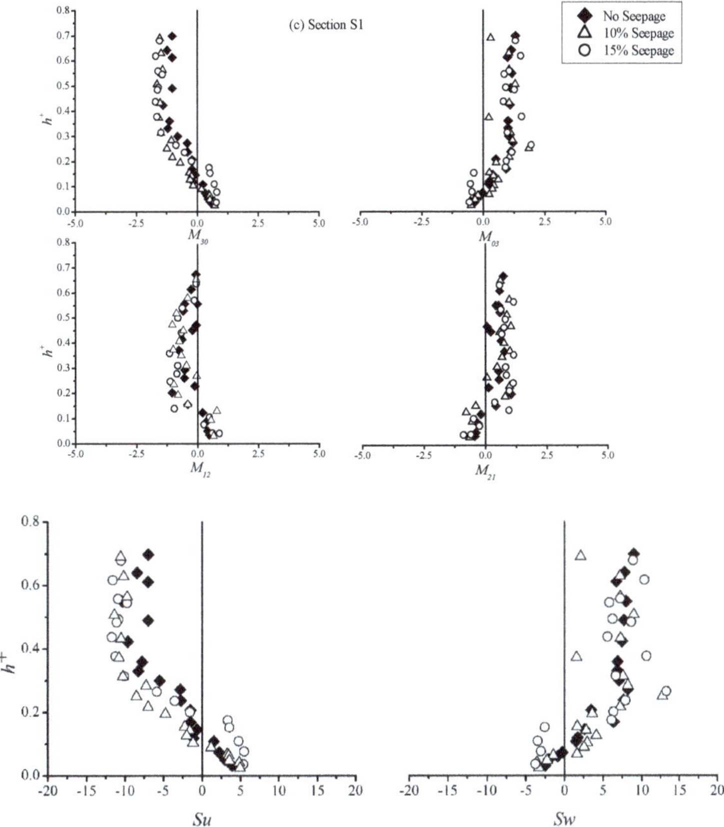

Figure 3. Non-dimensional distribution of third order moment and skewness factors in streamwise and vertical direction for without seepage (NS), 10% seepage (S), and 15% S runs in the following sections: (**a**) upstream the pier (U); (**b**) downward the pier (D); (**c**) laterally to the pier (S1).

3.1.3. Turbulent Kinetic Energy Flux (TKE-Flux)

TKE-flux significantly describes the bed material movement in a sandy channel. The streamwise and vertical TKE-flux are calculated as follows [34]:

$$f_{ku} = 0.75(\overline{u\prime u\prime u\prime} + \overline{u\prime w\prime w\prime})$$
$$f_{kw} = 0.75(\overline{u\prime u\prime w\prime} + \overline{w\prime w\prime w\prime}) \quad (1)$$

and are scaled with the shear velocity u^*, thus obtaining ($Fku = fku/u^*3$) and ($Fkw = fkw/u^*3$), respectively. The vertical profiles of the TKE-flux around the pier, for all of the experimental conditions, are shown in Figure 4. In particular, upstream of the pier (U), near the bed (i.e., $h+ < 0.2$), Fku is positive and Fkw is negative, in agreement with the bed erosion as a result of the flow separation upstream of the pier. However, in the case of seepage runs, the absolute values of Fku and Fkw decrease, because of the lower erosive capacity of the reversal flow. Downward the pier (D), near the bed, FKu and FKw show slightly negative and positive values, respectively, with absolute values decreasing with the seepage percentage. This trend shows the hindered flow and the effects of low-speed incoming fluid particles. Moving towards the free surface, approximately at the edge of the scour hole ($h+ \approx 1$), FKu and FKw become positive and negative, respectively.

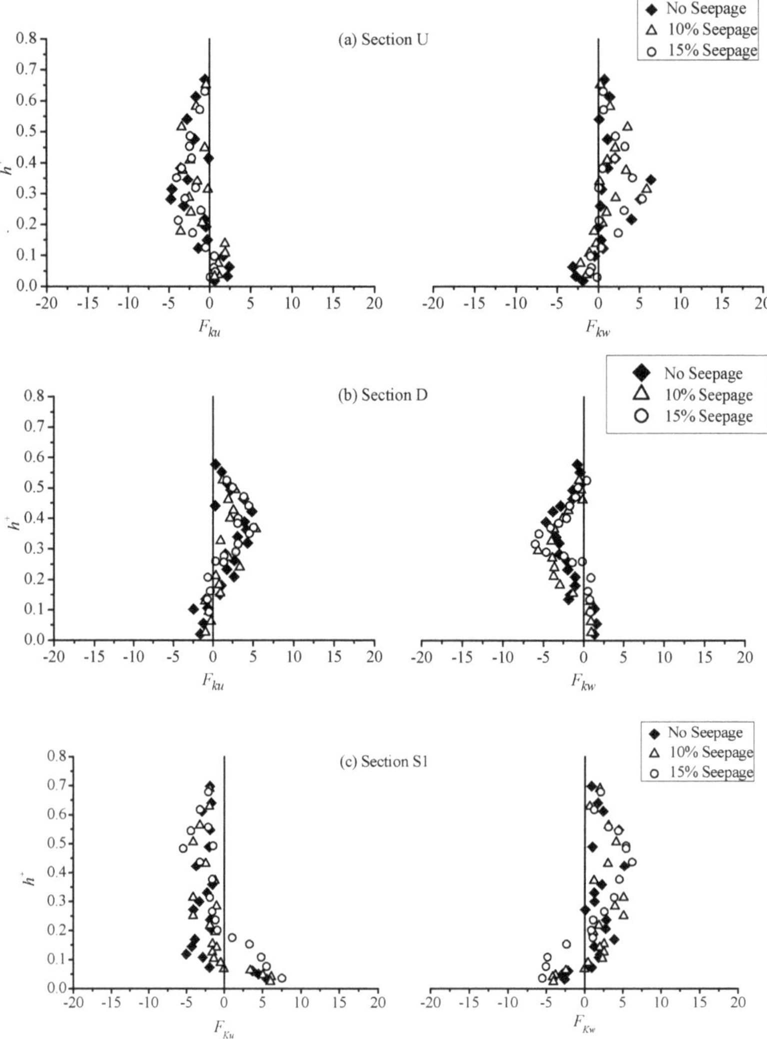

Figure 4. Non-dimensional distributions of turbulent kinetic energy flux (TKE-flux) in the following sections: (**a**) U; (**b**) D; (**c**) S1.

Laterally to the pier (S1)1, near the bed, the increased positive and negative values of *Fku* and *Fkw* with a downward seepage show the enhanced erosion of the bed material in respect to the other sections.

3.1.4. Turbulence Production, Dissipation, and Diffusion

To deeply understand the trend of the velocity fluctuations, it is essential to consider the production, dissipation, and diffusion of the turbulent kinetic energy, expressed as follows:

$$\begin{array}{ll} \text{Turbulence production} & t_p = -\overline{u\prime w\prime}\frac{\partial \overline{U}}{\partial z} \\ \text{Turbulence dissipation} & \varepsilon = \frac{15\vartheta}{\overline{U}^2}\left(\frac{\partial u\prime}{\partial t}\right) \\ \text{Turbulence diffusion} & t_D = \frac{\partial f_{kw}}{\partial z} \end{array} \quad (2)$$

where t_p, ε, and t_D were scaled with $h/u_*{}^2$, and are denoted by T_P, E_D, and T_D, respectively. In Figure 5, the vertical profiles of the production (T_P), diffusion (E_D), and dissipation (T_D) of the turbulent kinetic energy around the pier, for all of the experimental conditions, are shown.

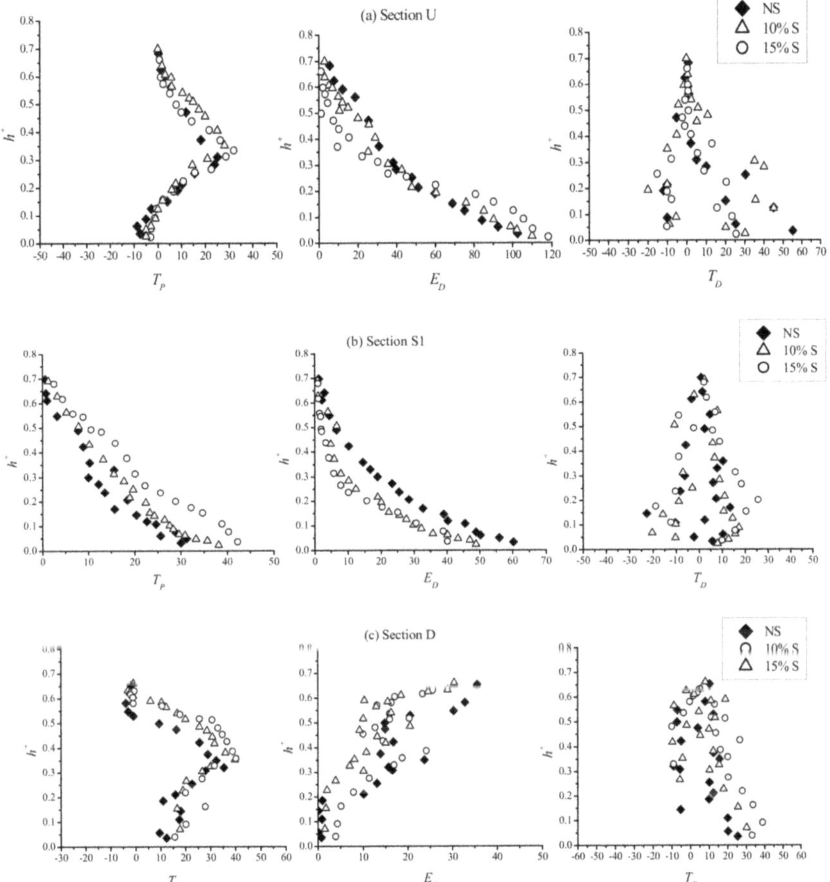

Figure 5. Non-dimensional distributions of turbulent production (T_P), turbulent kinetic energy dissipation (E_D), and diffusion (T_D) for NS, 10% S, and 15% S in the following sections: (**a**) U; (**b**) S1; (**c**) D.

The turbulence production (TP) comes from the exchange of energy between the mean flow velocity and the velocity fluctuations. Therefore, the positive TP expresses the energy flowing from the mean flow velocity to the velocity fluctuations, while, on the contrary, the negative TP shows the energy flowing from the velocity fluctuations to the mean flow velocity. In Figure 5, it is possible to observe that in the section of U, near the bed, in the seepage runs, the decreased TP, with respect to sections D and S1, is evidence of lower turbulence intensities and moment transfer, as the reversal flow is hindered by the seepage flow. On the contrary, in sections D and S1, when the downward seepage is applied, a higher TP shows the enhanced turbulence level in the sections, because of more energy being converted into the turbulent fluctuations. Consequently, in section U, near the bed, in the seepage runs, the ED is higher than in sections D and S1. Moreover, in sections D and S1, near the bed, in the seepage runs, lower TD values correspond to a gain in TP, and a consequent reduction in ED.

3.2. Scour Hole Characteristics

Marine structures such as bridges, docks, and so on, are supported by piers embedded into the sand. The piers obstruct the flow of water, and result in a distortion of the flow pattern and local scour. Such scour can have detrimental effects on the stability of the structure, and hence it is essential to evaluate scour hole characteristics, such as the depth, length, width, area, and volume, so as to prevent the structure from the adverse effects of bed erosion. Obstruction to the flowing stream by piers develops a vortex system in the flow field around them. A horseshoe vortex, upstream of the pier and near the bed, is dominant in the vortex system, and is responsible for sweeping out the bed material around the pier, producing a local scour hole. Moreover, a wake vortices system arises behind the piers, which is shed from their sides as a result of flow separation. Because of the vertical axes, the vortices suck the sand into their cores as they move downstream, and increase the scouring action along their line of movement. The variation in the flow pattern as a result of the piers results in the erosion of the bed material around the piers in both longitudinal and transverse directions. The lateral flow through the channel boundaries, in the form of downward seepage, leads to an enhancement of mass and momentum transfer.

Hence, in the presence of downward seepage, additional forces may be exerted, and this results in enhancing the rate of sediment transport. In this section, the effect of downward seepage on scour hole characteristics is studied.

3.2.1. Geometry of Scour Hole

The experimental measurements carried out in the flume showed that scour starts upstream of the pier, and that the scour hole increases in size over time (Figure 1 shows a snapshot of the scour hole after an experimental run of 24 h, with various dimensions of scour hole). The slope of the scour hole is slightly milder downstream than upstream of the pier. In Figure 6, the length of the scour hole in a longitudinal direction along the centerline is shown, for different flow rates and increasing in seepage percentage. While the scour depth is decreasing, the scour hole length is increasing, and, moreover, the scour hole length is slightly shifting downstream.

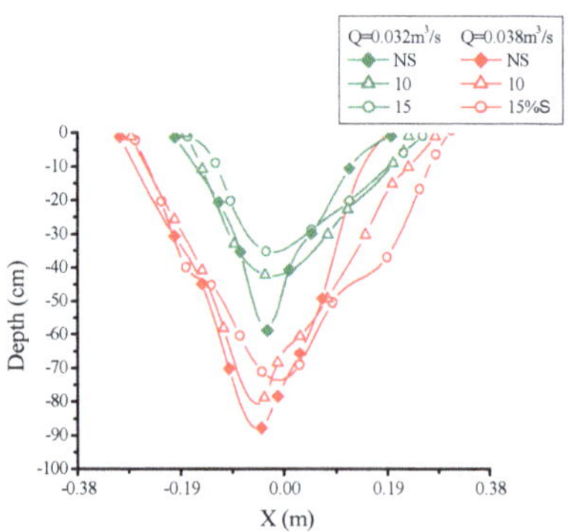

Figure 6. Longitudinal bed profile along the centerline.

Upstream of the pier, in the seepage runs, the reversal flow hindered by the lateral flow through the channel boundaries results in a reduction of the scour depth and the scour length, while downstream the pier, the increased momentum transfer results in the erosion of the bed material and an increase in the scour length. Similarly, from Figure 7, it is possible to observe that the scour hole is wider in the seepage runs than in no seepage run.

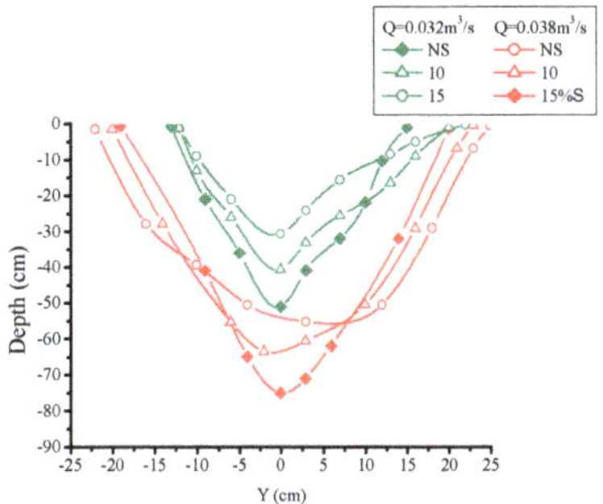

Figure 7. Lateral bed profile upstream from the piers.

This phenomenon depends on the turbulent characteristics of the flow—increasing the turbulence near the bed, laterally to the pier, in the seepage runs, enhances the bed erosion, and, therefore, increases the scour hole width.

3.2.2. Dimensional Analysis

The scour process is mainly influenced by the flow and fluid characteristics, the geometry of the piers, and the bed material properties [15]. However, the lateral flow through the river boundaries also plays significant role in modifying the channel geometry. So, it is necessary to include the downward seepage parameters for the prediction of scour hole characteristics. In this study, the seepage Reynolds number (Re_s = *Vsd50/υ*) is used as a dimensionless seepage parameter. Hence, in the analysis of the scour hole characteristics, such as the length (*ls*) and width (*ws*) of the pier, the following independent parameters can be considered to affect the scour process:

$$l_s, w_s, = \text{f}\,(d_{50}, \sigma_g, g, \upsilon, h, U, Vs, R) \quad (3)$$

where, d_{50} (mm) is the median sand diameter, $\sigma_g\left(\sqrt{d_{84}/d_{16}}\right)$ is the standard deviation, g (m/s^2) is the gravitation acceleration, υ (m^2/s) is the kinematic viscosity, h (m) is the flow depth, U (m/s) is the depth average velocity, Vs (m/s) is the seepage velocity, and R is the hydraulic radius (m). Applying and suitably modifying Equation (1), it can be rearranged as Equation (2).

$$L_s,\ W_s = f\left\{\frac{U}{\sqrt{gR}}, \frac{h}{d_{50}}, \frac{V_s d_{50}}{\vartheta}, \sigma_g\right\} \quad (4)$$

where *Ls* (*ls/D*) is the dimensionless scour length and *Ws* (*ws/D*) is the dimensionless scour hole width. The dimensionless characteristics of the scour hole are a function of the following dimensionless parameters: the Froude number is F_d ($U/\sqrt{gR}$); *h/d50* is aspect ratio, $V_s d_{50}/\upsilon$ is the Seepage Reynolds number (Re_s), and σ_g is the standard deviation. Starting from Equation (2), the length and width of the scour hole can be expressed as follows:

$$L_s = a(F_d)^b\,(h/d_{50})^c\,exp(Re_s)^d\left(\sigma_g\right)^e \quad (5)$$

$$W_s = a(F_d)^b\,(h/d_{50})^c\,exp(Re_s)^d\left(\sigma_g\right)^e \quad (6)$$

where a, b, c, d and e, are the empirical parameters evaluated through experimental measurements with both sands.

The seepage Reynolds number is stated in an exponential, form in order to fulfil the condition without seepage. The final forms of Equations (3) and (4) are the following ones:

$$L_s = 0.06(F_d)^{2.69}\,(h/d_{50})^{1.54}\,exp(Re_s)^{0.703}(\sigma_g)^{-0.644} \quad (7)$$

$$W_s = 0.02(F_d)^{4.23}\,(h/d_{50})^{2.05}\,exp(Re_s)^{0.707}(\sigma_g)^{-0.705} \quad (8)$$

The area and the volume of the scour hole are the function of the scour depth. Hence, Equation (2) can be modified as follows:

$$A_s,\ V_s = f\left\{\frac{U}{\sqrt{gR}},\ \frac{h}{d_{50}},\ \frac{V_s d_{50}}{\vartheta},\ \frac{d_s}{D}\right\} \quad (9)$$

where, $As = as/ap$, and as is the scour hole area and ap is the pier area, and $Vs = vs/vp$, where vs is the scour hole volume and vp is the pier volume.

From the dimensional analysis, the following expressions for the dimensionless area and volume of the scour hole result in the following:

$$A_s = 0.49(F_d)^{4.007}\,(h/d_{50})^{1.18}\,exp(Re_s)^{0.324}\left(\frac{d_s}{D}\right)^{1.69} \quad (10)$$

$$V_s = 0.81(F_d)^{3.39}\,(h/d_{50})^{1.05}\,exp(Re_s)^{3.15}\left(\frac{d_s}{D}\right)^{2.17} \quad (11)$$

Scour hole characteristics, such as the length, width, area, and volume, have been calculated from Equations (5), (6), (8) and (9), and the results have been compared with the measured values, as shown in Figure 8. The overall agreement between Equations (5), (6), (8) and (9), and the measured values is good, with R^2 ranging between 0.88 and 0.92.

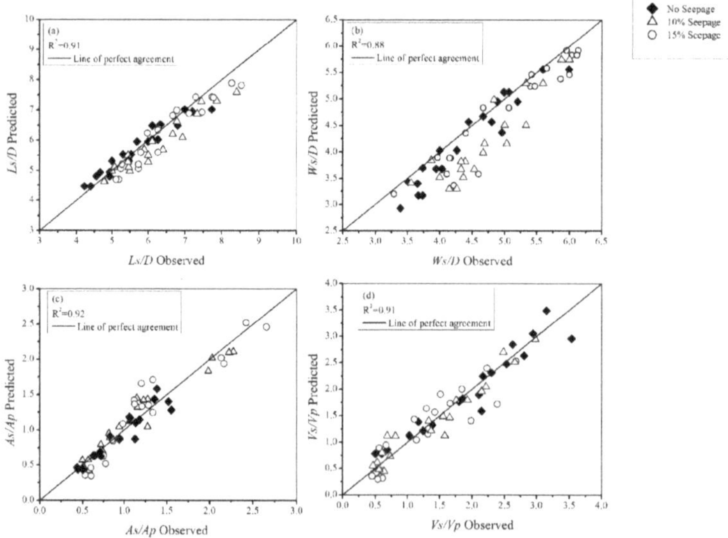

Figure 8. Comparison between the predicted and observed values of the dimensionless (**a**) scour length (Equation (5)), (**b**) scour width (Equation (6)), (**c**) scour area (Equation (8)), and (**d**) scour volume (Equation (9)).

4. Summary and Conclusions

In order to deepen the impact of the downward seepage and the turbulent flow characteristics on the scour geometry around a pier, an experimental measurement of the instantaneous velocity with different percentages of with seepage (10% S and 15% S) and without seepage (NS) have been carried out in four different sections around a pier, for two different diameters (75 and 90 mm).

Upstream of the pier, the streamwise turbulence intensities, with respect of zero seepage, are reduced by 15% in case of the 10% seepage, and by 22% in case of the 15% seepage, respectively. Downstream of the pier, the higher values of vertical turbulence intensities show the presence of wake vortices. Laterally to the pier, the streamwise and vertical turbulence intensities are higher in the seepage runs than without seepage. Upstream of the pier, the positivity of *M30* and *M12* decreases, thus increasing the seepage percentage. Downstream of the pier, decreasing the negativity of *M30* and the positivity of *M03*, respectively, with seepage, shows ejections. Laterally to the pier, the streamwise and vertical turbulence intensities increase by 20%–35% in the seepage runs, resulting in an enhanced in sediment transport.

Upstream of the pier, in the seepage runs, a reduction in TKE-flux confirms the lower erosive capability of the flow. However, the erosive capacity is found to be increased laterally to the pier, by the application of downward seepage. A pronounced effect of seepage has been found on TKE budget. Upstream of the pier, a reduction in the turbulence production, in the case of seepage runs, is in agreement with a lower turbulence level in the same section. From the turbulence statistics of the flow around the pier, in the seepage runs, it has been observed that they decrease upstream of the pier, which results in a lesser scour depth, while they increase laterally the pier, which results in a wider scour hole and shifting towards the downstream of the scour hole length.

The primary characteristics of the scour hole, like the length and width, were physically measured; and the area and volume were measured by using commercial software Surfer®. Finally, with the help of the experimental results obtained from the laboratory flume, empirical relationships for the prediction of the scour hole characteristics, like length, width, area, and volume, including the downward seepage parameter in terms of the seepage Reynolds number, have been developed. Considering that the laboratory flume is a distorted model of a river, the obtained expression showed a good fit with the estimated values.

Author Contributions: B.K. proposed the idea. R.C. did the experiments. R.C., B.K., and P.G. analyzed the results and drafted the manuscript; all of the authors contributed to reviewing and editing the manuscript.

Funding: R.C., B.K. and P.G. acknowledge the support by the UNINA-DICEA research project "Experimental and Numerical Fluids Dynamics".

Conflicts of Interest: The authors declare no conflict of interest.

References

1. Chiew, Y.M. Local Scour at Bridge Piers. Ph.D. Dissertation, The University of Auckland, Auckland, NJ, USA, 1984.
2. Melville, B.W. Local Scour at Bridge Sites. Ph.D. Dissertation, The University of Auckland, Auckland, NJ, USA, 1975.
3. Ettema, R. Scour at Bridge Piers. Ph.D. Dissertation, The University of Auckland, Auckland, NJ, USA, 1980.
4. Melville, B.W.; Coleman, S.E. *Bridge Scour*; Water Resources Publication: Littleton, CO, USA, 2000.
5. Richardson, E.V.; Davis, S.R. Evaluating Scour at Bridges, Hydraulic Engineering Circular No. 18 (hec-18). Publication No. FHWA NHI, 01-001. Available online: http://www.engr.colostate.edu/CIVE510/Manuals/HEC-18%204th%20Ed.(2001)%20-%20Evaluating%20Scour%20at%20Bridges.pdf (accessed on 27 July 2019).
6. Istiarto, I. Flow Around a Cylinder in a Scoured Channel Bed. Ph.D. Thesis, Gadjah Mada University, Yogyakarta, Indonesia, 2001.
7. Izadinia, E.; Heidarpour, M.; Schleiss, A.J. Investigation of turbulence flow and sediment entrainment around a bridge pier. *Stoch. Environ. Res. Risk Assess.* **2013**, *27*, 1303–1314. [CrossRef]

8. Chavan, R.; Sharma, A.; Kumar, B. Effect of downward seepage on turbulent flow characteristics and bed morphology around bridge piers. *J. Mar. Sci. Appl.* **2017**, *16*, 60–72. [CrossRef]
9. Breusers, H.N.C.; Nicollet, G.; Shen, H. Local Scour Around Cylindrical Piers. *J. Hydraul. Res.* **1977**, *15*, 211–252. [CrossRef]
10. Ben Meftah, M.; Mossa, M. Scour holes downstream of bed sills in low-gradient channels. *J. Hydraul. Res.* **2006**, *44*, 497–509. [CrossRef]
11. Guan, D.; Melville, B.W.; Friedrich, H. Flow Patterns and Turbulence Structures in a Scour Hole Downstream of a Submerged Weir. *J. Hydraul. Eng.* **2014**, *140*, 68–76. [CrossRef]
12. Raudkivi, A.J.; Ettema, R. Clear-Water Scour at Cylindrical Piers. *J. Hydraul. Eng.* **1983**, *109*, 338–350. [CrossRef]
13. Chiew, Y.M.; Melville, B.W. Local scour around bridge piers. *J. Hydraul. Res.* **1987**, *25*, 15–26. [CrossRef]
14. Chavan, R.; Kumar, B. Prediction of scour depth and dune morphology around circular bridge piers in seepage affected alluvial channels. *Environ. Fluid Mech.* **2018**, *18*, 923–945. [CrossRef]
15. Richardson, J.R.; Abt, S.R.; Richardson, E.V. Inflow Seepage Influence on Straight Alluvial Channels. *J. Hydraul. Eng.* **1985**, *111*, 1133–1147. [CrossRef]
16. Shukla, M.K.; Mishra, G.C. Canal discharge and seepage relationship. In Proceedings of the 6th National symposium on Hydro, Shillong, India, 6–7 March 1994; pp. 263–274.
17. Tanji, K.K.; Kielen, N.C. *Agricultural Drainage Water Management in Arid and Semi-Arid Areas*; Food and Agriculture Organization: Rome, Italy, 2002.
18. Kinzli, K.D.; Martinez, M.; Oad, R.; Prior, A.; Gensler, D. Using an ADCP to determine canal seepage loss in an irrigation district. *Agric. Water Manag.* **2010**, *97*, 801–810. [CrossRef]
19. Martin, C.A.; Gates, T.K. Uncertainty of canal seepage losses estimated using flowing water balance with acoustic Doppler devices. *J. Hydrol.* **2014**, *517*, 746–761. [CrossRef]
20. Lu, Y.; Chiew, Y.-M.; Cheng, N.-S. Review of seepage effects on turbulent open-channel flow and sediment entrainment. *J. Hydraul. Res.* **2008**, *46*, 476–488. [CrossRef]
21. Rao, A.R.; Sreenivasulu, G.; Kumar, B. Geometry of sand-bed channels with seepage. *Geomorphology* **2011**, *128*, 171–177. [CrossRef]
22. Cao, D.; Chiew, Y.M. Suction effects on sediment transport in closed-conduit flows. *J. Hydraul. Eng.* **2013**, *140*. [CrossRef]
23. MacLean, A.G. Open channel velocity profiles over a zone of rapid infiltration. *J. Hydraul. Res.* **1991**, *29*, 15–27. [CrossRef]
24. Chen, X.; Chiew, Y.M. Velocity Distribution of Turbulent Open-Channel Flow with Bed Suction. *J. Hydraul. Eng.* **2004**, *130*, 140–148. [CrossRef]
25. Singh, A.; Al Faruque, M.A.; Balachandar, R. Vortices and large-scale structures in a rough open-channel flow subjected to bed suction and injection. *J. Eng. Mech.* **2011**, *138*, 491–501. [CrossRef]
26. Devi, T.B.; Sharma, A.; Kumar, B. Turbulence Characteristics of Vegetated Channel with Downward Seepage. *J. Fluids Eng.* **2016**, *138*, 121102. [CrossRef]
27. Marsh, N.A.; Western, A.W.; Grayson, R.B. Comparison of Methods for Predicting Incipient Motion for Sand Beds. *J. Hydraul. Eng.* **2004**, *130*, 616–621. [CrossRef]
28. Dey, S.; Sarkar, S.; Ballio, F. Double-averaging turbulence characteristics in seeping rough-bed streams. *J. Geophys. Res. Space Phys.* **2011**, *116*, F03020. [CrossRef]
29. Kumar, V.; Raju, K.G.R.; Vittal, N. Reduction of Local Scour around Bridge Piers Using Slots and Collars. *J. Hydraul. Eng.* **1999**, *125*, 1302–1305. [CrossRef]
30. Goring, D.G.; Nikora, V.I. Despiking Acoustic Doppler Velocimeter Data. *J. Hydraul. Eng.* **2002**, *128*, 117–126. [CrossRef]
31. Chavan, R.; Kumar, B. Experimental investigation on flow and scour characteristics around tandem piers in sandy channel with downward seepage. *J. Mar. Sci. Appl.* **2017**, *16*, 313–322. [CrossRef]
32. Bandyopadhyay, P.R.; Gad-El-Hak, M. Reynolds Number Effects in Wall-Bounded Turbulent Flows. *Reynolds Number. Eff. Wall-Bounded Turbul. Flows* **1994**, *47*, 307–365.
33. Raupach, M.R. Conditional statistics of Reynolds stress in rough-wall and smooth-wall turbulent boundary layers. *J. Fluid Mech.* **1981**, *108*, 363. [CrossRef]

34. Krogstadt, P.Å.; Antonia, R. Surface roughness effects in turbulent boundary layers. *Exp. Fluids* **1999**, *27*, 450–460. [CrossRef]

Article

Two-Phase Flow Simulation of Tunnel and Lee-Wake Erosion of Scour below a Submarine Pipeline

Antoine Mathieu *, Julien Chauchat, Cyrille Bonamy and Tim Nagel

Laboratoire des Ecoulements Géophysiques et Industriels (LEGI), Centre National de la Recherche Scientifique (CNRS), University of Grenoble Alpes, Grenoble-INP, F-38000 Grenoble, France

* Correspondence: antoine.mathieu@univ-grenoble-alpes.fr

Received: 29 July 2019; Accepted: 13 August 2019; Published: 19 August 2019

Abstract: This paper presents a numerical investigation of the scour phenomenon around a submarine pipeline. The numerical simulations are performed using SedFoam, a two-phase flow model for sediment transport implemented in the open source Computational Fluid Dynamics (CFD) toolbox OpenFOAM. The paper focuses on the sensitivity of the granular stress model and the turbulence model with respect to the predictive capability of the two-phase flow model. The quality of the simulation results is estimated using a statistical estimator: the Brier Skill Score. The numerical results show no sensitivity to the granular stress model. However, the results strongly depend on the choice of the turbulence model, especially through the different implementations of the cross-diffusion term in the dissipation equation between the $k-\varepsilon$ and the $k-\omega2006$ models. The influence of the cross-diffusion term tends to indicate that the sediment transport layer behaves more as a shear layer than as a boundary layer, for which the $k-\varepsilon$ model is more suitable.

Keywords: two-phase flow; scour; pipeline; numerical modeling; turbulence modeling

1. Introduction

The scour phenomenon is a major cause of the rupture or self-burying of submarine pipelines. This phenomenon has to be accurately predicted during the design process of these submarine structures. An important amount of experimental and numerical studies has been published in the literature to characterize both the shape and the time development of the scour hole. The erosion under a submarine pipeline can be decomposed into three steps: (1) the onset; when the current around the cylinder is strong enough, it generates a pressure drop, which liquefies the sediments underneath the cylinder; (2) the tunneling stage; when a breach is formed between the cylinder and the sediment bed, it expands due to the strong current in the breach; (3) the lee-wake erosion stage; when the gap is large enough, vortices are shed in the wake of the cylinder, leading to erosion downstream of the scour hole [1]. Mao (1986) [2] performed extensive experiments for which the shape of the sediment bed around the cylinder was recorded for various flow conditions. The data collected by Mao (1986) [2] have been widely used as a benchmark for sediment transport models applied to the scour phenomenon.

The numerical simulation of scour under a pipeline has been extensively investigated during the last few decades. The first numerical simulations were based on single-phase models and differed mainly by the way turbulence was modeled. Models based on potential flow theory like Chiew (1991) [3] or Li and Cheng (1999) [4] tend to predict correctly the final shape of the upstream part of the scour hole and its maximum depth. However, they fail to reproduce the downstream part because, according to Sumer (2002) [1], the erosion in this region is dominated by the wake of the cylinder and vortex shedding in the sediment bed. Other single-phase flow models using $k-\varepsilon$ turbulence models like Leeuwenstein and Wind (1984) [5] were used to determine the shape of the scour hole, but also failed to reproduce the equilibrium stage. Indeed, in early numerical models, the role played by the suspended sediment was underestimated, and only the bed load transport component of sediment transport was taken into account. Clear water cases were better predicted than live-bed cases because suspension plays a more important role in live-bed conditions. Liang et al. (2004) [6] presented a numerical model able to simulate the time development of the scour hole in live-bed and clear-water conditions. The authors' model took into account bed load and suspended load transport, and a $k-\varepsilon$ turbulence model was used. Furthermore, the influence of the local bed slope on the critical shear stress was incorporated. This means that the threshold Shields parameter θ_c was adjusted to a higher value for sediments moving up slope and to lower value for sediments moving down slope. The mathematical model presented by Liang et al. (2004) [6] showed fairly good agreement with the experimental data from Mao (1986) [2], but the accumulation of sediment behind the pipeline was over-predicted for both live-bed and clear-water scour cases. Liang et al. (2004) [6] argued that this over-prediction was caused by the choice of the $k-\varepsilon$ turbulence model. Indeed, the $k-\varepsilon$ model tends to smooth out the fluctuations in the wake of the pipeline, and therefore, the interaction with the sediment bed can be altered.

The mutual interactions between the fluid and the sediments are more complex than a simple local shear stress relation, as is classically assumed in single-phase flow models. During the past two decades, a new modeling approach has emerged, two-phase flow models [7–9]. One of the main advantages over classical models is that the two-phase flow approach does not require the use of the empirical sediment transport rate and erosion-deposition formulas. The physical grounds on which this new generation of sediment transport models is based should allow improving scour modeling. The two-phase flow approach has already been applied to the scour below a submarine pipeline configuration. Zhao and Fernando (2007) [10] used a two-phase model implemented in the CFD software FLUENT with a $k-\varepsilon$ turbulence model. The temporal evolution of the maximum scour hole was captured in clear-water conditions, but they found that sediments were still in motion in the "fixed" sediment bed layer. It was the first time that a two-phase flow model was applied to the case of scour under a pipeline. At the time, their results were encouraging given the complexity of the phenomenon.

Bakhtiary et al. (2011) [11] also simulated scour under a pipeline in clear-water conditions with a $k-\varepsilon$ turbulence model. The tunneling stage of scour was correctly reproduced, and the shape of the sediment bed corresponded to the experimental data from Mao (1986) [2]. Nevertheless, the upstream part of the scour hole seemed to be better predicted than the downstream part.

More recently, Lee et al. (2016) [12] used a two-phase flow model with a $k-\varepsilon$ turbulence model and the $\mu(I)$ rheology for the granular phase [13]. The authors performed simulations of scour under a pipeline in live-bed conditions. The undisturbed Shields number was higher than in the previous simulations, and no comparison can be made between this model and the other models cited previously. Nevertheless, the time evolution of the scour hole corresponded to the experimental data. They found a strong influence of the turbulence model parameters on the final morphology. Two-phase flow models reached the level of performance of the classical models. Lee et al. (2016) [12] also pointed out the limitation of the $k-\varepsilon$ model and the necessity to use a $k-\omega$-type turbulence model for prediction of the final shape of the scour hole downstream of the pipeline.

The revisited $k-\omega$ turbulence model from Wilcox (2006) [14] adapted for two-phase flow by Nagel (2019) [15] (referred to as $k-\omega 2006$ in this paper) can reproduce vortex-shedding phenomenon (see Appendix A). It should therefore be able to simulate the lee-wake erosion stage.

In Section 2, the two-phase flow model is presented. The numerical setup is detailed in Section 2.4, and the results are presented and discussed in Section 3.

2. Materials and Methods

2.1. Governing Equations

The two-phase flow model for sediment transport, SedFoam, developed by Cheng et al. (2017) [16] and Chauchat et al. (2017) [17], was used as a starting point. SedFoam is an open-source and multi-version solver implemented in the open-source CFD toolbox OpenFOAM and proposes several granular stress and turbulence models (https://github.com/SedFoam/sedfoam). SedFoam is designed to study three-dimensional sediment transport.

In the two-phase flow formulation, both solid and fluid phases are described by Eulerian equations. The fluid and solid mass conservation equations are written as:

$$\frac{\partial \phi}{\partial t} + \frac{\partial \phi u_i^s}{\partial x_i} = 0 \tag{1}$$

$$\frac{\partial (1-\phi)}{\partial t} + \frac{\partial (1-\phi) u_i^f}{\partial x_i} = 0 \tag{2}$$

where ϕ is the sediment phase concentration, u_i^s and u_i^f are the sediment and fluid phase velocities, respectively, and $i = 1, 2$ are the stream-wise and vertical components.

The momentum equations for the solid and fluid phases are given by:

$$\begin{aligned} \frac{\partial \rho^s \phi u_i^s}{\partial t} + \frac{\partial \rho^s \phi u_i^s u_j^s}{\partial x_j} = -\phi \frac{\partial p}{\partial x_i} - \frac{\partial p^s}{\partial x_i} + \frac{\partial \tau_{ij}^s}{\partial x_j} + \phi \rho^s g_i + \phi(1-\phi) K (u_i^f - u_i^s) \\ -(1-\phi) \frac{1}{\sigma_c} K \nu_t^f \frac{\partial \phi}{\partial x_i} \end{aligned} \tag{3}$$

$$\begin{aligned} \frac{\partial \rho^f (1-\phi) u_i^f}{\partial t} + \frac{\partial \rho^f (1-\phi) u_i^f u_j^f}{\partial x_j} = -(1-\phi) \frac{\partial p}{\partial x_i} + \frac{\partial \tau_{ij}^f}{\partial x_j} + (1-\phi) \rho^f g_i \\ -\phi(1-\phi) K (u_i^f - u_i^s) + (1-\phi) \frac{1}{\sigma_c} K \nu_t^f \frac{\partial \phi}{\partial x_i} \end{aligned} \tag{4}$$

with ρ^s and ρ^f the solid and fluid density, g_i the acceleration of gravity, p the fluid pressure, p^s the solid phase normal stress, and τ_{ij}^s and τ_{ij}^f the solid and fluid phase shear stresses. The solid phase shear stress closure model is detailed in Section 2.2, and the fluid phase shear stress is expressed as:

$$\tau_{ij}^f = \rho^f (1-\phi) \left[2 \nu_{Eff} S_{ij}^f - \frac{2}{3} k \delta_{ij} \right]. \tag{5}$$

$S_{ij}^f = 1/2 \left(\partial u_i^f / \partial x_j + \partial u_j^f / \partial x_i \right) - 1/3 \left(\partial u_k^f / \partial u_k^f \right)$ is the deviatoric part of the fluid strain rate tensor; k is the turbulent kinetic energy (TKE); and ν_{Eff} is the effective velocity defined by $\nu_{Eff} = \nu_t^f + \nu_{mix}$ with ν_t^f the eddy viscosity calculated by a turbulence closure model (see Section 2.3) and ν_{mix} the mixture viscosity following the model proposed by Boyer et al. (2011) [18]:

$$\frac{\nu_{mix}}{\nu^f} = 1 + 2.5 \phi \left(1 - \frac{\phi}{\phi_{max}} \right)^{-1}, \tag{6}$$

where $\phi_{max} = 0.635$ is the maximum value for the solid phase concentration and ν^f is the fluid kinematic viscosity.

The last two terms of the right-hand side of both momentum equations represent the drag force coupling the two phases. σ_c is the Schmidt number, and K is the drag parameter modeled according to Richardson and Zaki (1954) [19]:

$$K = 0.75 C_d \frac{\rho^s}{d} \parallel u^f - u^s \parallel (1-\phi)^{-h_{Exp}}. \tag{7}$$

d is the particles' diameter; h_{Exp} is the hindrance exponent controlling the drag increase with increasing solid concentration; and C_d is the drag coefficient calculated by the empirical formula given by Schiller and Naumann (1933) [20]:

$$C_d = \begin{cases} \frac{24}{Re_p}(1 + 0.15Re_p^{0.687}), & Re_p \leq 1000 \\ 0.44, & Re_p > 1000 \end{cases}, \tag{8}$$

where Re_p is the particulate Reynolds number defined by: $Re_p = (1 - \phi) \parallel u^f - u^s \parallel d/\nu^f$.

2.2. Granular Stress Models

The particle pressure is the sum of the the pressure induced by collision p^s, calculated differently depending on the solid phase stress closure model and the pressure induced by the permanent contact between the particles p^{ff} defined as:

$$p^{ff} = \begin{cases} 0, \phi < \phi_{min}^{Fric} \\ Fr\frac{(\phi - \phi_{min}^{Fric})^{\eta_0}}{(\phi_{max} - \phi)^{\eta_1}}, \phi \geq \phi_{min}^{Fric}, \end{cases} \tag{9}$$

where $\phi_{min}^{Fric} = 0.57$, $Fr = 0.05$, $\eta_0 = 3$, and $\eta_1 = 5$ are empirical coefficients. In the present work, numerical simulations are conducted using two granular stress models: the dense granular flow rheology ($\mu(I)$ rheology) and the kinetic theory for granular flows (KT).

2.2.1. $\mu(I)$ Rheology

The pressure induced by collisions and frictional interactions is modeled following Chauchat et al. (2017) [21]:

$$p^s = \left(\frac{B_\phi \, \phi}{\phi_{max} - \phi}\right)^2 \rho^s d^2 \parallel S^s \parallel^2 \tag{10}$$

where $B_\phi = 1/3$ is a parameter of the dilatancy law [22] and $\parallel S^s \parallel$ is the norm of the deviatoric part of the solid phase strain rate tensor S_{ij}^s defined as $\parallel S^s \parallel = \sqrt{2S_{ij}^s S_{ij}^s}$.

Following Jop et al. (2006) [23], the particle shear stress is proportional to the particle pressure following a frictional law:

$$\tau_{ij}^s = \mu(I)(p^s + p^{ff})\frac{S_{ij}^s}{\parallel S^s \parallel}. \tag{11}$$

According to GDRmidi(2004) [13], the friction coefficient $\mu(I)$ is given by:

$$\mu(I) = \mu_s + \frac{\mu_2 - \mu_s}{I_0/I + 1}, \tag{12}$$

where $I = \parallel S^s \parallel d\sqrt{\rho^s/\tilde{p}^s}$ is the inertial number, $\mu_s = 0.63$ is the static friction coefficient for sand, and $\mu_2 = 1.13$ and $I_0 = 0.6$ are empirical coefficients.

A frictional shear viscosity is introduced to be consistent with the fluid phase momentum equation, and the particle shear stress is written as $\tau_{ij}^s = \nu_{Fr}^s S_{ij}^s$, with the frictional shear viscosity ν_{Fr}^s written as:

$$\nu_{Fr}^s = min\left(\frac{\mu(I)(p^s + p^{ff})}{\rho^s\left(\parallel S^s \parallel^2 + D_{small}^2\right)^{1/2}}, \nu_{max}\right). \tag{13}$$

D_{small} is regularization parameter from Chauchat and Médale (2014) [24], and ν_{max} is a viscosity limiter set to $\nu_{max} = 10$

2.2.2. Kinetic Theory for Granular Flows

The model adopted was suggested by Ding and Gidaspow (1990) [25]. The particulate pressure is a function of the particle velocity fluctuations represented by the granular temperature Θ following:

$$p^s = \rho^s \phi [1 + 2(1+e)\phi g_{s0}]\Theta, \tag{14}$$

with e the coefficient of restitution during the collision and $g_{s0} = (2-\phi)/2(1-\phi)^3$ a radial distribution function from Carnahan and Starling (1969) [26] introduced to describe the crowdedness of particles.

The particle shear stress τ^s_{ij} is decomposed into the sum of a frictional and a collisional stress component:

$$\tau^s_{ij} = \tau^{ff}_{ij} + \tilde{\tau}^s_{ij}. \tag{15}$$

The frictional components allow reproducing the immobile sediment bed behavior and are defined as $\tau^{ff}_{ij} = 2\rho^s \nu^s_{Fr} S^s_{ij}$, with ν^s_{Fr} calculated as:

$$\nu^s_{Fr} = \frac{p^{ff} sin(\theta_f)}{\rho^s \left(|| S^s ||^2 + D^2_{small} \right)^{1/2}}, \tag{16}$$

using a constant friction angle $\theta_f = 32°$.

The particle collisional stress is calculated as:

$$\tilde{\tau}^s_{ij} = 2\mu^s S^s_{ij} + \lambda \frac{\partial u^s_k}{\partial x_k} \delta_{ij}. \tag{17}$$

The particle shear viscosity μ^s and bulk viscosity λ are functions of the granular temperature and the radial distribution function following:

$$\mu^s = \rho^s d\sqrt{\Theta} \left[\frac{4\phi^2 g_{s0}(1+e)}{5\sqrt{\pi}} + \frac{\sqrt{\pi} g_{s0}(1+e)(3e-1)\phi^2}{15(3-e)} + \frac{\sqrt{\pi}\phi}{6(3-e)} \right] \tag{18}$$

and:

$$\lambda = \frac{4}{3}\phi^2 \rho^s d g_{s0}(1+e)\sqrt{\frac{\Theta}{\pi}}. \tag{19}$$

The balance equation for the granular temperature is written as:

$$\frac{3}{2}\left[\frac{\partial \phi \rho^s \Theta}{\partial t} + \frac{\partial \phi \rho^s u^s_j \Theta}{\partial x_j} \right] = (-p^s \delta_{ij} + \tilde{\tau}^s_{ij}) \frac{\partial u^s_i}{\partial x_j} - \frac{\partial q_j}{\partial x_j} - \gamma + J_{int}, \tag{20}$$

where the granular temperature flux q_j is modeled following Fourier's law of conduction:

$$q_j = -D_\Theta \frac{\partial \Theta}{\partial x_j} \tag{21}$$

with D_Θ the conductivity calculated as:

$$D_\Theta = \rho^s d\sqrt{\Theta} \left[\frac{2\phi^2 g_{s0}(1+e)}{\sqrt{\pi}} + \frac{9\sqrt{\pi} g_{s0}(1+e)^2(2e-1)\phi^2}{2(49-33e)} + \frac{5\sqrt{\pi}\phi}{2(49-33e)} \right]. \tag{22}$$

The expression of the dissipation rate of granular temperature γ is modeled following Ding and Gidaspow (1990) [25]:

$$\gamma = 3(1-e^2)\phi^2\rho^s g_{s0}\Theta \left[\frac{4}{d}\sqrt{\frac{\Theta}{\pi}} - \frac{\partial u_j^s}{\partial x_j}\right]. \tag{23}$$

Finally, the fluid particle interaction term J_{int} is expressed as:

$$J_{int} = \phi K(2\alpha k - 3\Theta), \tag{24}$$

where α characterizes the degree of correlation between particles and fluid velocity fluctuations following the expression $\alpha = e^{-BSt}$, where B is an empirical coefficient and St is the Stokes number defined as the ratio between the particle response time $t_p = \rho^s/(1-\phi)K$ and the characteristic time scale of the most energetic eddies $t_l = k/(6\varepsilon)$, with ε the dissipation rate of TKE.

2.3. Turbulence Models

Three two-phase flow versions of Reynolds-averaged Navier–Stokes (RANS) turbulence models were used in the simulations: a $k-\varepsilon$ model [9,16], a $k-\omega 2006$ model [15], and a $k-\varepsilon$ turbulence model written in terms of the specific dissipation rate of TKE ω, denoted hereafter as the modified $k-\varepsilon$ model. The last model was only used to compare the $k-\varepsilon$ and $k-\omega 2006$ behaviors.

In the framework of two-equation RANS turbulence models, transport equations for the dissipation rate and for the TKE need to be solved to compute the turbulent viscosity ν_t^f. The general expression for the TKE transport equation is given by:

$$\begin{aligned}\frac{\partial k}{\partial t} + u_j^f\frac{\partial k}{\partial x_j} = \frac{R_{ij}}{\rho^f}\frac{\partial u_i^f}{\partial x_j} + \frac{\partial}{\partial x_j}\left[\left(\nu^f + \sigma_k \nu_t^f\right)\frac{\partial k}{\partial x_j}\right] - \varepsilon \\ - \frac{2K(1-\alpha)\phi k}{\rho^f} - \frac{\nu_t^f}{\sigma_c(1-\phi)}\frac{\partial \phi}{\partial x_j}(s-1)g_i\end{aligned} \tag{25}$$

with R_{ij}^f the Reynolds stress tensor and σ_k an empirical coefficient.

The transport equation for the dissipation rate and the expression of the turbulent viscosity differ for the different turbulence models.

2.3.1. $k-\varepsilon$ Model

For the $k-\varepsilon$ model, the turbulent viscosity ν_t^f is calculated as:

$$\nu_t^f = C_\mu \frac{k^2}{\varepsilon}, \tag{26}$$

and the following transport equation for the dissipation rate ε is solved:

$$\begin{aligned}\frac{\partial \varepsilon}{\partial t} + u_j^f\frac{\partial \varepsilon}{\partial x_j} = C_{1\varepsilon}\frac{\varepsilon}{k}\frac{R_{ij}}{\rho^f}\frac{\partial u_i^f}{\partial x_j} + \frac{\partial}{\partial x_j}\left[\left(\nu^f + \sigma_\varepsilon \nu_t^f\right)\frac{\partial \varepsilon}{\partial x_j}\right] - C_{2\varepsilon}\frac{\varepsilon^2}{k} \\ - C_{3\varepsilon}\frac{\varepsilon}{k}\frac{2K(1-\alpha)\phi k}{\rho^f} - C_{4\varepsilon}\frac{\varepsilon}{k}\frac{\nu_t^f}{\sigma_c(1-\phi)}\frac{\partial \phi}{\partial x_j}(s-1)g_i.\end{aligned} \tag{27}$$

The values of the empirical coefficients σ_k, σ_ε, $C_{1\varepsilon}$, $C_{2\varepsilon}$, $C_{3\varepsilon}$, $C_{4\varepsilon}$, and C_μ are listed in Table 1.

Table 1. Empirical coefficients for the $k-\varepsilon$ turbulence model from Chauchat et al. (2017) [17].

σ_k	σ_ε	$C_{1\varepsilon}$	$C_{2\varepsilon}$	$C_{3\varepsilon}$	$C_{4\varepsilon}$	C_μ
1.0	0.77	1.44	1.92	1.2	1.0	0.09

2.3.2. $k-\omega2006$ Model

The dissipation rate ε can be expressed in term of specific dissipation rate ω following the expression $\varepsilon = C_\mu k\omega$. The $k-\omega2006$ turbulence model uses ω and the norm of the deviatoric part of the strain rate tensor $\| S^f \|$ to compute the eddy viscosity:

$$\nu_t^f = \frac{k}{max\left[\omega,\ C_{lim}\frac{\| S^f \|}{\sqrt{C_\mu}}\right]} \tag{28}$$

Compared with the $k-\varepsilon$ or the standard $k-\omega$ model, a stress limiter is incorporated and adjusted by the coefficient C_{lim} [14].

The transport equation for the specific dissipation rate ω reads:

$$\begin{aligned}\frac{\partial \omega}{\partial t} + u_j^f \frac{\partial \omega}{\partial x_j} = C_{1\omega}\frac{\omega}{k}\frac{R_{ij}}{\rho^f}\frac{\partial u_i^f}{\partial x_j} + \frac{\partial}{\partial x_j}\left[\left(\nu^f + \sigma_\omega \nu_t^f\right)\frac{\partial \omega}{\partial x_j}\right] - C_{2\omega}\omega^2 - C_{3\omega}\omega\frac{2K(1-\alpha)\phi}{\rho^f} \\ -C_{4\omega}\frac{\omega}{k}\frac{\nu_t^f}{\sigma_c(1-\phi)}\frac{\partial \phi}{\partial x_j}(s-1)g_i + \sigma_d\frac{1}{\omega}\frac{\partial k}{\partial x_j}\frac{\partial \omega}{\partial x_j}.\end{aligned} \tag{29}$$

The empirical coefficients for this turbulence model are presented in Table 2, and the coefficient before the cross-diffusion term σ_d is given by:

$$\sigma_d = \begin{cases} 0 & \text{for} \quad \frac{\partial k}{\partial x_j}\frac{\partial \omega}{\partial x_j} < 0 \\ \frac{1}{8} & \text{for} \quad \frac{\partial k}{\partial x_j}\frac{\partial \omega}{\partial x_j} \geq 0. \end{cases} \tag{30}$$

Table 2. Empirical coefficients for the $k-\omega2006$ turbulence model.

σ_k	σ_ω	$C_{1\omega}$	$C_{2\omega}$	$C_{3\omega}$	$C_{4\varepsilon}$	C_μ	C_{lim}
0.6	0.5	0.52	0.0708	0.35	1.0	0.09	0.875

2.3.3. Modified $k-\varepsilon$ Model

The modified $k-\varepsilon$ model is obtained by substituting the dissipation rate ε by $C_\mu k\omega$ in Equations (26) and (27). The new expression for the eddy viscosity is given by:

$$\nu_t^f = \frac{k}{\omega} \tag{31}$$

and the new dissipation rate transport equation is the same as Equation (29) from the $k-\omega2006$ model with different coefficients (see the coefficients in Table 3). The coefficient σ_d in front of the cross-diffusion term allows switching from the $k-\varepsilon$ and the $k-\omega2006$ models and investigating the sensitivity of the model to this cross-diffusion term.

$$\sigma_d = \begin{cases} 0 & \text{for} \quad \frac{\partial k}{\partial x_j}\frac{\partial \omega}{\partial x_j} < 0 \\ 1.712 & \text{for} \quad \frac{\partial k}{\partial x_j}\frac{\partial \omega}{\partial x_j} \geq 0. \end{cases} \tag{32}$$

Table 3. Empirical coefficients for the modified $k-\varepsilon$ turbulence model.

σ_k	σ_ω	$C_{1\omega}$	$C_{2\omega}$	$C_{3\omega}$	$C_{4\varepsilon}$	σ_d
1.0	0.856	0.44	0.0828	0.35	1.0	1.712 or Equation (32)

2.4. Numerical Setup

Following Lee et al. (2016) [12], the configuration of Mao (1986) [2] was used as a benchmark to study the time evolution of the bed morphology. A pipeline having a diameter $D = 5$ cm was placed just above a sediment bed made of medium sand (median diameter $d_{50} = 360$ µm and density $\rho^s = 2600$ kg·m^{-3}). The incoming current had a mean velocity $\overline{U} = 0.87$ m·s^{-1} corresponding to a Shields number $\theta_\infty = 0.33$. Initially, the pipeline was just laid on the sediment bed with no embedment. Mao (1986) [2] measured the sediment bed profile at different times until scour equilibrium was reached. For the simulations, the dynamics was resolved up to 30 s to ensure that the small perturbations of the sediment bed coming from the inlet did not reach the scour hole.

2.4.1. General Setup

The numerical domain dimensions presented in Figure 1 are similar to the ones used by Lee et al. (2016) [12]. For both configurations, the cylinder was placed 15 diameters away from the inlet. The overall domain dimensions were 35 diameters long and 6.1 diameters high. The top boundary condition was a symmetry plane. For the reduced pressure, the outlet boundary condition was a homogeneous Dirichlet condition ($p - \rho^f gy = 0$ Pa). For the outlet velocity, a homogeneous Neumann boundary condition was used for outgoing flows, and a homogeneous Dirichlet boundary condition was used for incoming flows. The inlet was decomposed into two parts. From the bottom to $y = 1.5D$, a wall-type boundary condition was applied. From $y = 1.5D$ to the top, a rough wall log law velocity profile was used following the expression:

$$u_1^f(y) = \frac{u_*}{\kappa} ln\left(\frac{30y}{k_s}\right), \tag{33}$$

where $\kappa = 0.41$ is the von Karman constant, u_* is the friction velocity, and $k_s = 2.5d$ is the Nikuradse roughness length. The different boundary conditions can be found in the test case 2DPipelineScour, publicly available on GitHub (after the paper is accepted). Second order schemes (Gauss linearUpwind) and the default preconditioned biconjugate gradient pressure solver were used for all the simulations presented in this paper.

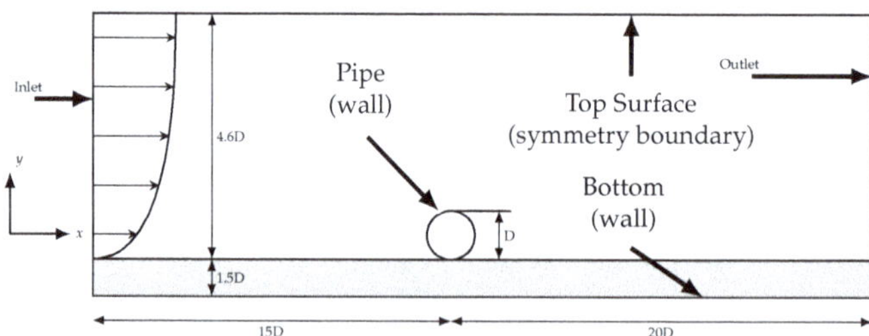

Figure 1. Sketch of the geometry and the boundary conditions used for the computational domain.

For the two configurations, following Chauchat et al. (2017) [17], the turbulent parameter B was set to $B = 1$. According to Van Rijn (1984) [27], the value of σ_c depends on the suspension number w_s/u_* with w_s the particles' fall velocity:

$$\frac{1}{\sigma_c} = 1 + 2\left[\frac{w_s}{u_*}\right]^2, \; 0.1 < \frac{w_s}{u_*} < 1 \tag{34}$$

Therefore, σ_c is bounded between 1/3 and one. For the present configuration, the sediment transport was intense; the suspension number was small; and following Lee et al. (2016) [12], σ_c was set to one.

The mesh was generated using the OpenFOAM utility snappyHexMesh. Cells were refined in the sediment bed region. Non-refined cells were squares having 3×10^{-3}m sides, and refined cells were squares having 7.5×10^{-4} m sides. The different turbulence models required a specific near-wall resolution. Therefore, cells' refinement close to the cylinder and turbulent boundary conditions depended on the choice of the turbulence model. More details are available in the test case 2DPipelineScour.

2.4.2. Simulations with the $k - \varepsilon$ Turbulence Model

The first cells near the cylinder were 6×10^{-4} m thick, giving a dimensionless near-wall cell thickness equal to the required $y^+ = 30$ with y^+ the dimensionless wall distance defined as $y^+ = u^* y/\nu$. A homogeneous Dirichlet boundary condition of 1×10^{-10} m^2 s^{-2} was applied on the cylinder surface for k, and a homogeneous Neumann boundary condition was applied for the rate of dissipation of the TKE ε. Similar to Lee et al. (2016) [12], inlet values for turbulent quantities were set following Ferziger (2002) [28]: $k = 10^{-4}\bar{U}$ and $\varepsilon = k^{3/2}/0.1$ h, with h the distance from the bed to the top boundary.

2.4.3. Simulations with the $k - \omega 2006$ and the Modified $k - \varepsilon$ Turbulence Models

For the simulations using the $k - \omega 2006$ and the modified $k - \varepsilon$ turbulence models, cells near the cylinder were 2×10^{-5} m thick, giving a near-wall cell thickness equal to $y^+ = 1$. Wall functions for smooth walls were applied on the cylinder surface for k and ω. Inlet values for k and ω were calculated similarly to Section 2.3.1 following Ferziger (2002) [28] with $\omega = \varepsilon / C_\mu k$, except that the dissipation was enhanced at the inlet by two orders of magnitude to reduce the incoming TKE, susceptible to damping the vortex-shedding.

2.5. Brier Skill Score

In order to estimate the prediction capability of the different combinations of models objectively (granular stress and turbulence models), the Brier Skill Score (BSS) was used. It is a statistical approach used to measure the quality of an agreement between simulation results and experimental data. This statistical tool has been extensively used in the coastal engineering community [29–31]. It provides an estimation of a model's performance and is defined following the expression:

$$BSS = 1 - \frac{\sum_i^n |y_i^s - y_i^e|^2}{\sum_i^n |y_i^0 - y_i^e(x)|^2}. \tag{35}$$

The BSS compares the sum of the squared difference between the bed elevation y^s from simulations and y^e from experiments at point i (from 0–n, the total number of experimental points) with the mean squared difference between the initial bed elevation y^0 and y^e also at point i. The bed elevation y^s in the simulations is the line of isoconcentration $\phi = 0.5$. A BSS equal to one expresses a perfect agreement between simulation and experimental results. The agreement quality decreases with the BSS, whereas a negative BSS expresses simulation results further away from the experimental results than the initial bed elevation.

3. Results

3.1. Solid Phase Stress Model Sensitivity

Two simulations were conducted to evaluate the sensitivity of the model results to the choice of the granular stress model: the kinetic theory for granular flows and the $\mu(I)$ rheology. For both simulations, the $k-\varepsilon$ turbulence model was used. The time evolution of the maximum scour depth and the shape of the sediment bed are compared with the experimental data from Mao (1986) [2] and numerical simulation results from Lee et al. (2016) [12] in Figures 2 and 3, respectively.

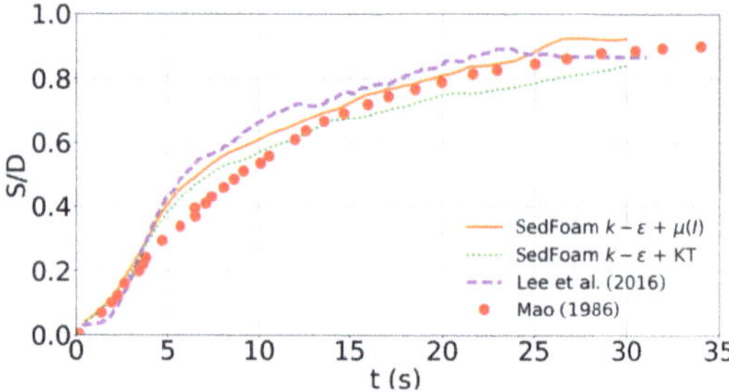

Figure 2. Time evolution of the maximum scour depth for simulations with the $k-\varepsilon$ turbulence model using $\mu(I)$ rheology (orange line) and kinetic theory (green dotted line) compared with the experimental data from Mao (1986) [2] (red dots) and numerical data from Lee et al. (2016) [12] (purple dashed line).

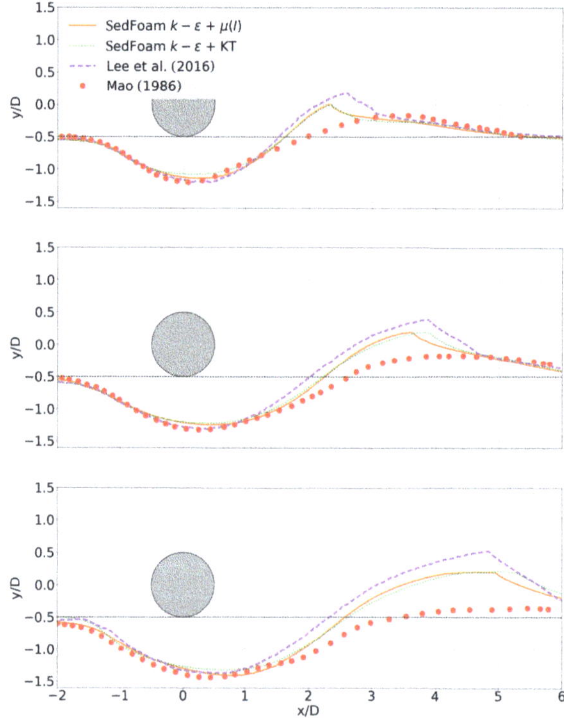

Figure 3. Bed profiles from simulations with the $k-\varepsilon$ turbulence model using $\mu(I)$ rheology (orange line) and kinetic theory (green dotted line) at 11 s (top), 18 s (middle), and 25 s (bottom) compared with experimental data from Mao (1986) [2] (red dots) and numerical data from Lee et al. (2016) [12] (purple dashed line).

It clearly appears that the shape of the sediment bed and its time evolution were not very sensitive to the granular stress model. For both models, the temporal evolution of the maximum scour depth (Figure 2) and the upstream part of the sediment bed (Figure 3) agreed reasonably well with the experimental data from Mao (1986) [2] and the numerical results from Lee et al. (2016) [12]. At 25 s, the BSS calculated from the simulations using the $\mu(I)$ rheology ($BSS = 0.731$) and the KT ($BSS = 0.721$) were very close. The quality of the results was therefore equivalent for both granular stress models.

In both simulations, the sediments tended to accumulate at the downstream side of the pipeline, generating a sand dune. According to Lee et al. (2016) [12], this accretion phenomenon can be explained by the inability of the $k-\varepsilon$ model to reproduce the oscillatory wake behind the cylinder (responsible for the lee-wake erosion stage).

The present results demonstrated that the sediment accumulation observed at the lee side of the cylinder was not due to the solid stress model. Since the granular stress models provided similar results, in the following, only the $\mu(I)$ rheology will be used, as it was more computationally efficient.

3.2. Turbulence Model Sensitivity

In this subsection, the influence of the turbulence model on the shape of the sediment bed is investigated. The results using the $k-\varepsilon$ and the $k-\omega2006$ turbulence models are compared in Figures 4 and 5 in term of the time evolution of the maximum scour depth and the shape of the sediment bed.

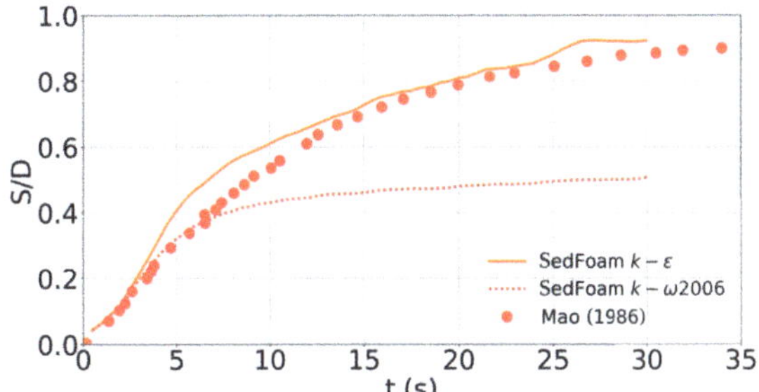

Figure 4. Time evolution of the maximum scour depth from simulations with the $\mu(I)$ rheology using the $k-\varepsilon$ (orange line) and $k-\omega2006$ (red dotted line) turbulence models compared with the experimental data from Mao (1986) [2] (red dots).

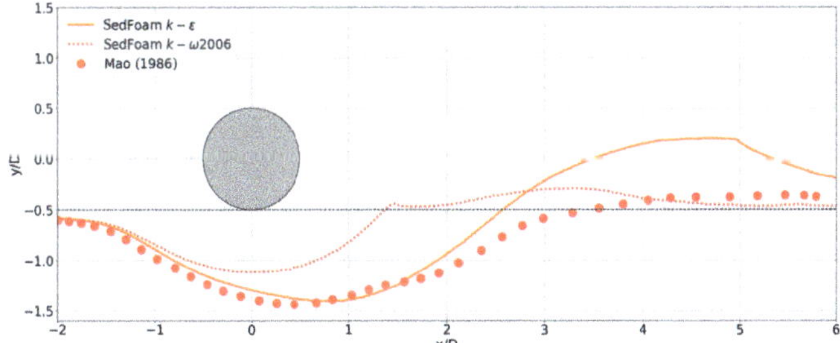

Figure 5. Bed profiles at 25 s from simulations with the $\mu(I)$ rheology using the $k-\varepsilon$ (orange line) and $k-\omega2006$ (red dotted line) turbulence models compared with the experimental data from Mao (1986) [2] (red dots).

The time evolution and the equilibrium depth of the scour hole were significantly underestimated when using the $k-\omega2006$ turbulence model. This turbulence model did not provide quantitative results in this configuration. However, the vortex shedding phenomenon was predicted, and the sediment bed downstream of the pipeline was eroded (see bed interface between x/D = 4 and x/D = 6).

The snapshot provided in Figure 6 confirms that the erosion was caused by the vortices in the wake of the cylinder. A strong sediment flux was associated with a vortex reaching the sand dune downstream of the pipeline. Therefore, the accretion of sediment visible using the $k-\varepsilon$ model was no longer present when using the $k-\omega2006$ turbulence model.

The BSS at 25 s using the $k-\omega2006$ was 0.569, which was significantly lower than the one obtained with the $k-\varepsilon$ model. The lee-wake erosion of the sand dune did not compensate the underestimation of the scour depth in the BSS.

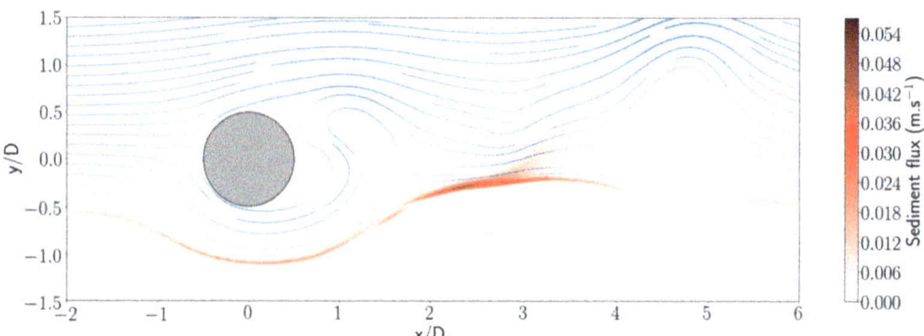

Figure 6. Streamlines and sediment volumetric flux at 25 s for the simulation using the $k-\omega/2006$ turbulence model.

A sensitivity analysis on the cross-diffusion term appearing in the dissipation equation through the coefficient σ_d was performed to identify the main differences between the two aforementioned models. The time evolution of the maximum scour depths from simulations using the $k-\omega2006$, the modified $k-\varepsilon$, and the modified $k-\varepsilon$ taking only the positive contribution of the cross-diffusion term is presented in Figure 7. It appears that removing the negative contribution of the cross-diffusion term in the modified $k-\varepsilon$ turbulence model provided results closer to the ones obtained using the $k-\omega2006$ turbulence model with an equilibrium erosion depth largely underestimated. The definition of the turbulent viscosity in the $k-\omega2006$ turbulence model mainly affected the dilute regions, but the differences in terms of bed elevation visible in Figures 4 and 7 came from the cross-diffusion term.

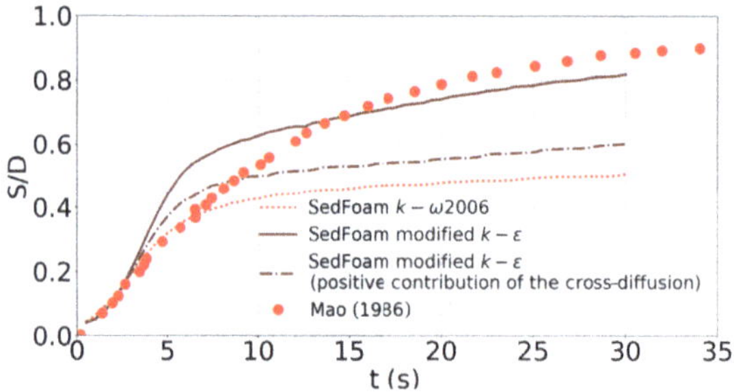

Figure 7. Time evolution of the maximum scour depth from simulations with $\mu(I)$ rheology using the $k-\omega2006$ (red dotted line) and modified $k-\epsilon$ turbulence model with (brown line) and without (brown dashed dotted line) the negative contribution of the cross-diffusion term compared with the experimental data from Mao (1986) [2] (red dots).

The cross-diffusion term was responsible for the behavior of $k-\varepsilon$. It became positive in free shear flows where the $k-\varepsilon$ model is known to provide better predictions and became negative near solid boundaries where the classical $k-\omega$ model provides better predictions. In the $k-\omega2006$ turbulence model, from Equation (30), only the positive contribution of the cross-diffusion term was incorporated with a coefficient one order of magnitude smaller than in the $k - \varepsilon$ model. The idea behind the $k - \omega 2006$ model is to incorporate the cross-diffusion term in free shear flows (positive contribution) to have a $k - \varepsilon$ behavior and suppress it near solid boundaries (negative contribution) to have a classical $k - \omega$ behavior.

4. Discussion

Close to the sediment bed, the cross-diffusion term became negative. Its contribution was incorporated in the $k - \varepsilon$ model, but not in the $k - \omega 2006$ turbulence model. However, from Figure 7, the negative contribution of the cross-diffusion term played an important role in the time development of the scour hole depth.

The negative contribution of the cross-diffusion term seemed to be necessary to reproduce quantitatively the time development of the scour hole. Typical TKE (k) and the specific dissipation rate of the TKE (ω) profiles for free shear, boundary layer, and sediment transport flows are presented in Figure 8. For free shear flows, the peak value of k corresponds to the peak value of ω. The cross-diffusion term was always positive, and the $k - \omega 2006$ turbulence model behaved like a $k - \varepsilon$. For boundary layer flows, the peak value of ω was located at the wall, whereas the peak value of k was located further away. The gradient of k changed sign toward the boundary, as did the cross-diffusion term, which became negative. The negative cross-diffusion contribution was suppressed in the $k - \omega 2006$ model, having the effect of relaminarizing the flow close to the wall. When sediment transport was involved, the peak values of k and ω were offset, so that the cross-diffusion term became negative between the two peaks. Using the $k - \omega 2006$ model in this configuration suppressed the influence of the negative contribution of the cross-diffusion term, and the flow was relaminarized close to the sediment bed. This phenomenon was not physical and was responsible for the underestimation of the sediment erosion observed using the $k - \omega 2006$ turbulence model. Finally, our numerical results suggested that sediment transport shares more similarities with a free shear flow than with boundary layer flows. The negative contribution of the cross-diffusion term should therefore be incorporated to behave like a $k - \varepsilon$ model near the sediment bed, while suppressed far from the bed to behave like the $k - \omega 2006$ model and allow vortex-shedding to develop.

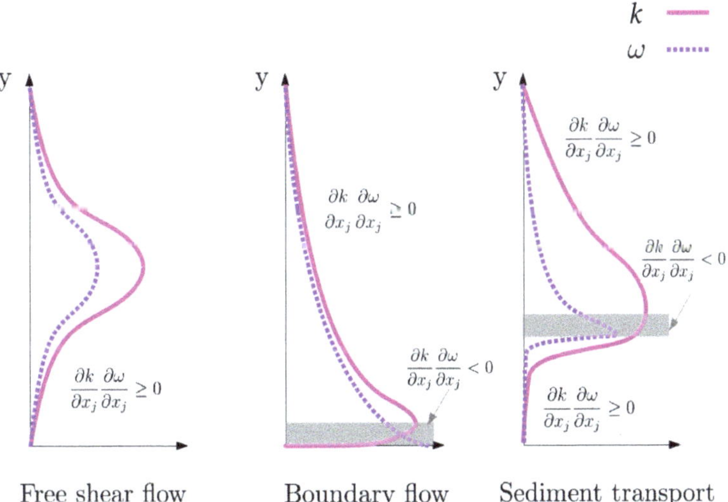

Figure 8. Typical turbulent kinetic energy (k) and specific dissipation rate (ω) profiles for free shear flows, boundary flows, and sediment transport configurations.

Finally, even though the two-phase flow model relied on a more theoretical background than the classical single-phase flow models, empirical expressions are still needed, especially for the granular stress and turbulence models. However, these models are at a lower level of approximation in the sense that they have been developed and validated on other fluid and granular flow configurations. In this respect, they are more general and better describe the complex physics at work in sediment transport. From Section 3.1, the empiricism in the granular stress model did not seem to be a limitation since the dense granular flow rheology and the kinetic theory of granular flows provided accurate results. However, the available two-phase turbulence models did not fully take into account the complex interactions between the granular phase and the fluid turbulence. For this type of configuration, the coupling between the fluid turbulence and the sediment dynamics was crucial, and Reynolds averaged two-phase flow models showed their limitations.

5. Conclusions

This paper presented a numerical investigation of the scour phenomenon below a submarine pipeline. SedFoam, a two-phase flow model for sediment transport applications, was used to study the sensitivity of the scour hole formation and of the bed morphology to the granular stress and the turbulence closure. The quality of the different simulations was measured using the Brier Skill Score. The granular stress model was not sensitive, and similar results were obtained between simulations using $\mu(I)$ rheology and the kinetic theory for granular flows. Both models provided a quantitative time evolution of the erosion depth and of the bed morphology when coupled with the $k-\varepsilon$ turbulence model.

The turbulence model however had a significant influence on the bed morphology. On the one hand, the $k-\varepsilon$ model provided the right equilibrium maximum erosion depth, but overestimated the bed elevation downstream of the pipeline. This accretion phenomenon was explained by the incapacity of the $k-\varepsilon$ model to reproduce the vortex-shedding phenomenon and the lee-wake erosion stage of scour. Therefore, a turbulence model able to reproduce vortex shedding should be used. The $k-\omega 2006$ model, which can reproduce the vortex-shedding, strongly underestimated the erosion depth, but allowed qualitatively reproducing the lee-wake stage of scour.

An in-depth analysis of the $k-\varepsilon$ and the $k-\omega 2006$ revealed the importance of the cross-diffusion term responsible for the behavior of $k-\varepsilon$. The negative and positive contribution of the cross-diffusion term were incorporated in the $k-\varepsilon$ model, whereas only the positive contribution was incorporated in $k-\omega 2006$. The numerical results showed that the negative contribution of the cross-diffusion term was required near the sediment bed to reproduce quantitatively the time development of the scour hole.

An improved URANS two-phase flow turbulence model should have a $k-\omega 2006$ behavior in the outer regions and a $k-\varepsilon$ behavior near the sediment bed. Such a turbulence model would allow providing accurate results in conditions where the interactions between the fluid vortices and the sediment bed are important.

The coupling between the sediments dynamics and the turbulence is a very complex phenomenon, and it should be investigated in detail using large eddy simulations. It would allow better understanding the interactions between the turbulent wake of the cylinder and the sediment bed downstream of the pipeline. This is beyond the scope of the present paper, and it is left for future research.

Author Contributions: Funding acquisition, writing—review, project administration and supervision, J.C.; methodology, software, validation, A.M., T.N., C.B. and J.C.; writing—original draft preparation A.M. and J.C.

Funding: The work presented in this manuscript was financially supported by the French national research agency ANR projects SegSed (ANR-16-CE01-0005-03) and SheetFlow (ANR-18-CE01-0003) and the European Community's Horizon 2020 Program through the Integrated Infrastructure Initiative HYDRALAB + FREEDATA (654110).

Acknowledgments: We would like to thank Cheng-Hsien Lee, Tian-Jian Hsu, and Zhen Cheng for fruitful discussions around multiphase flow modeling of scour. Most of the computations presented in this paper were performed using the GENCI infrastructure under Allocation A0060107567 and the GRICAD infrastructure. We are also grateful to the developers involved in OpenFOAM.

Conflicts of Interest: The authors declare no conflict of interest.

Appendix A. Hydrodynamic Simulations

Two numerical simulations without solid phase ($\phi = 0$) were performed to study the behavior of the turbulence models in the case of a cylinder placed in a steady current. The first simulation used the $k - \omega 2006$ turbulence model presented in Section 2.3.2, and the second used the standard $k - \varepsilon$ turbulence model presented in Section 2.3.1.

Appendix A.1. Numerical Setup

The mesh used in both simulations was a 0.41 m (8.2D)-wide and 1.2 m (24D)-long rectangle with a cylinder of diameter $D = 0.05$ cm placed 0.2 m (4D) from the inlet (Figure A1). The top and bottom boundary conditions were symmetry-type boundary conditions. The outlet boundary condition for the reduced pressure was a homogeneous Dirichlet condition ($p - \rho^f gy = 0$ Pa). For the outlet velocity, a homogeneous Neumann boundary condition was used for outgoing flows, and a homogeneous Dirichlet boundary condition was used otherwise. A fixed value of $U = 0.87$ m·s^{-1} was given for the inlet velocity, giving a Reynolds number $Re = UD/\nu = 4.35 \times 10^4$.

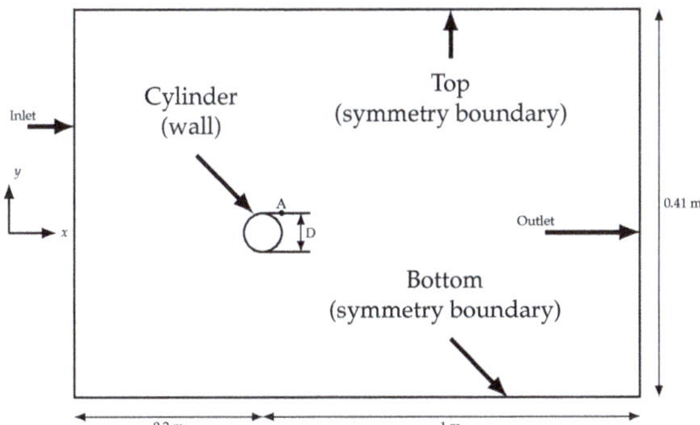

Figure A1. Sketch of the geometry and the boundary conditions used for the computational domain.

For the simulation using a $k - \omega 2006$ turbulence model, the cells near the cylinder were 2×10^{-5} m thick, giving a near-wall $y^+ \approx 1$. Wall functions for smooth walls were applied on the cylinder surface.

For the simulations using the standard $k - \varepsilon$, the cells near the cylinder were 6×10^{-4} m thick, giving a near-wall $y^+ \approx 30$. A homogeneous Dirichlet boundary condition of 1×10^{-10} m^2·s^{-2} was applied on the cylinder surface for k, and a homogeneous Neumann boundary condition was applied for ε.

For both simulations, inlet boundary conditions for turbulent quantities were set following Table A1 with a turbulence intensity $I = 2\%$ and a turbulence length scale $l = 0.07D$.

Table A1. Inlet boundary conditions for turbulent quantities.

	Turbulence Kinetic Energy	Dissipation
$k - \varepsilon$	$k_{inlet} = \frac{3}{2}(UI)^2$	$\varepsilon_{inlet} = C_\mu \frac{k^{1.5}}{l}$
$k - \omega 2006$	$k_{inlet} = \frac{3}{2}(UI)^2$	$\omega_{inlet} = \frac{\sqrt{k}}{l}$

Second order differentiation schemes and a time step equal to 2×10^{-4} s were used. A probe placed on the top right of the cylinder (Point A in Figure A1) recorded the velocity signal along the simulations.

Appendix A.2. Results

The Strouhal number is the dimensionless frequency defined by $St = fD/U$, with f the frequency at which the vortices are shed in the wake of the cylinder. A fast Fourier transform was performed on the velocity signal, and the result is presented in Figure A2. For each simulation, the velocity signal showed one peak value. However, the peak value was not the same for the two turbulence models. Periodic structures were shed in the wake of the cylinder for both simulations, but the frequency differed depending on the turbulence model.

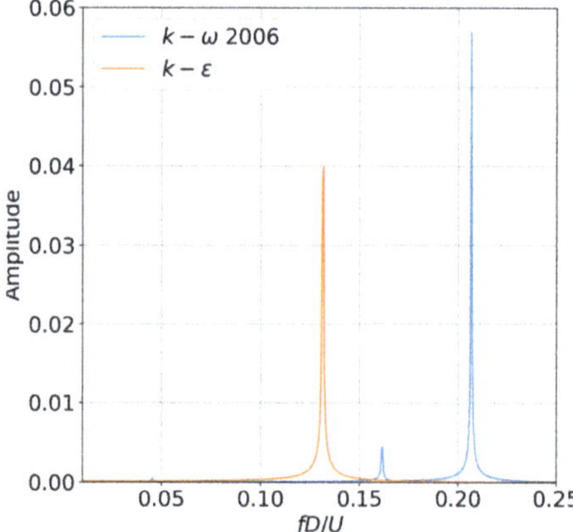

Figure A2. Spectrum of the velocity signal.

A typical value of the Strouhal number for cylinder at Reynolds number $Re = 4.35 \times 10^4$ is 0.2 [32]. This value corresponded to the peak value found for the $k - \omega 2006$ turbulence model. The oscillatory behavior of the flow downstream of the pipeline was consistent with the results from the literature. On the contrary, the Strouhal number found with the $k - \varepsilon$ model was largely underestimated. The unsteady turbulent wake behind the cylinder was not properly captured. In conclusion, the $k - \omega 2006$ turbulence model was more suitable to reproduce the lee-wake erosion stage of scour.

References

1. Sumer, B.M.; Fredsoe, J. *The Mechanics of Scour in the Marine Environment*; World Scientific Publishing Company: Singapore, 2002; Volume 17.
2. Mao, Y. *The Interaction Between a Pipeline and an Erodible Bed*; Institute of Hydrodynamics and Hydraulic Engineering København: Series Paper; Institute of Hydrodynamics and Hydraulic Engineering, Technical University of Denmark: Lyngby, Denmark, 1986.
3. Chiew, Y. Prediction of Maximum Scour Depth at Submarine Pipelines. *J. Hydraul. Eng.* **1991**, *117*, 452–466. [CrossRef]
4. Li, F.; Cheng, L. Numerical Model for Local Scour under Offshore Pipelines. *J. Hydraul. Eng.* **1999**, *125*, 400–406. [CrossRef]
5. Leeuwenstein, W.; Wind, H.G. The Computation of Bed Shear in a Numerical Model. In Proceedings of the 19th International Conference on Coastal Engineering, Houston, TX, USA, 3–7 September 1985.
6. Liang, D.; Cheng, L.; Li, F. Numerical modeling of flow and scour below a pipeline in currents: Part II. Scour simulation. *Coast. Eng.* **2005**, *52*, 43–62, doi:10.1016/j.coastaleng.2004.09.001. [CrossRef]
7. Jenkins, J.T.; Hanes, D.M. Collisional sheet flows of sediment driven by a turbulent fluid. *J. Fluid Mech.* **1998**, *370*, 29–52. [CrossRef]
8. Revil-Baudard, T.; Chauchat, J. A two-phase model for sheet flow regime based on dense granular flow rheology. *J. Geophys. Res. Ocean.* **2013**, *118*, 619–634. [CrossRef]

9. Hsu, T.J.; Liu, P.L.F. Toward modeling turbulent suspension of sand in the nearshore. *J. Geophys. Res. Ocean.* **2004**, *109*. [CrossRef]
10. Zhao, Z.; Fernando, H.J.S. Numerical simulation of scour around pipelines using an Euler-Euler coupled two-phase model. *Environ. Fluid Mech.* **2007**, *7*, 121–142, doi:10.1007/s10652-007-9017-8. [CrossRef]
11. Yeganeh-Bakhtiary, A.; Kazeminezhad, M.H.; Etemad-Shahidi, A.; Baas, J.H.; Cheng, L. Euler-Euler two-phase flow simulation of tunnel erosion beneath marine pipelines. *Appl. Ocean Res.* **2011**, *33*, 137–146. [CrossRef]
12. Lee, C.H.; Low, Y.M.; Chiew, Y.M. Multi-dimensional rheology-based two-phase model for sediment transport and applications to sheet flow and pipeline scour. *Phys. Fluids* **2016**, *28*, 053305. [CrossRef]
13. MiDi, G.D.R. On dense granular flows. *Eur. Phys. J. E* **2004**, *14*, 341–365. [CrossRef]
14. Wilcox, D.C. *Turbulence Modeling for CFD*; DCW Industries: La Canada, CA, USA, 2006.
15. Nagel, T. *Three-Dimensional Scour Simulations with a Two-Phase Flow Model*; 2019; under review.
16. Cheng, Z.; Hsu, T.J.; Calantoni, J. SedFoam: A multi-dimensional Eulerian two-phase model for sediment transport and its application to momentary bed failure. *Coast. Eng.* **2017**, *119*, 32–50, doi:10.1016/j.coastaleng.2016.08.007. [CrossRef]
17. Chauchat, J.; Cheng, Z.; Nagel, T.; Bonamy, C.; Hsu, T.J. SedFoam-2.0: A 3D two-phase flow numerical model for sediment transport. *Geosci. Model Dev. Discuss.* **2017**, *10*, 4367–4392. [CrossRef]
18. Boyer, F.; Guazzelli, E.; Pouliquen, O. Unifying Suspension and Granular Rheology. *Phys. Rev. Lett.* **2011**, *107*, 188301, doi:10.1103/PhysRevLett.107.188301. [CrossRef]
19. Richardson, J.F.; Zaki, W. Sedimentation and Fluidization: Part I. *Chem. Eng. Res. Des.* **1954**, *32*, 35–53.
20. Schiller, L.; Naumann, A.Z. Über die grundlegenden Berechnungen bei der Schwerkraftaufbereitung. *Ver. Deut. Ing.* **1933**, *77*, 318–320.
21. Chauchat, J. A comprehensive two-phase flow model for unidirectional sheet-flows. *J. Hydraul. Res.* **2017**, *2017*, 1289260. [CrossRef]
22. Maurin, R.; Chauchat, J.; Frey, P. Dense granular flow rheology in turbulent bedload transport. *J. Fluid Mech.* **2016**, *804*, 490–512. [CrossRef]
23. Jop, P.; Forterre, Y.; Pouliquen, O. A constitutive law for dense granular flows. *Nature* **2006**, *441*, 727–730. [CrossRef]
24. Chauchat, J.; Marc, M. A three-dimensional numerical model for incompressible two-phase flow of a granular bed submitted to a laminar shearing flow. *Comput. Methods Appl. Mech. Eng.* **2009**, *199*, 439–449. [CrossRef]
25. Ding, J.; Gidaspow, D. A bubbling fluidization model using kinetic theory of granular flow. *AIChE J.* **1990**, *36*, 523–538. [CrossRef]
26. Carnahan, N.F.; Starling, K.E. Equation of State for Nonattracting Rigid Spheres. *J. Chem. Phys.* **1969**, *51*, 635–636. [CrossRef]
27. Van Rijn, L. Sediment Transport, Part II: Suspended Load Transport. *J. Hydraul. Eng.* **1984**, *110*, 1613–1641. [CrossRef]
28. Ferziger, J.H.; Peric, M. *Computational Methods for Fluid Dynamics*; Springer Science & Business Media: Berlin/Heidelberg, Germany, 2002.
29. Sutherland, J.; Hall, L.; Chesher, T. *Evaluation of the Coastal Area Model PISCES at Teignmouth (UK)*; HR Wallingford Report TR125; 2001. [CrossRef]
30. Brady, A.; Sutherland, J. *COSMOS Modelling of COAST3D Egmond Main Experiment*; HR Wallingford Report TR115; 2001. [CrossRef]
31. Van Rijn, L.; Ruessink, B.; Mulder, J. *Coast3D-Egmond: The Behaviour of a Straight Sandy Coast on the Time Scale of Storms and Seasons*; Aqua Publications: Locust Valley, NY, USA, 2002.
32. Sumer, B.; Fredsoe, J. *Hydrodynamics Around Cylindrical Structures*; World Scientific: Singapore, 2006; Volume 12.

Article

Numerical Study of the Hydrodynamics of Waves and Currents and Their Effects in Pier Scouring

Matias Quezada [1,*], Aldo Tamburrino [1,2] and Yarko Niño [1,2,*]

[1] Department of Civil Engineering, Faculty of Physics and Mathematical Sciences, University of Chile, Santiago 8370449, Chile; atamburr@ing.uchile.cl

[2] Advanced Mining Technology Center, Faculty of Physics and Mathematical Sciences, University of Chile, Santiago 8370451, Chile

* Correspondence: matias.quezada@ing.uchile.cl (M.Q.); ynino@ing.uchile.cl (Y.N.); Tel.: +56-229-784-400 (M.Q. & Y.N.)

Received: 31 August 2019; Accepted: 22 October 2019; Published: 28 October 2019

Abstract: The scour around cylindrical piles due to codirectional and opposite waves and currents is studied with Reynolds Averaged Navier–Stokes (RANS) equations via REEF3D numeric modeling. First, a calibration process was made through a comparison with the experimental data available in the literature. Subsequently, not only the hydrodynamics, but also the expected scour for a set of scenarios, which were defined by the relative velocity of the current (U_{CW}), were studied numerically. The results obtained show that the hydrodynamics around the pile for codirectional or opposite waves and currents not have significant differences when analyzed in terms of their velocities, vorticities and mean shear stresses, since the currents proved to be more relevant compared to the net flow. The equilibrium scour, estimated by the extrapolation of the numerical data with the equation by Sheppard, enabled us to estimate values close to those described in the literature. From this extrapolation, it was verified that the dimensionless scour would be less when the waves and currents are from opposite directions. The U_{CW} parameter is an indicator used to adequately measure the interactions between the currents and waves under conditions of codirectional flow. Nevertheless, it is recommended to modify this parameter for currents and waves in opposite directions, and an equation is proposed for this case.

Keywords: horseshoe vortex; pile scour; waves and currents; computational fluid dynamics; REEF3D

1. Introduction

The hydrodynamics of the coastal environment usually correspond to the result of the interaction of several force, such as waves, tides, and winds, that act at different spatial and temporal scales, thereby modulating circulation. Meanwhile, rivers run mainly due to the gravitational action that moves the waters resulting from snowmelt or rain, which flow into the alluvial channel that transports to the ocean by runoff. The convergence of coastal and fluvial environments is known as an estuary zone, and the resulting currents correspond to a complex interaction between tides, waves, winds, and river flow.

The hydrodynamics of environments where waves and currents interact have been previously studied by various authors both co-directionally [1–3] and perpendicularly [4–6].

The research developed by Umeyama [1–3], sought to analyze the behavior of Reynolds stress and velocity vertical distributions [1], the changes induced by the combined wave currents over the turbulent flow structures [2], and the surface elevation and particle velocities [3].

In the case of waves orthogonal to currents, the experimental research developed by Feraci et al. [4], Lim and Madsen [5] and Feraci et al. [6] allows us to understand the effects of joint action on the behavior of the resulting velocity of the fluid. For example, Feraci et al. [4] experimentally demonstrated the joint action of orthogonal waves and currents, which speeds up the evolution process of a sandpit. Lim and Madsen [5] analyzed, via an experimental study the effects of the roughness in an experimental study on the velocity distribution in a wave-current interaction. A complete statistical analysis of the near bed velocity behavior due to waves and a current acting perpendicularly was developed by Feraci et al. [6]. They concluded that the probability distribution of near-bed velocity follows a Gaussian distribution in a flow field generated by a current alone. In the presence of waves, the distribution changes and another peak over a Gaussian distribution appears.

However, all the studies in the previous paragraphs do not include any type of obstruction to the flow, which generates additional modifications to the hydrodynamic characteristics of the flow field.

It is well known that when placing a circular pile in an environment that has a specific current (that can be produced by waves/tides, river flow, or both), a hydrodynamic modification will be produced around it and, therefore, vortexes will be produced (a horseshoe vortex and vortex shedding), which are the main elements responsible for the scour around the pile.

Through time, different authors have studied pile scour due to a uniform flow. Among these authors it is worth mentioning Hjorth [7], Melville [8], Ettema [9], Chiew and Melville [10], Melville and Chiew [11], Oliveto and Hager [12], Link et al. [13], Diab et al. [14], and Link et al. [15], who focused their interests mainly on the scour around bridge piles. When it comes to scour by waves, the number of studies is limited. On this subject, the authors of this paper consider the contributions of Sumer et al. [16], Sumer et al. [17] and Sumer and Fredsøe [18] to be fundamental to our understanding of multiple hydrodynamic processes responsible for the movement of sediments near the pile.

Experimental studies on the scour around piles under a flow associated with the combined action of waves and currents have been carried out by different authors [19–26], who have contributed, through their laboratory tests, to our understanding of the scour phenomenon in this type of environments. The following is a brief bibliographic description.

Eadie and Hernich [19] studied a physical model with the purpose of evaluating the effects that the combined action of two co-directional forces, waves (random) and currents, have over the scour around cylindric piles. The main results of Eadie and Hernich [19] indicate that the scour process due to waves and currents together is faster and reaches higher equilibrium compared to currents acting alone. Similar results were determined by Kawata and Tsichiya [20], who characterized the scour process for clear-water and live-beds in a similar manner to Eadie and Hernich [19].

Raaijmakers and Rudolph [15] studied the temporal dependency of the scour around a pile due to the combined action of waves and currents with the purpose of analyzing the equilibrium scour, the temporal scales needed to reach such depths and the backfilling process. As part of their results, Raaijmakers and Rudolph [21] propose an equation to determine the scour as a function of time and additionally concluded that the equilibrium scour is of a higher magnitude in cases of currents acting alone compared to the conditions reached for the combined action of waves and currents. The equilibrium scour equation as a function of time, presented by Raaijmakers and Rudolph [21], was validated through the comparison of field data, as presented by Rudolph et al. [22].

Zanke et al. [23], through an analysis of data gathered by other authors, proposed a unified equation to determine the scour depth due to the actions of waves and currents, through the incorporation of a transition function (x_{rel}) defined by the effective scour (x_{eff}). Similarly, Ong et al. [24] developed a stochastic method by which the maximum equilibrium scour could be determined in piles exposed to long-crested and short-crested nonlinear random waves plus a current. They validated their approach by comparison with the experimental data provided by Sumer and Fredsøe [25].

The contribution carried out by Sumer and Fredsøe [25] to understand the process of scour is significant, since through its dimensional analysis, their model is able to represent the dimensionless scour (S) over the pile diameter (D) as a function of relative flow velocity (U_{cw}), as defined by Equation (1), where U_c corresponds to the undisturbed current velocity at the transverse distance $z = D/2$ and

U_m is the maximum value of the undisturbed orbital velocity at sea bottom just above the wave's boundary layer:

$$U_{cw} = \frac{U_c}{U_c + U_m}. \quad (1)$$

Evidently, the relative flow velocity will have values close to zero when the environment is dominated by waves, but it will approach one if currents are the main flow mechanism.

The main conclusions presented by Sumer and Fredsøe [25] indicate that in a wave environment, the scour increases significantly in the presence of a current, even if the current is mild. This current, is mainly associated with a strong horseshoe vortex in front of the pile, even in the case of a mild vortex. In addition, the current apparently dominates the pile's scour when $U_{cw} \geq 0.7$; the scour approaches this value due to the current acting alone.

Even though the articles mentioned above have studied the scour around cylindrical piles due to the combined action of waves and currents, they considered forcing to act co-directionally. Qi and Gao [26], in their experimental work, studied the scour around cylindrical piles under the combined action of co-directional and opposite waves and currents, for different pile diameters, waves conditions and currents. The main conclusion they reached was that the scour in the combined flows of waves and currents is a nonlinear process, and the time required to reach scour equilibrium is much lower than that required for waves or currents acting independently. Additionally, Qi and Gao [26] mentioned that the maximum flow velocity in waves and co-directional currents is much higher than that in waves and currents from opposite directions, thereby affecting the maximum scour magnitude, which is lower in opposite flows.

While there is a number (albeit limited) of experimental articles related to the study of scour caused by combined waves and currents, investigations based on numerical models are even more scarce. It is only possible to find simulations of scour acting separately around piles due to currents or oscillatory flows. A literature review on this subject is available on Quezada et al. [27].

The application of Reynolds-averaged Navier–Stokes equations (RANS) in simulated environments, in which waves and currents coexist, has been demonstrated by several authors [28–30], who nonetheless fail to include the vertical pile in the flow. Ahmad et al. [31] recently developed a numerical study based on the REEF3D model in order to study scour on a horizontal pile (pipeline) caused by combined waves and currents. This study is relevant to the research presented in this article, since the same numerical model used by Ahmad et al. [31] was applied.

Based on the above, the main objective of this article is to study, through numerical models, scour's hydrodynamics around cylindrical piles where waves and currents coexist, both co-directional and opposite to the wave direction as well.

2. Materials and Methods

2.1. Numerical Model: REEF3D

The numerical modelling developed in this research was conducted using the numerical model known as REEF3D [32,33], which is a computational fluid dynamic (CFD) tool used to solve Reynolds-averaged Navier–Stokes equations (RANS). This tool is able to simulate hydraulic, coastal, and estuarine phenomena for both compressible and incompressible fluids.

The REEF3D model provides a three-dimensional solution for governing equations composed of Equation (2), which corresponds to the continuity of an incompressible flow, and Equation (3), which is the momentum conservation.

$$\frac{\partial u_i}{\partial x_i} = 0 \quad (2)$$

$$\frac{\partial u_i}{\partial t} + u_j \frac{\partial u_i}{\partial x_j} = -\frac{1}{\rho}\frac{\partial p}{\partial x_i} + \frac{\partial}{\partial x_j}\left[(v + v_T)\frac{\partial u_i}{\partial x_j}\right] + g_i, \quad (3)$$

where i, j = 1,2,3, u_i is the mean velocity vectoral component, x_i is the spatial vectoral component, t is time, ρ is water density, p is pressure, υ is kinematic viscosity, υ_T is kinematic eddy viscosity, and g is gravity.

The turbulence closure model used was the well-known $k-\omega$ model [34], in which υ_T is defined according to Equation (4), and both the turbulent kinetic energy (k) and the kinetic energy for the specific turbulent dissipation per unit of turbulence (ω) are determined with a transport equation, according to Equations (5) and (6), respectively. The constant values in $k-\omega$ were taken according to β_k = 9/100, α = 5/9, β = 3/40, σ_ω = 1/2, and σ_k = 1; and P_k is the turbulent production rate defined by Equation (7).

$$\upsilon_T = \frac{k}{\omega} \tag{4}$$

$$\frac{\partial k}{\partial t} + u_j \frac{\partial k}{\partial x_j} = \frac{\partial}{\partial x_j}\left[\left(\upsilon + \frac{\upsilon_T}{\sigma_k}\right)\frac{\partial k}{\partial x_j}\right] + P_k - \beta_k k\omega \tag{5}$$

$$\frac{\partial \omega}{\partial t} + u_j \frac{\partial \omega}{\partial x_j} = \frac{\partial}{\partial x_j}\left[\left(\upsilon + \frac{\upsilon_T}{\sigma_\omega}\right)\frac{\partial \omega}{\partial x_j}\right] + \frac{\omega}{k}\alpha P_k - \beta\omega^2 \tag{6}$$

$$P_k = \upsilon_T \frac{\partial u_i}{\partial x_j}\left[\frac{\partial u_i}{\partial x_j} + \frac{\partial u_j}{\partial x_i}\right]. \tag{7}$$

To solve the flow around complex structures, the ghost-cell method [35,36] was applied to impose the boundary conditions when a pile is included in the numerical domain. This approach uses fictional cells that are incorporated in the domain (specifically on the obstacles) and corresponds to a particular case of the immersed boundary method [37].

The numerical treatment of the governing equations was based on the second order of the Runge–Kutta method for temporal discretization. Meanwhile, the convective terms were solved by the applied weight essentially non-oscillatory (WENO) scheme [38]. The velocities and pressures were determined under a staggered grid, using the Semi-Implicit Method for Pressure Linked Equations (SIMPLE) [39].

The model configuration to represent waves in the numerical domain was adopted similar to the method presented by Ahmad et al. [31], using a Dirichlet boundary condition and the second-order Stokes waves theory [40]. We defined the surface elevation (η) and the incident velocity (u, w) according to Equations (8) and (9), respectively.

$$\eta(x,z,t) = \frac{H}{2}\cos\vartheta + \frac{H^2 K}{16}\frac{\cos h(Kh)}{\sin h^3(Kh)}(2+\cos h(2Kh))\cos(2\vartheta) \tag{8}$$

$$u(x,z,t) = \frac{\partial \Phi}{\partial x},\ w(x,z,t) = -\frac{\partial \Phi}{\partial z} \tag{9}$$

$$\Phi(x,z,t) = \frac{Hg}{2\varpi}\frac{\cos h[K(h+z)]}{\cos h(Kh)}\sin\vartheta + \frac{3}{32}H^2\varpi\frac{\cos h[2K(h+z)]}{\sin h^2(Kh)}\sin(2\vartheta), \tag{10}$$

where H is wave height, K is the wave number, h is the water depth, ϑ is the wave phase, ϖ is the angular frequency, and Φ is the velocity potential.

In order to avoid the wave reflection, active wave absorption (AWA), was used in the outlet according to the method described by Schäffer and Klopman [41]. In this methodology, the waves that reach the outlet cancel out the reflected waves, prescribing the velocity as in [31]:

$$u_0 = -\sqrt{\frac{g}{h}}\eta_r \tag{11}$$

$$\eta_r = \eta_m - h, \tag{12}$$

where η_r is the reflected wave amplitude and η_m is the actual elevation of the free surface.

Complementary information on the numerical model´s hydrodynamic configuration is shown in Table 1.

Table 1. Complementary information for the REEF3D hydrodynamic simulation.

Configuration	Definition
Boundary condition	Non-slip for velocities Non-slip for k and ω Logarithmic profile for inlet Fix pressure at inlet Zero-gradient outflow Active wave absorption at outlet (waves)
Initialization	Potential flow for velocities Hydrostatic for pressure

In order to ascertain the sediment transport and changes in the bed, both the bed load and the suspended load were considered, for the morphodynamics evolution, the equation for the conservation of sediment initially proposed by Exner [42] and later generalized by Paola and Voller [43], was used.

The bed load sediment transport was computed by using the Meyer–Peter formula [44] according to Equation (13), where q_b^* is the dimensionless bed load, τ^* is the dimensionless shear stress on the bed computed by Equation (14), τ_c^* is the dimensionless critical shear stress defined according to the Equation (15), τ is the bed shear stress, τ_c is the critical bed shear stress, ρ_s the sediment density, and d is the sediment diameter. The real magnitude of the bed load (q_b) can be obtained from Equation (16).

$$q_b^* = 8(\tau^* - \tau_c^*)^{\frac{3}{2}} \tag{13}$$

$$\tau^* = \frac{\tau}{(\rho_s - \rho)gd} \tag{14}$$

$$\tau_c^* = \frac{\tau_c}{(\rho_s - \rho)gd} \tag{15}$$

$$q_b^* = \frac{q_b}{\sqrt{\frac{(\rho_s-\rho)gd^3}{\rho}}}. \tag{16}$$

The suspended load was computed using the advection diffusion equation shown in Equation (17), where c is the suspended load concentration, and w_s is the fall velocity of the sediment. To solve Equation (17) two boundary conditions were applied. The first is the zero vertical flux on the surface and the second is the bed load concentration (c_b) according to Van Rijn [45], defined by Equation (18), where Υ is the relative bed shear stress defined by Equation (19), and D_* is the particle parameter computed according to Equation (20).

$$\frac{\partial c}{\partial t} + u_j \frac{\partial c}{\partial x_j} + w_s \frac{\partial c}{\partial z} = \frac{\partial}{\partial x_j}\left(v_T \frac{\partial c}{\partial x_j}\right) \tag{17}$$

$$c_b = 0.015 \frac{d}{a}\left(\frac{\Upsilon^{1.5}}{D_*^{0.3}}\right) \tag{18}$$

$$\Upsilon = \frac{(\tau - \tau_c)}{\tau_c} \tag{19}$$

$$D_* = d\left[\frac{(s-1)g}{v^2}\right]^{\frac{1}{3}} \tag{20}$$

In Equation (18), a is the reference level for computing the suspended load that has been determined according to the methodology proposed by Rouse [46].

The critical bed shear stress was defined according to Shields [47] and parameterized according to Yalin [48], including the slope correction proposed by Dey [49] and, in addition, applying the incipient transport relaxation factor extensively described by Quezada et al. [27].

The morpho-dynamic evolution was determined by the sediment volume conservation equation described in Equation (21), where z_b is the bed elevation, and $q_{b,x}$ and $q_{b,y}$ are the sediment transport for the bed load in the x and y directions, respectively. E is the sediment entrainment rate from the bed load to the suspended-load, and D is the sediment deposition rate from the suspended load onto the bed.

$$\frac{\partial z_b}{\partial t} + \frac{1}{(1-n)}\left[\frac{\partial q_{b,x}}{\partial x} + \frac{\partial q_{b,y}}{\partial y}\right] + E - D = 0. \tag{21}$$

The difference between E and D is defined by REEF3D according to Wu et al. [50]. Meanwhile, a more extensive description of the estimation of $q_{b,x}$ and $q_{b,y}$ can be found in Quezada et al. [27].

In order to develop a numerical study of the present investigation, the model was developed as indicated in Table 2, where L is the flume length, W is the width, h_t is the height, and Δx_i is the cell dimension used in the model; N° Cells is the total number of elements comprising the numerical domain, and t_{test} is the simulation time of the numerical model. The W (W1 to W3) and WC series (WC1 to WC3) correspond to a hydrodynamic calibration stage for waves acting alone and waves plus current, respectively, without a pile placed on the flume. Detailed information can be found in Table 3.

Table 2. Summary of the REEF3D model configuration applied to each simulated cases.

Case	L (m)	W (m)	h_t (m)	Δx_i (m)	N° Cells	t_{test} (min)
W1	24.00	0.70	1.00	0.01	16,768,400	5.00
W2	24.00	0.70	1.00	0.01	16,768,400	5.00
W3	24.00	0.70	1.00	0.01	16,768,400	5.00
WC1	24.00	0.70	1.00	0.01	16,768,400	5.00
WC2	24.00	0.70	1.00	0.01	16,768,400	5.00
WC3	24.00	0.70	1.00	0.01	16,768,400	5.00
C01	24.00	1.00	1.00	0.01	23,968,400	7.00
C02	24.00	1.00	1.00	0.01	23,968,400	7.00
C03	24.00	1.00	1.00	0.01	23,968,400	7.00
C04	24.00	24.00	1.00	0.01	575,968,400	1.00
E01	24.00	1.00	1.00	0.01	23,968,400	30.00
E02	24.00	1.00	1.00	0.01	23,968,400	30.00
E03	24.00	1.00	1.00	0.01	23,968,400	30.00
E04	24.00	1.00	1.00	0.01	23,968,400	30.00
E05	24.00	1.00	1.00	0.01	23,968,400	30.00
E06	24.00	1.00	1.00	0.01	23,968,400	30.00
E07	24.00	1.00	1.00	0.01	23,968,400	30.00
E08	24.00	1.00	1.00	0.01	23,968,400	30.00

Δx_i is the dimension for the x, y, and z axis, due to the model using regular element definitions.

Table 3. Hydrodynamic calibration of waves and currents coexisting without a pile.

Case	h (m)	H (cm)	T (s)	U_C (cm/s)
W1	0.30	1.03	1.00	0.00
W2	0.30	2.34	1.00	0.00
W3	0.30	3.61	1.00	0.00
WC1	0.30	0.91	1.00	8.00
WC2	0.30	2.02	1.00	8.00
WC3	0.30	3.09	1.00	8.00

Series C (C01 to C04) corresponds to a hydrodynamic stage for waves plus current with a pile. Detailed information for each test can be found in Table 4. Finally, series E (E01 to E08) corresponds to the simulated cases for hydrodynamic analysis due to the combined action of waves and currents; their detailed information can be found in Table 5.

Table 4. Hydrodynamic calibration of waves and currents coexisting with pile.

Case	h (m)	D (m)	H (cm)	T (s)	U_C (m/s)	Direction
C01	0.50	0.20	2.60	1.40	0.23	Codirectional
C02	0.50	0.20	5.20	1.40	0.23	Codirectional
C03	0.50	0.20	8.50	1.40	0.23	Codirectional
C04	0.50	0.20	4.00	1.25	0.25	Perpendicular

Table 5. Simulated cases for the analysis of hydrodynamics due to the combined action of waves and currents.

Case	h (m)	D (m)	H (m)	T (s)	U_C (m/s)	U_m (m/s)	U_{cw}	KC	Direction
E01	0.50	0.20	0.085	1.40	0.23	0.12	0.65	0.86	Codirectional
E02	0.50	0.20	0.085	1.40	0.23	0.12	0.65	0.86	Opposite
E03	0.50	0.20	0.129	1.40	0.22	0.19	0.54	1.31	Codirectional
E04	0.50	0.20	0.129	1.40	0.22	0.19	0.54	1.31	Opposite
E05	0.50	0.20	0.150	2.00	0.24	0.28	0.47	2.76	Codirectional
E06	0.50	0.20	0.150	2.00	0.24	0.28	0.47	2.76	Opposite
E07	0.50	0.20	0.150	3.00	0.10	0.31	0.25	4.61	Codirectional
E08	0.50	0.20	0.150	3.00	0.10	0.31	0.25	4.61	Opposite

The following sections provide more exstensive information on the simulated scenarios and the data processing used.

2.2. Hydrodynamics Calibration Test

Prior to executing a numerical simulation of scour due to the combined waves and currents around a circular pile, the process of hydrodynamically calibrating the model REEF3D was conducted by comparing the numerical results with the experimental results obtained by other authors.

The calibration process was carried out in two phases. The first phase to verified the numerical model's capacity to represent a flow of waves and currents combined, without the pile in the flume, while the second calibration phase included the presence of a vertical pile in the center of the flume.

In Table 3, cases executed in the hydrodynamic calibration process without considering the pile are described. The experimental information for each of the cases was obtained from Umeyama [3] and these data correspond to the unevenness data for the water surface and the vertical profiles of flow horizontal velocity for different times steps, both for the flow caused only by the actions of waves (case W1 to W3) and by waves and currents (cases WC1 to WC3).

The variables in Table 3 are defined as follows: h is water depth, H is wave height, T is the wave periodic time, and U_C is the undisturbed current velocity defined according to Sumer and Fredsøe [25].

In order to compare the results obtained by the numerical model and those provided by Umeyama's [3] experimental work, a virtual sensor was established in the middle of the numerical domain, from which the vertical distribution (from the total velocity), and the surface elevation were extracted. The simulation time for all scenarios of the situation without pile (W1 to WC3) was five minutes (300 waves).

Simulated cases for the hydrodynamic calibration of the numerical model, including one pile in a flow due to waves and the combined action of currents, are summarized in Table 4, which considers simulated cases for codirectional (C01 to C03) and perpendicular (C04) waves and currents.

The numerical results obtained by the simulation for cases C01 to C03 (codirectional) contrasted with the experimental data provided by Qi and Gao [26], where the flow velocity was the comparison variable. For the case of the perpendicular waves and currents (C04), the experimental information provided by Miles et al. [51] was used to verify the numerical results, where the contrasted variable was the average vertical profile of the total flow velocity.

The comparison of the numerical and experimental data obtained by Qi and Gao [26] (C01 to C03) was carried out by obtaining the time series of the total velocity in a virtual monitoring station located at the horizontal 20D and 1D relative to the bed. The total time of the simulation was seven minutes (300 waves).

For case C04, the comparison of numerical and experimental data published by Miles et al. [51] was conducted via the vertical distribution of the longitudinal velocity averaged in eight points around the pile, which were denominated P1 to P8 and distributed as shown in Figure 1. The monitoring stations were located 0.75D from the center of the pile, while their angular separation was 45°.

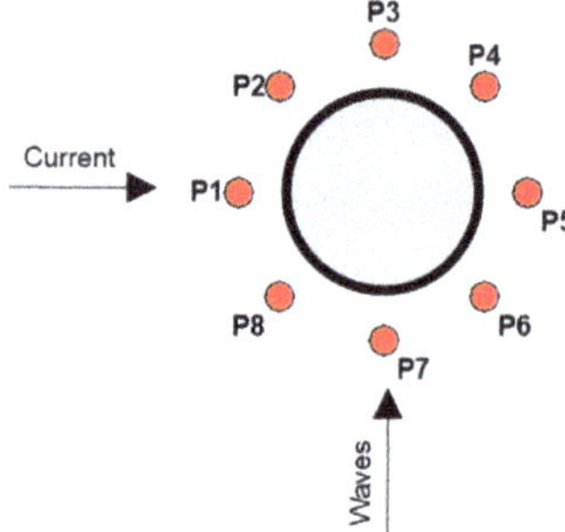

Figure 1. Identification of the comparison points of the average velocity profiles obtained from the numerical modeling and those published by Miles et al. [51].

The velocities were nondimensionalized according to the characteristic velocity (C) described in Equation (22) for all compared cases between the modeled and experimental data (Umeyama [3], Qi and Gao [20], and Miles et al. [51]).

$$C = \sqrt{gh} \tag{22}$$

2.3. Hydrodynamics Behavior of the Flow around a Cylindrical Pile

To numerically determine the hydrodynamic behavior around a cylindrical pile subjected to the combined action of waves and currents, a set of 8 simulations were developed, as described in Table 5. The cases were defined to cover a wide range of waves and current interactions according to the flow relative velocity (U_{cw}) proposed by Sumer and Fredsøe [25]. Thus, we formed scenarios dominated by currents (E01 and E02), waves and currents but with a tendency toward current domination (E03 and E04), waves and currents but with a tendency toward wave domination (E05 and E06), and environments dominated by waves (E07 and E08).

The estimation of the flow relative velocity (U_{cw}) was conducted by considering the maximum value of the undisturbed orbital velocity at the sea bottom just above the wave boundary layer (U_m), according to Equation (23), while the undisturbed current velocity at the transverse distance $z = D/2$ (U_c) was defined as an edge condition for each of the modeling scenarios. Additionally, the Keulegan-Carpenter number was estimated as indicated in Equation (24), in order to identify the influence that waves have over maximum scour.

$$U_m = \frac{\pi H}{T \sin h(Kh)} \tag{23}$$

$$KC = \frac{U_m T}{D}. \tag{24}$$

All simulations conducted (E01 to E08) were set to solve the hydrodynamics model within 30 minutes throughout the entire numerical domain, using the potential flow and a hydrostatic distribution of pressures as the initial condition (see Table 1).

To analyze the velocities and vortexes associated with the flow, two main vertical planes of the channel were designed. The first of these plains corresponds to the longitudinal axis (flow development) passing through the center of the pile from the beginning of the channel to its end. The second is associated with the axis perpendicular to the channel, as shown in Figure 2.

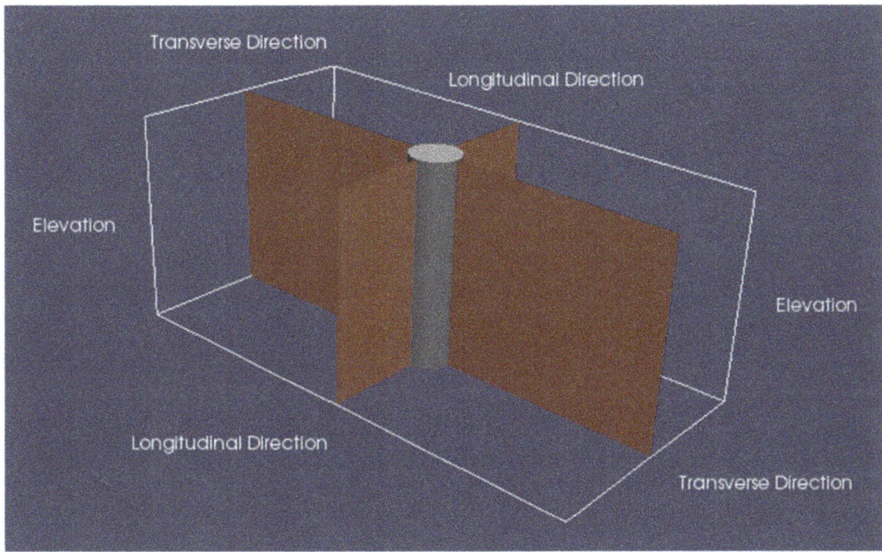

Figure 2. Analysis of the hydrodynamic planes around the cylindrical pile.

As the first stage of the analysis conducted on the results obtained from the numerical model, the velocity fields along the channel were inspected in order to identify patterns in spatiotemporal flow performance. Subsequently, for each of the planes traced around the cylindrical pile (see Figure 2), the vorticity's average performance was determined and the associated streamlines were traced to study and analyze the average performance of the horseshoe vortex. Additionally, eight monitoring stations were defined to obtain the vertical distribution of the flow velocity Figure 1 indicates the distribution of these stations, which is concordant with the methodological approach of Miles et al. [51].

Moreover, the amplification of the shear stress (α_τ) around the pile was determined, associated with the average flow conditions and (on a Cartesian plane) based on the numerical domain bed, prior to scour. For such purposes, the proportion of the bed shear stress (τ_0) and the undisturbed bed shear stress (τ_∞) according to Equation (25) were considered. To determine τ_0 and τ_∞, the shear velocity was computed by the numerical model in the bed's nearest cell for two locations: around the pile $\left(u_{*T,pile}\right)$ for the calculation of τ_0 and 2 meters downstream of the inlet ($u_{*T,inlet}$) for τ_∞.

The relation employed for the bed shear stress estimations corresponds to the conventional hydraulic definition described in Equation (26) for τ_0 and Equation (27) for τ_∞, where $u_{*T,inlet}$ or $u_{*T,pile}$ corresponds to the total bed velocity, determined from the longitudinal and transverse velocity vector magnitude.

$$\alpha_\tau = \frac{|\tau_0|}{\tau_\infty} \tag{25}$$

$$\tau_0 = \rho u_{*T,\ pile}^2 \tag{26}$$

$$\tau_{\infty} = \rho u_{*T,\ inlet}^{2} \quad (27)$$

2.4. Scour around a Cylindrycal Pile

Using a simulated scenario for the study of hydrodynamics, the scour around the pile was estimated by considering a bed composed of spherical sediments with diameter of 0.38 mm (d), 2650 Kg/m^3 in density, a 30° angle of repose, and a dimensionless critical shear stress (τ_{cr}^{*}) equal to 0.036, in order to make the results obtained in this investigation for E01 and E02 comparable to those previously presented by Qi and Gao [26].

The scour estimation using the numerical model was activated during the last 25 minutes of the hydrodynamic modelling (S_1), to obtain enough information at the beginning of the simulation for the immobile bed condition and, subsequently, the associated condition for the mobile bed.

The features of the numerical tests conducted on the scour around the modeled cylindrical pile are summarized in Table 6. The parameters associated with the incipient transport of sediments for the combined regimen of waves and currents have been estimated according to the methodology extensively described in Soulsby [52]. The fundamental equations for estimating shear stress are described below in summary.

Table 6. General characteristics of numerical tests of scour modeling.

Case	τ_c^*	τ_w^*	τ_{wc}^*	$\frac{\tau_{wc}^*}{\tau_{cr}^*}$	Regimen
E01	0.023	0.003	0.023	0.63	Clear water
E02	0.023	0.003	0.023	0.63	Clear water
E03	0.021	0.005	0.021	0.58	Clear water
E04	0.021	0.005	0.021	0.58	Clear water
E05	0.025	0.008	0.025	0.69	Clear water
E06	0.025	0.008	0.025	0.69	Clear water
E07	0.004	0.008	0.004	0.15	Clear water
E08	0.004	0.008	0.004	0.15	Clear water

Bed shear stress due to the combined action of waves and currents was estimated according to Equation (28), where τ_c is shear stress considering only the action of currents, while τ_w corresponds to the shear stress for waves alone. The shear velocity (u_*) was determined based on a resistance law according to Equation (31), where U_B is the bulk velocity of the flow due to the current. The wave boundary layer velocity (U_{fm}) is defined in Equation (32), where f_w is the friction factor, which was determined according to Fredsøe and Deigaard [53] (page 25).

$$\tau_{wc} = \tau_c \left[1 + 1.2 \left(\frac{\tau_w}{\tau_c + \tau_w} \right)^{3.2} \right] \quad (28)$$

$$\tau_c = \rho u_*^2 \quad (29)$$

$$\tau_w = \frac{1}{2} \rho f_w U_{fm}^2 \quad (30)$$

$$\frac{u_*}{U_B} = \frac{1}{7} \left(\frac{d}{h} \right)^{\frac{1}{7}} \quad (31)$$

$$U_{fm} = \sqrt{\frac{f_w}{2}} U_m. \quad (32)$$

The shear stress values associated with currents, waves and the combined action of both were nondimensionalized according to Equation (14). The results are presented in Table 6.

Considering that the numerical modeling duration of the scour was 25 minutes and that in this time scale the equilibrium condition was not reached, a projection was made via the equation proposed by Sheppard et al. [54], which corresponds to a four-parameter exponential function for the extrapolation of the equilibrium scour depth (S_t), which is presented in Equation (33) where a_i corresponds to the adjustment coefficient i of the equation by Sheppard et al. [54]. This approach was also used by Qi and Gao [20], with experimental data.

$$S_t = a_1[1 - \exp(-a_2 t)] + a_3[1 - \exp(-a_4 t)]. \tag{33}$$

The equilibrium scour was compared with the results presented by Qi and Gao [26], Sumer and Fredsøe [25], Raaijmakers and Rudolph [15], Sumer et al. [55], and Mostafa and Agamy [56] to check whether the numerical results obtained are concordant with the experimental data developed by other authors.

3. Results

3.1. Hydrodynamics Calibration Test

The hydrodynamic calibration process for the numerical model REEF3D is as follows. Figure 3 shows the results obtained by numerical modeling and those reported by Umeyama [3], based on experimental data, for a domain forced by waves and a mixed domain of wave and currents. The comparison variable corresponds to the instant surface elevation (η) nondimensionalized with the wave height (H) as a nondimensional time function ($\frac{t}{T}$), with T as the period. The physical sense of the variable $\frac{\eta}{H}$ corresponds to the fraction of the increase or decrease in the water surface and, evidently, the maximum and minimum fractions are the descriptors of wave asymmetry. The variable $\frac{t}{T}$ corresponds to an indicator of the wave time fraction being simulated.

A comparison between the experimental data for cases where waves are acting alone (W1 to W3, illustrated with circles) and those for waves and currents (WC1 to WC3, illustrated with triangles) are shown in Figure 3, boxes 1.1 to 1.3. Here, it can be deduced that a codirectional current modifies the wavelength and amplitude of the wave. The numerical results are found in Figure 3, boxes 2.1 to 2.3, both for a flow with waves (blue line) and for waves and currents (red line). Thus, it can be verified that the wavelength and amplitude are modified between both flows, as shown in the experimental data (Figure 3, box 1.1 to 1.3). The results of the comparison between the numerical modeling REEF3D and the experimental data provided by Umeyama [3] are included in boxes 3.1 to 3.3 for waves acting alone, while waves and currents acting together are shown in boxes 4.1 to 4.2.

When analyzing numerical and experimental data associated to waves acting alone (W1 to W3), all model cases were able to adequately reproduce the maximum and minimum wave amplitude, as well as its temporal evolution within a wave time coinciding with the necessary time to reach the peak, the zero crossing time and the time to reach the minimum. Taking into account the difference in the estimation of the crest and trough, for the waves and currents (WC1 to WC3), it can be observed that the numerical model slightly underestimates the experimental data, but at a magnitude of less than one millimeter (around 2% error).

The vertical profiles for the instant velocity associated with each of the simulated cases in the present investigation (which were experimentally registered by Umeyama [3]) are compared in Figure 4 for waves acting alone and in Figure 5 for waves and currents combined. Both figures illustrate four instants of time for each simulated case, ordered from left to right and corresponding to t/T = 0.00, 0.25, 0.50, 0.75, and 1.00.

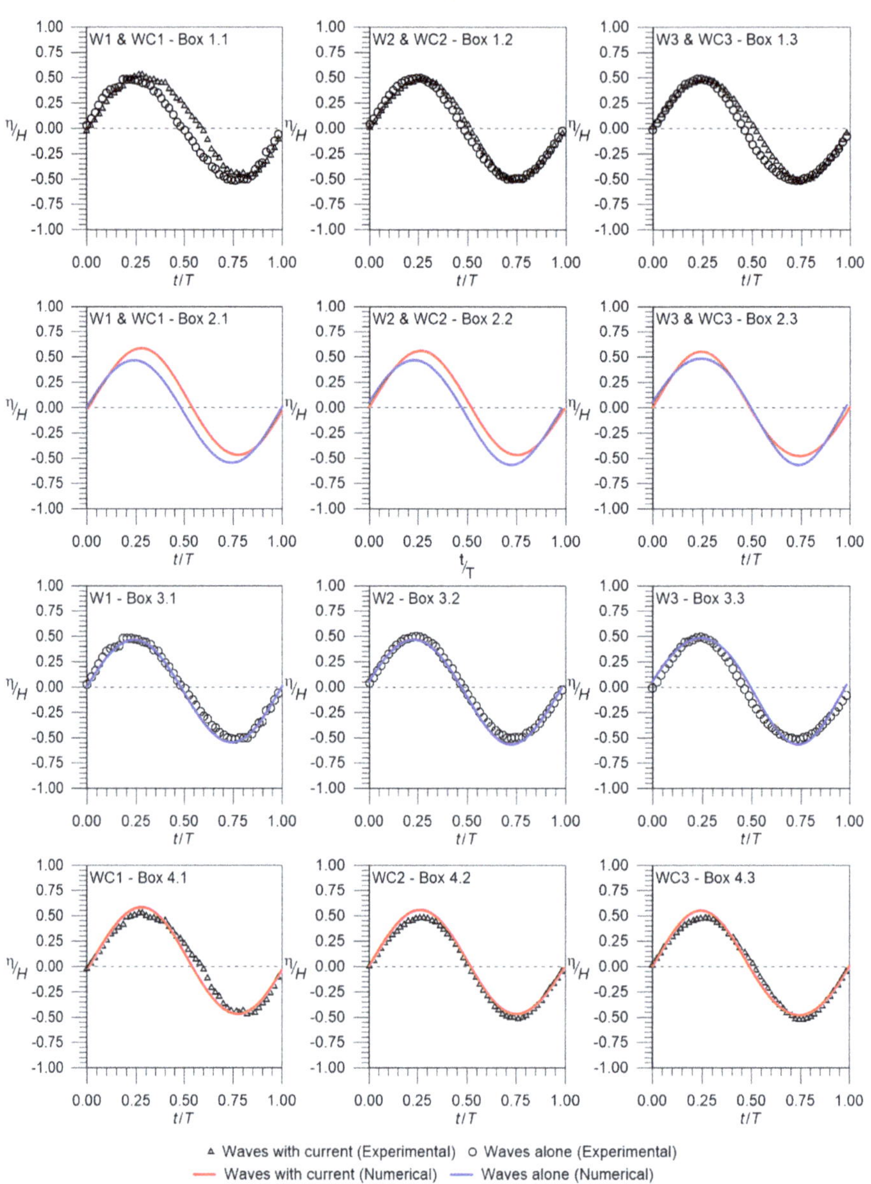

Figure 3. Phase-average surface displacements comparison between experimental data (Umeyama [3]) and numerical results.

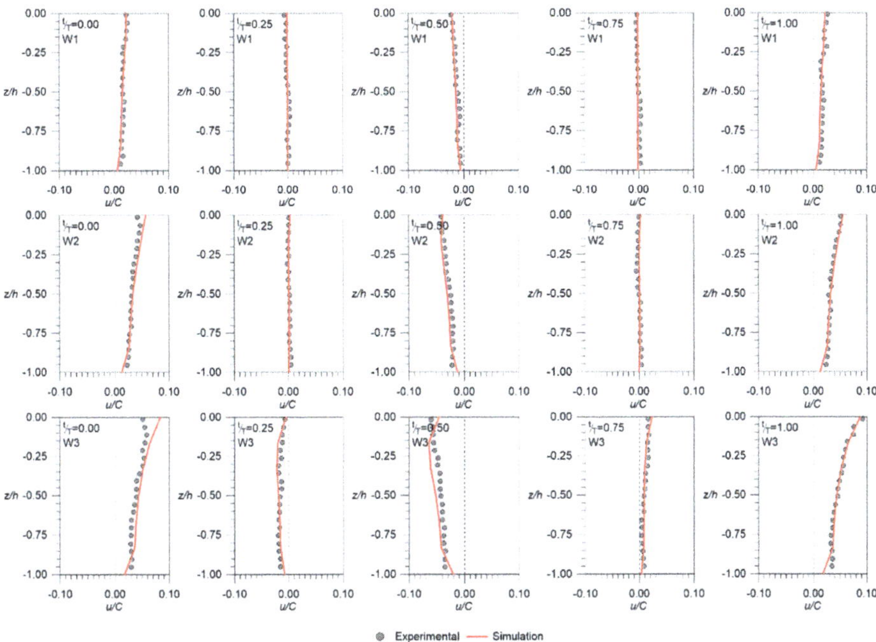

Figure 4. Instantaneous horizontal velocity profile comparison between the experimental data (Umeyama [3]) and numerical results for different time steps, for cases with waves alone.

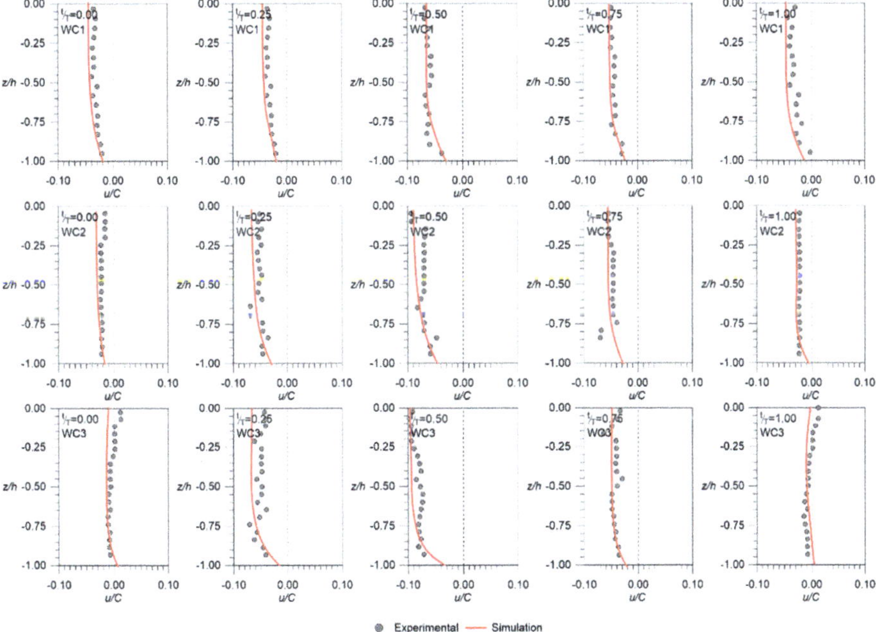

Figure 5. Instantaneous horizontal velocity profile comparison between the experimental data (Umeyama [3]) and numerical results for different time steps, both for waves alone and for current cases.

The results of the comparison of numerical and experimental data considered in the presence of waves alone (Figure 4) reflect a high consistency between vertical and temporal behavior, as the model is capable of adequately representing the flood direction (negative u/C) and the ebb direction (positive u/C). The experimental and numerical data show equivalent vertical structures for the velocity profile, with a slight increase toward the surface, which is evidenced in a greater proportion when analyzing the case with the highest wave (W3).

Near the bed, the current magnitudes determined based on the numerical model coincide with those determined by experimental means, showing slight differences between the simulated and instrumental data.

The detected differences for $t/T = 0.25$ and 0.75 show a low magnitude. However, the flood and ebb conditions reached different magnitudes described, as follows. For the W1 case, the maximum difference in the nondimensional velocity (u/C) obtained by the ebb direction was 0.004 ($t/T = 1.00$), while for the flood direction it was −0.001. For the W2 case, the differences fluctuated between −0.005 and −0.011.

Greater differences between the numerical model and experimental data were found for the W3 case for $t/T = 0.00$, which is produced at the water surface. Meanwhile, near the bed, the greatest difference in u/C was 0.015 for $t/T = 1.00$.

In Figure 4, the maximum and minimum velocities are not at 0.25 and 0.75 t/T respectively, because the wave phase effect on the initial condition was adjusted to represent the same oscillatory flow that Umeyama [3] reported in his research.

When incorporating a codirectional current to waves, Figure 5 shows that the vertical velocity distribution along the channel only shows the flood direction because the currents control the hydrodynamics. This can be confirmed by calculating the flow relative velocity proposed by Sumer and Fredsøe [26] (U_{cw}), which offers results equal to 0.84, 0.70, and 0.60, for WC1, WC2, and WC3, respectively.

In general terms, the numerical model adequately captured the behavior of the velocity profile, showing a greater similarity between the currents near the bed and those obtained toward the free surface. Compared to the hydrodynamic scenarios, where only waves were present, the differences found for the dimensionless velocity (u/C) have a greater magnitude when the current is incorporated in the centre, reaching 0.010 for WC1, −0.017 for WC2, and −0.037 for WC3.

The previous results correspond to the scenarios of interactions between waves and currents without the incorporation of a circular pile blocking the flow. However, they allow us to confirm that the numerical model is capable of representing the complex hydrodynamics resulting from the combined action of both components (waves and currents). To strengthen this analysis, the results of the model for a circular pile in the flow are shown next.

The results of the comparison of the numerical model with the experimental data obtained from Qi and Gao [26] are presented in Figure 6, which considers the total flow velocity ($U_c + U_w$) non-dimensionalized with the characteristic velocity (C). Based on this comparison, it was observed that the numerical model was able to represent the dynamic behavior of the combined flow velocity, for the different characteristics of the simulated waves and currents. In test case C01, it can be observed that the numerical model predicted slightly higher velocities in the trough located between $1.00 < t/T < 1.50$, $(U_c + U_w)/C$ (equal to 0.02). This result, however, was not observed in cases C02 and C03. Hence this case does not correspond to the numerical model configuration.

It is important to emphasize that the total flow velocity measurements experimentally obtained by Qi and Gao [26] were registered upstream from the pile at a distance of $20D$. Ergo, the effects of the opaque structure on the velocities field would not be shown in their behavior. Therefore, this comparison (Figure 6) complements that previously shown in Figure 3 for the instant surface elevation. Thus, the numerical model is capable of representing the wave and current interactions in a freestream.

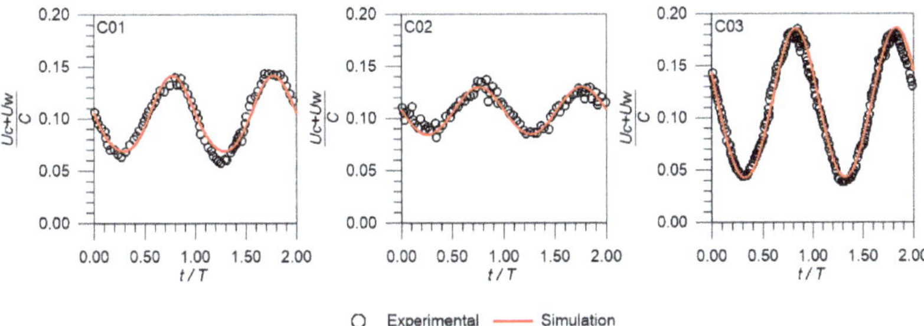

Figure 6. Total flow velocity comparison between the experimental data (Qi and Gao [26]) and numerical results for waves alone and for current cases in the presence of a cylindrical pile.

The effects on the velocity fields caused by the cylindrical pile in a flow field with waves perpendicular to the currents obtained from the numerical model were compared with data provided by Miles et al. [51]. The results for which waves and currents come from perpendicular directions are shown in Figure 7 (associates with case C04). When analyzing monitoring station P1, a high consistency is observed between the velocity profiles modeled (red line) and experimentally obtained data (circles), highlighting that the model is capable of representing the mean velocity near the bed and in the vertical direction as well. Equivalent performance was also verified for P2 and P8, which corresponds to the monitoring stations exposed to the current direction.

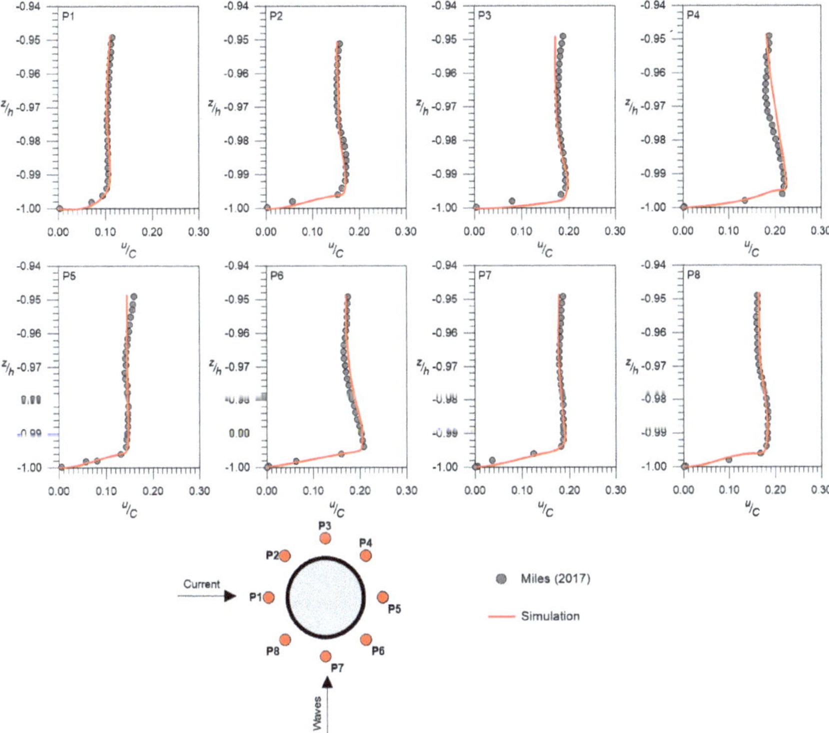

Figure 7. Mean velocity profiles comparison between the experimental data (Miles et al. [51]) and numerical results associated with case C04, for waves perpendicular to the current cases in presence of a cylindrical pile.

In station P4, it is noted that the numerical model may have slight differences in its vertical velocity distributions for elevation range, here z/h is equal to -0.98 to -0.96, which is not clearly presented in the other velocity monitoring stations. This performance detected in station P4 may be caused by the combined wake effect due to currents and waves, since geometrically this location is completely downstream of both forcings (associates with case C04).

3.2. Hydrodynamics Behaviour of the Flow Around a Cylindrycal Pile

The results obtained for the spatio–temporal evolution of the velocity and vorticity fields for case E07 are illustrated in Figure 8. These results consider four instants in time that allow us to visualize the interaction between the flow and the pile. All charted variables have been nondimensionalized by test characteristic scales. For example, the longitudinal and the vertical axes have been made dimensionless with the pile diameter, and, the beginning of the coordinated system has been placed at the center of the pile. The velocities (u and w) have been made dimensionless with the characteristic velocity (C). Vorticity (ω_x, ω_y, and ω_z), on the other hand, was nondimensionalized by the integral temporal scale of the experiment and corresponds to the proportion D/C.

In Figure 8, in the first instant of time (Time 1 in Figure 8), it can be observed that the free surface is on a trough nearby the pile where the longitudinal velocity is mainly an ebb type, which usually form a horseshoe vortex when interacting with the incoming boundary layer. Descending velocities are developed in the vertical axis and near the pile, which may be associated with the down-flow.

As time progresses and the crest approaches the pile (Time 2 in Figure 8), the longitudinal velocity starts to diminish its magnitude (nearing zero), thus reflecting the oscillating effect, since this hydrodynamic behavior is an indicator that both ebb and flow can be present based on the phase of the wave. Contrary to the previous instant, the vertical velocity near the pile shows a positive magnitude, which would not produce down-flow and would consequently modify the horseshoe vortex behavior, as follows.

At time instant 3 (time 3 in Figure 8), the wave crest interacts directly with the pile, developing longitudinal velocities mainly oriented downstream the pile, while the vertical component of the velocity is once again oriented toward the bed, thus producing a down-flow. This is maintained for as long as the wave interacts with the pile during the crest phase (time 4 in Figure 8), while the longitudinal velocity changes to an ebb direction.

The association of the main component of the vorticity with the cross vorticity ($\omega_y D/C$) is obtained from the vorticity development near the pile and the adjacent bed for the four illustrated instants of time in Figure 8. Thus, that the minimum intensities occur under the trough, while in the crest there is a general tendency to maximize these intensities. When the crest is approaching the pile, it is observed that in the bed (upstream), a cross vorticity structure with a positive magnitude approaches the pile, and, conversely, the cross vorticity considerably diminishes its magnitude when the trough moves through the pile. The geometrical characteristics of this cross vorticity structure mainly present a greater longitudinal than vertical development.

The vorticity behavior upstream from the pile (described above) corresponds to the horseshoe vortex, and its intermittence would be conditioned by direction changes of the vertical and longitudinal velocity.

The downstream sector under the pile showed a vorticity behavior ($\omega_x D/C$ and $\omega_z D/C$) with alternate structures (positives and negatives) and a greater development in the vertical axis than in the horizontal axis, compared to the cross vorticity. This spatio–temporal development could be associated with vortex shedding, resulting from structure-fluid interactions.

Based on a general analysis of the information obtained from the numerical modeling (a visual inspection of the results), there are no significant differences in the mean spatio–temporal behavior of the streamlines, vorticities, and velocity field; hence, these element can be broadly described by the specification of one of the eight case simulations, with scenario E01 selected for this effect.

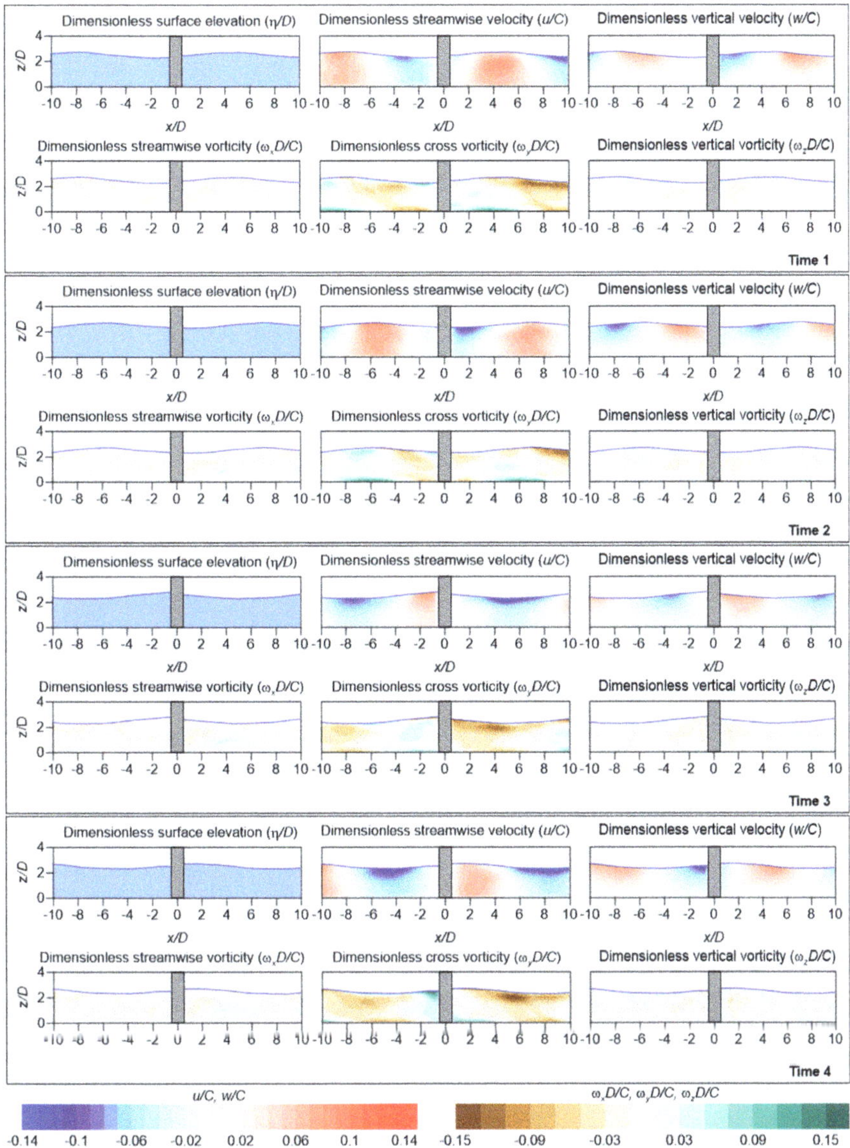

Figure 8. Temporal behavior for velocities and vorticities near to the pile, waves and current are coming from the left to the right.

The characteristic results of scenario E01 are presented in Figure 9. Unlike Figure 8, which illustrates the behavior from the free surface to the bed, in this section, the analysis focused on the bottom to obtain the characteristics of the hydrodynamics that might be responsible for sediments transport and, consequently the scour around the pile.

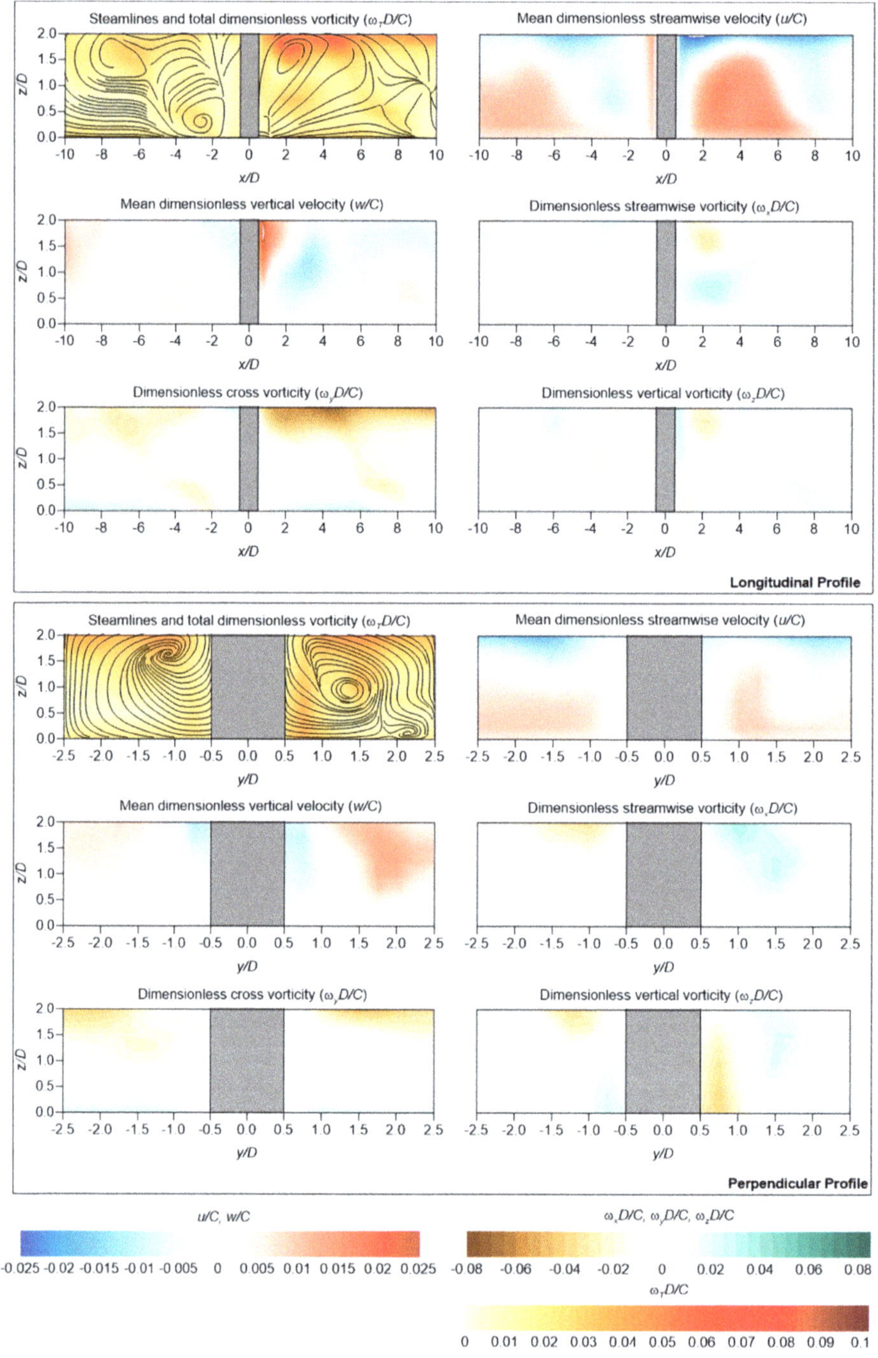

Figure 9. General description of the mean velocities and vorticities near the pile for the longitudinal and cross profiles (associated to the scenario E01). Waves and currents are move from the left to the right.

Figure 9 is divided into two main boxes. The upper box illustrates the flow characteristics in the longitudinal profile of the channel, while the lower box illustrates the transverse section. Both have included the streamlines and total vorticity (vector addition, $\omega_T D/C$), the mean velocity along the channel (u/C), the vertical mean velocity (w/C), the mean vorticity along the channel ($\omega_x D/C$), the transversal mean vorticity ($\omega_y D/C$), and the vertical mean vorticity ($\omega_z D/C$) in a nondimensionalized manner.

The total vorticity in the longitudinal profile and flow lines reflect the presence of a horseshow vortex as the main structure upstream the pile ($x/D < 0$), this structure approximately centered at $x/D = -2.6$ and $z/D = 0.3$. This vorticity system also reflects the average behavior of the flow longitudinal velocity, as in the zone equivalent to the horseshoe vortex location, u/C has negative values that are driven by the vortex counter clockwise rotation. Additionally, w/C also reflects the horseshoe vortex effects, since negative magnitudes of the vertical velocity (known as the down-flow) can be revealed near the pile. Meanwhile, at a distance lower than $x/D = -2.6$, the down-flow becomes positive (mainly driven by the counter clock turn of the vortex).

The pile downstream area ($x/D > 0$) in the mean streamline condition showed a rotational centered structure at around $x/D = 2.4$ and $z/D = 1.6$, which is related to the vortex shedding and could produce alterations in the mean field of the vertical and longitudinal velocities. Longitudinally, near the pile, a flow could be produced toward it, mainly caused by the momentum balance that would be developed at an approximate distance of $x/D = 1.2$. Then, a positive direction of the flow velocity up to a distance of $x/D = 6$ would be present, to subsequently re-adopt a negative velocity; since this behavior is associated with vortex shedding, as stated previously.

Vertical velocities in the downstream area show that, nearby the pile, the flow moves upwards. Meanwhile, an $x/D = 2.7$ distance would make the vertical velocity negative. This and the flow line behavior could be caused by clockwise rotation and related to the formation of vortex shedding.

Analyzing the transverse section of the flow in Figure 9, clear that in the mean streamlines, a horseshoe vortex system develops around the pile, with a longitudinal component of mainly positive velocities near the pile. In the case of a vertical component, these velocities would be negative. This means that: a downstream flow component is present once the horseshoe vortex surrounds the pile, and it remains near the bed pushed by the flow vertical velocities.

The presence of structures concordant with the horseshoe vortex around the pile is shown by the transverse vorticity ($\omega_y D/C$). Thus, the vertical development of these structures is limited to an approximately height of $z/D = 0.5$ from the bed. The mean vorticity fields obtained for vertical and longitudinal components are congruent with the classical description of the flow around the cylindrical piles for a permanent flow, which has been widely addressed in the literature [1–4].

The mean velocity vertical profiles for monitoring stations P1 to P8 are presented in Figure 10 for each modeled scenario. From this, it can be observed that the mean characteristics of the codirectional currents and wave velocity profiles are not significantly different under opposite flow conditions in either of the proposed scenarios (those controlled by currents or those controlled by waves).

Clear evidence of this phenomenon is shown in P5 (Figure 10), which corresponds to a downward station for codirectional currents and waves and an upstream station for waves in opposite flow scenarios. In each case, the station presented the smallest flow magnitudes, indicating that for all the simulated scenarios of station P5, currents dominate the flow, according to the flow relative velocity (U_{CW}) proposed by Sumer and Fredsøe [25], this result indicates that a combined regimen is present. This aspect will be further addressed in the analysis of the results.

It is important to note that the currents are symmetrical based on the pile geometrical characteristics (cylinder) and the studied flow. Figure 10, shows that the velocity profiles of station pairs P2 with P8; P3 and P7; and P4 and P6 are concordant not only in their magnitudes but also in their vertical axis. Thus, describing only one of the stations associated in a pair is sufficient to understand the flow around the pile.

The velocity profiles for both codirectional and opposite cases reached their maximum value in station P3 (P7), mainly due to the contraction of flow lines producing accelerations, thereby increasing the velocities from station P1 towards P3 (P7), followed by a gradual decrease from station P3 (P7) towards P5.

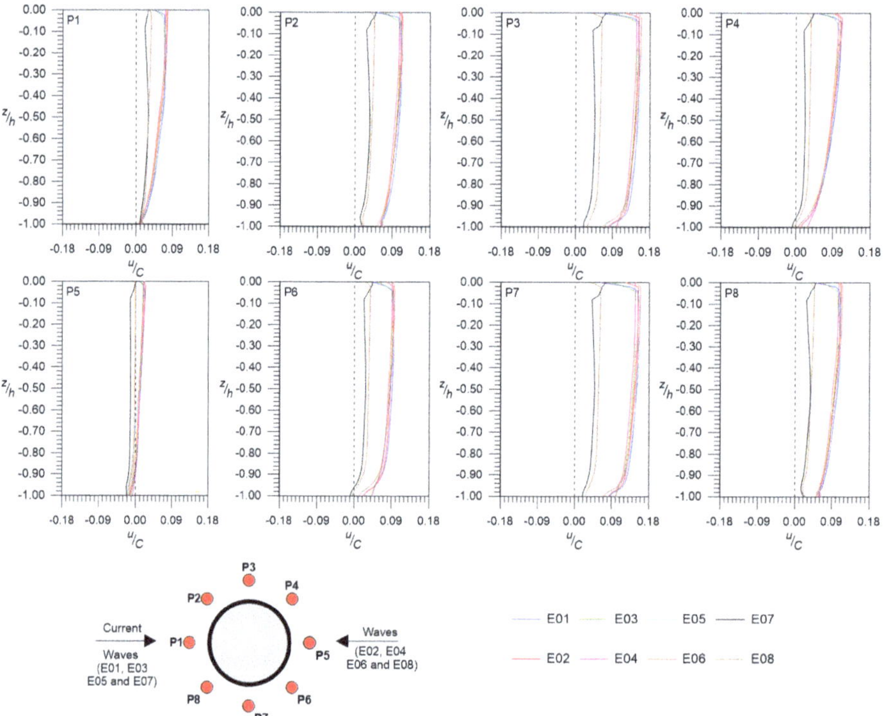

Figure 10. Mean streamwise velocity vertical profile for all the simulated cases.

In light of the results described above, and for the simulated scenarios, the hydrodynamics around a pile for combined flow of waves and currents (codirectional and opposite) should behave like a normal flow around a cylinder due to its steady state flow, which is widely described in the literature.

The amplification of the mean shear stresses (α_τ) made dimensionless with an undisturbed bed shear stress around a cylindrical pile is shown in Figure 11 for each of the simulated cases, in which the left column shows codirectional currents and waves cases and the right column shows the opposite current and wave cases.

The maximum amplifications of the main bed shear stress were produced in the pile lateral edge and reached magnitudes of 2.5 for codirectional and opposite currents and waves cases. Nevertheless, when waves act on the current in an opposite direction, the amplification zone coverage is reduced (smaller area) compared to the codirectional current and wave cases. As a general trend, the amplification obtained shows that shear stress gradually decreases from hydrodynamic scenarios dominated by currents (E01 and E02) to those dominated by waves (E07 and E08).

Analyzing the results of E01 and E02, the differences found between the spatial distributions of the bed shear stress are not significant when the waves act codirectionally with currents or when they are opposite. No significant movements were noticed in the maximum amplification localization. This means no direct influence of the waves flow or ebb was found for the development of the shear stress in the bottom.

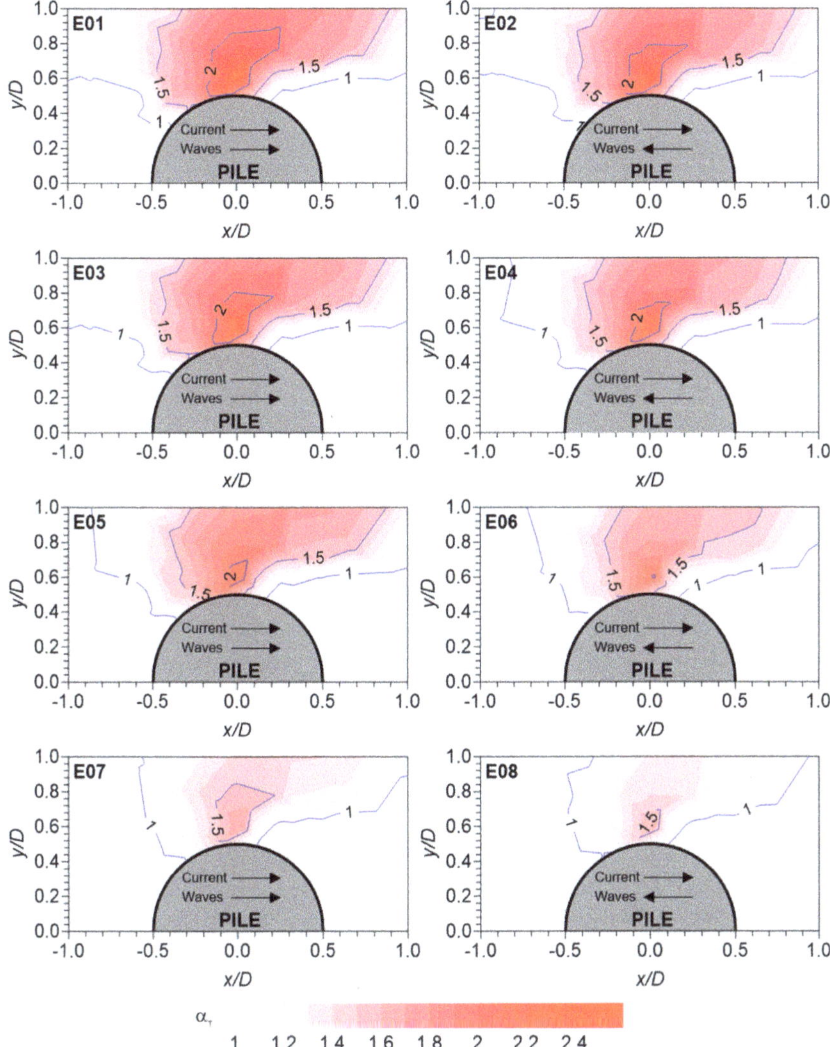

Figure 11. Bed shear stress made dimensionless with undisturbed bed shear stress amplification for each simulated case.

A behavior equivalent to that described for E01 and E02 was identified in E03, E04, E05, and E06, that is, no influence of the waves in the mean bed shear stress distribution was evidenced, even though according to U_{cw} waves domain. In general hydrodynamics, should become more significant when transiting from scenario E01 to E07 and E08. In these last scenarios (E07 and E08), the lowest amplifications of the mean bed shear stress were obtained, which were 1.5.

3.3. Scour around a Cylindrycal Pile

This section presents the analysis results of the scour around the pile. Figure 12 presents the dimensionless scour time series (S/D) obtained experimentally by Qi and Gao [26] and the series resulting from the numerical model, for codirectional (E01) and opposite (E04) waves and currents.

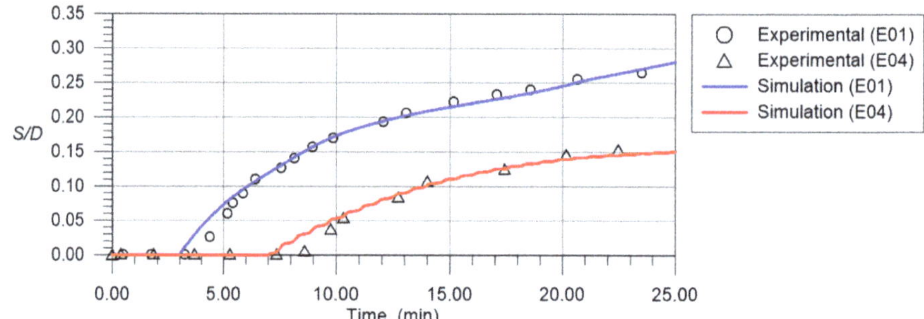

Figure 12. Maximum scour development comparison between the experimental data (Qi and Gao [26]) and numerical simulation results for cases E01 and E04.

The information for the codirectional flow (E01) illustrated in Figure 12 shows a strong agreement between the experimental and simulated data, in its temporal evolution (curve form) and in the magnitude reached by the dimensionless scour. For example, after 10 minutes of simulation the model reached a magnitude of $S/D = 0.173$, and the experimental data reached a magnitude of $S/D = 0.169$ (2.4% of the relative error).

The comparison of numerical data versus experimental data for the dimensionless scour with and opposite flow (E04) illustrated in Figure 12 (as the previously described for E01) showed a strong correspondence in both its the temporal evolution and the magnitude reached. For example, within 20 minutes, the experimental data shows the dimensionless scour would be 0.146, while the numerical model showed that the scour would be 0.139 (4.8% of the relative error).

The generality of the results obtained and presented in Figure 12 indicates that the scour would be greater when the flow acts codirectional to the waves and currents than in the opposite case. The latter is supported by the results summarized in Table 7 column S_1/D, which is the dimensionless scour obtained in the last time step of the numerical model. Additionally, Table 7 lists results for the adjustment coefficients of the equilibrium scour equation from Sheppard et al. [54] (a_1 to a_4), the equilibrium dimensionless scour (S_t/D), and the relative scour factor (S_t/S_1). A similar method was used by Qi and Gao [26] but with experimental data.

Table 7. Adjustment parameter for equilibrium scour estimate and the results obtained from the numerical simulation.

Case	a_1	a_2	a_3	a_4	$\frac{S_1}{D}$	$\frac{S_t}{D}$	$\frac{S_t}{S_1}$
E01	0.095	0.408	0.423	0.029	0.251	0.518	2.064
E02	0.330	0.078	0.006	1.078	0.161	0.336	2.083
E03	0.330	0.091	0.100	0.091	0.304	0.430	1.414
E04	7.663	0.230	−7.365	0.234	0.151	0.298	1.976
E05	2.553	0.068	−1.878	0.070	0.132	0.675	5.120
E06	0.755	0.002	0.149	0.089	0.110	0.626	5.711
E07	0.020	0.058	0.083	0.070	0.085	0.103	1.213
E08	0.081	0.067	−0.006	0.416	0.060	0.075	1.242

The relative scour factors indicated in Table 7 were greater in cases where the flow velocity defined at a distance of $z = D/2$ (U_c) was greater. This mean that the scour obtained from the numerical model of a 25 minutes sediment transport and resulting morphodynamics evolution of the bed, is very different from the equilibrium in cases whit greater flow magnitude.

The latter may be associated with major currents acting on the center, producing major scours and subsequently requiring greater action times in order to reach scour equilibrium [5]. Nonetheless, when magnitudes of the dimensionless equilibrium scour estimated by the equation developed by Sheppard et al. [54] and based on the data obtained here for the 25 minutes simulation are compared with the experimental data obtained by third parties, equivalent results and estimates within an acceptable range of experimental variability are obtained, as shown in Figure 13.

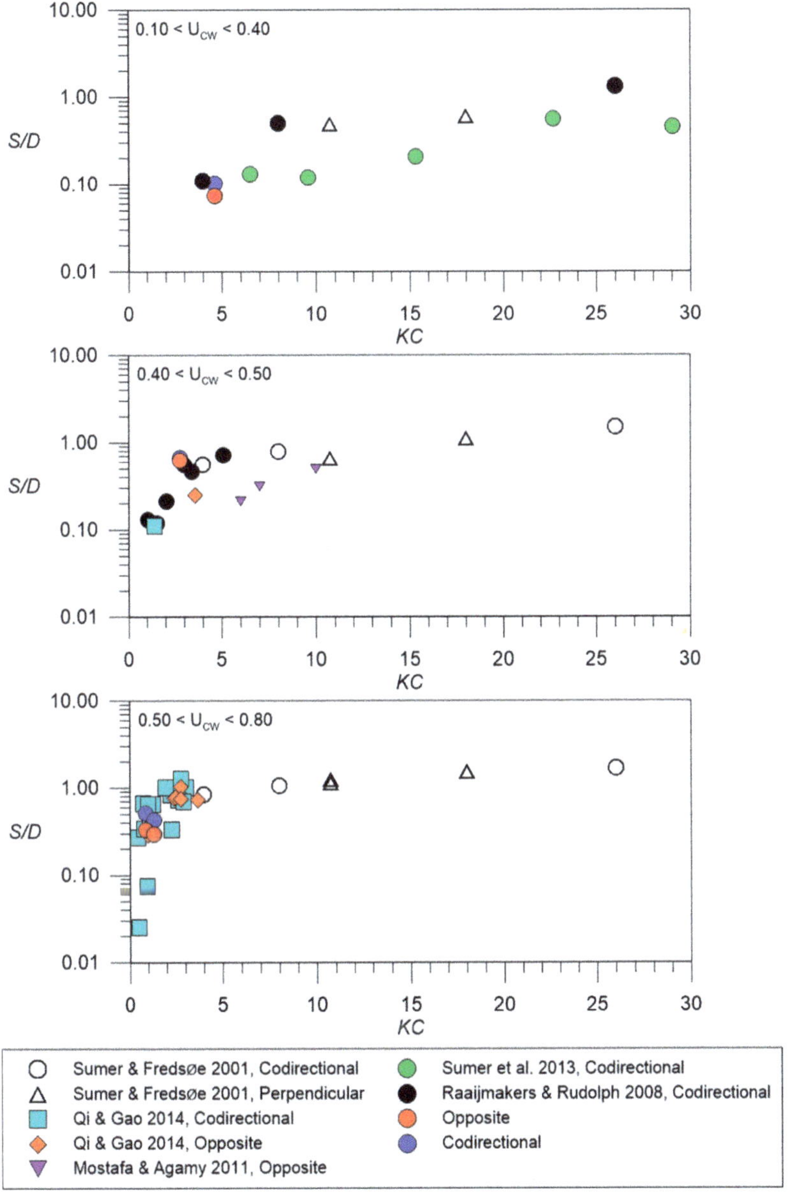

Figure 13. The equilibrium scour estimated from numerical data and its comparison with the equilibrium scour obtained by other authors.

Figure 13 presents a comparison between the numerical data obtained in this article (blue circles for codirectional scenarios and red circles for opposite flows), and experimental ones found in the literature [21,25,26,55,56]. This graph was formed similarly to that presented by Sumer and Fredsøe [25], i.e., the dimensionless equilibrium scour as a function of the Keulegan–Carpenter (*KC*) number for different relative velocity ranges between waves and currents (U_{CW}).

From the general analysis of Figure 13, it is observed that, for the range $0.10 < U_{CW} < 0.40$, the data obtained from the numerical model implemented in this article would be close to the data obtained by Raaijmakers and Rudolph [21] and Sumer et al. [55] for similar Keulegan–Carpenter numbers (around five) for both the codirectional flow condition and the opposite. This situation repeats in the comparison between ranges $0.40 < U_{CW} < 0.50$ and $0.50 < U_{CW} < 0.80$, meaning that the data obtained from the results projection of the numerical model toward the equilibrium scour based on the equation of Sheppard et al. [54] allows us to gather magnitudes that can be compared with the experimental records obtained by other researchers, for similar hydrodynamic characteristics.

4. Discussion

The experimental data with no scour provided by Umeyama [3] have been used to compare the numerical results of other authors, such as Zang et al. [30] and Ahmad et al. [31] who used the RANS approach to solve the hydrodynamics of waves and currents acting codirectionally over a grid featuring finite differences with a regular element ($\Delta x = \Delta y = \Delta z$), which correspond to the methodology applied in this investigation.

For the construction of the numerical domain, Zang et al. [30] utilized $\Delta x = 0.002$ m to solve the vertical domain in 150 layers, while the configuration applied by the authors of this study considered $\Delta x = 0.01$ m which determines the 30 layers in the vertical direction for the water flow adopted by Umeyama [3]. Despite of the coarser grid used in this research, the results are consistent with the experimental data for the vertical profile of velocities as well as for the instantaneous surface elevation of the water.

In order to model the scour around cylindrical piles, this study used the same element dimension as Ahmad et al. [31], who proved that the use of an element of 0.01 m is sufficient to estimate the scour under a pipeline [31]. This conclusion was reached by a grid analysis and time convergence study which analyzed the numerical behavior of REEF3D for element sizes of $\Delta x = 0.04$, 0.03, 0.02, 0.01 and 0.005 m.

The results obtained from the eight simulations (scenarios E01 to E08), showed a low variability of the mean velocity profile around the pile (stations P1 to P8), as illustrated in Figure 10, although these simulations were constructed to represent both, mixed and current or waves dominated environments, according to the criteria of Sumer and Fredsøe [26]. The results associated with the expected bed shear stresses for each of the eight scenarios (see Table 6), show that the effect of the waves on the first six scenarios (E01 to E06) is not significant in the bed dynamics, since the dimensionless shear stress due to waves (τ^*_w) is an order of magnitude less than the dimensionless shear stress due to currents (τ^*_c).

Based on the above, if it is considered that the shear stress is the hydraulic boundary condition to build at vertical profile of flow velocities, and the effect of the current dominates over the waves, it is expected that the first six scenarios present a high similarity for both a co-directional and opposed flow. On the other hand, in the remaining scenarios (E07 and E08), where the dimensionless shear stresses associated to waves and currents are of the same order of magnitude, greater effects on the velocity profile around the pile could be noticed (Figure 10), which would indicate that both the co-directional and opposed flow develop differences in the velocity mean behavior.

This difference between the shear stresses and the Sumer and Fredsøe criteria [26] seems to imply that the use of the dimensionless number called the relative velocity of the current (U_{CW}), does not fully describe the domain of the forcing over the total hydrodynamics, a discussion that is presented in the following paragraphs of this investigation.

From the results obtained, it is possible to verify that the presence of waves in the hydrodynamic behavior for scenarios E01 to E06 was less significant than for scenarios E07 and E08. This can be clearly seen in the mean profile analysis, since when the current dominated, the direction resulting from the flow agreed with the streamwise direction, these results were previously described by Feraci et al. [4], who, while performing wave and current test acting orthogonally and without the presence of a pile, obtained results comparable to those obtained in this investigation.

The vertical distribution of the mean velocity illustrated in Figure 10 is consistent with that described by Lim and Madsen [5], who mentioned that although the flow field can be mixed (waves and currents acting together and orthogonally), the velocity distribution can be simply modeled by a uniform return current. The foregoing analysis is also consistent with the results presented by Feraci et al. [6].

The equation by Sheppard et al. [54] was applied to estimate the equilibrium scour based on an extrapolation of the numerical model. These results were similar to those obtained experimentally by other authors, which indicates that the methodology applied as well as the configuration adopted by the numerical model are appropriate to describe the phenomenon under study, not only from the perspective of element size but also for the sediment transport equations applied.

An important aspect to highlight is that this investigation has used the relaxation factor previously applied by Quezada et al. [27], which also allowed the authors to correctly represent the scour for unsteady current and oscillatory flow. According to the results obtained in the current study, this value will also allow us to estimate the scour for uniform and oscillatory flow, not only codirectionally but also opposite. The relaxation coefficient permits in an auxiliary manner effects inherent to the structure of the fluid interaction produced around the pile, thereby improving the estimations of sediment transport and the resulting scour.

Sumer and Fredsøe [26] propose the relative velocity of the current (U_{CW}) as a dimensionless number relevant for the description of the scour due to codirectional or perpendicular waves and currents. This is defined in Equation (1), where the current magnitude (U_C) is estimated at a height of $D/2$ from the bed, while the velocity of the wave is considered as the maximum value of the undisturbed orbital velocity at the bottom, just above the wave boundary layer (U_m). By this dimensionless definition, Sumer and Fredsøe [26] established that values of U_{CW} higher than 0.7 indicate that the current dominates in the center, while waves have a significant effect when U_{CW} is close to zero (cases waves alone) and less than 0.4.

This dimensionless number considers that U_C and U_m are added, independently of the direction of incidence of the currents and waves. In this respect, Sumer and Fredsøe [26] consider a single value of U_{CW} for codirectional and perpendicular flow, if U_C and U_m are the same in magnitude but different in direction.

The above, according the authors of this paper, would not be appropriate as a general indicator of wave and current interaction, nor would their effects on the scour for cases in which the forcings are not codirectional, since when both flows face in the opposite direction, the wave would propagate with greater difficulty and modify the net velocity of the channel, such that the current present in the center would correspond to the residual value of both forcings.

Soulsby [52] and Van Rijn [57] indicates that current wave interactions must be treated in terms of the net current produced between the two forcing agents, which corresponds to an algebraic sum that is usually treated according to Equation (34), where ϖ is the angular frequency, U_B is the bulk velocity of the flow due to the current, K is the wave number, ϑ' is the angle between current and wave direction ($\vartheta' = 0$ for codirectional, and $\vartheta' = 180°$ for opposing), g is the gravity and h is the water depth:

$$\varpi - U_B K \cos(\vartheta') = \sqrt{gK \tan h(Kh)}. \tag{34}$$

The left term of Equation (34) correspond to the net velocity (defined according the relative velocity between the current and waves). Meanwhile, the right term corresponds to the dispersion relationship of the waves.

From Equation (34) it can be seen that in cases of co-directional or opposite waves, the $cos(\vartheta\prime)$ changes its sign and therefore, the system net velocity is the sum or subtraction of both forcings. Therefore, defining a dimensionless number that summarizes the wave and current interaction, must include a differentiation when the action is codirectional or when it is counter current.

An approximation to the description of scour due to opposite and codirectional currents and waves was conducted by Qi and Gao [58] who, using experimental data, obtained by the same authors in previous works (Qi and Gao [26]) and by third parties as well (Sumer and Fredsøe [18] and Sumer et al. [55]), propose the use of the Froude number (F_{ra}) defined in Equation (35), as a function of absolute velocity (U_a, defined by Equation (36)) and the pile diameter.

$$F_{ra} = \frac{U_a}{\sqrt{gD}} \tag{35}$$

$$U_a = U_C + \frac{2}{\pi} U_m. \tag{36}$$

From this analysis Qi and Gao [52] proposed a formula fitted to the experimental data for a dimensionless scour (S/D) which is presented in Equation (37) and is valid for the range $0.1 < F_{ra} < 1.1$ and $0.4 < KC < 2.6$.

$$log\left(\frac{S}{D}\right) = -0.8\exp\left(\frac{0.14}{F_{ra}}\right) + 1.11. \tag{37}$$

A comparison of the numerical results gathered in this paper, the experimental data and the equation proposed by Qi and Gao [58] are shown in Figure 14.

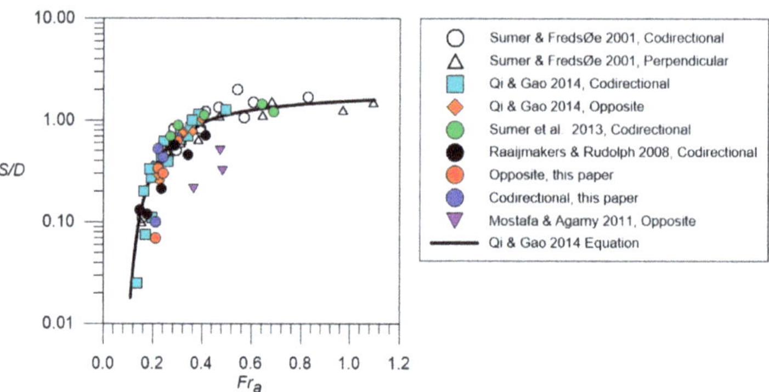

Figure 14. Equilibrium scour distribution according to absolute Froude number proposed by Qi and Gao [58].

In Figure 14, it is observed that the information available in the literature (experimental) and that generated in this study (numerical) are adequately concordant with the equation proposed by Qi and Gao [58], and such a description may be enough to collect information on the equilibrium scour around a dimensionless number that represents its behavior.

Notwithstanding the above, Qi and Gao [58], as well as Sumer and Fredsøe [26], consider current and wave actions added equally if they act in a codirectional or opposite manner, which, in general, is not consistent with a residual flow estimation that would be generated by the interaction. The foregoing disagrees with the results by Soulsby [52] and Van Rijn [57], who indicate that the sum of the forcing agents must be algebraic, respecting the angle between current and wave direction. Although the arguments are contradictory, both proposals (Qi and Gao [58], Sumer and Fredsøe [26]) compile reasonably well the scour information regardless of the direction. This should be analyzed in greater detail as proposed below.

Thus, the Froude number may be rewritten according to Equation (39) if the absolute velocity defined by Qi and Gao [58] is considered, albeit modified according to Equation (38), which is a proposal of the authors of this paper, and where C_φ is a coefficient to describe the flow direction, and where $C_\varphi = 1$ describes codirectional flows and $C_\varphi = -1$ describe opposite flows.

$$U'_a = U_C + \frac{2}{\pi} U_m C_\varphi \tag{38}$$

$$F'_{ra} = \frac{U'_a}{\sqrt{gD}}. \tag{39}$$

C_φ was included in order to incorporate the recommendations of Soulsby [52] and Van Rijn [57], in order to consider the effects of waves and currents directionality acting together on the pile.

Considering this proposal and collecting scour data (experimental and numerical from this paper), Figure 15 is obtained. The blue circles indicate codirectional cases and red circles indicate opposite flow cases. Two trends are found in two areas of the figure. The first trend corresponds to codirectional data, which are still represented by the equation proposed by Qi and Gao [58]. Nevertheless, the opposite flow cases are to the left of the codirectional data and apparently adjust to an equation different than that proposed by Qi and Gao [58], which, according to the available data set (experimental and numerical data from this paper) correspond to Equation (39).

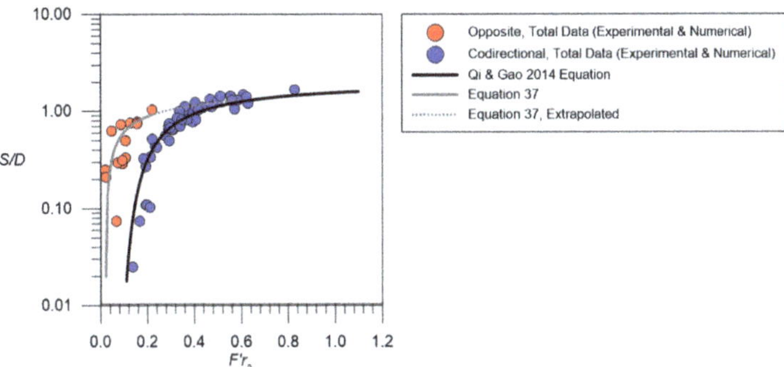

Figure 15. Equilibrium scour distribution according to the absolute Froude number (F'_{ra}) proposed in this research.

$$\frac{S}{D} = 0.39LN(F'_{ra}) + 1.51. \tag{40}$$

Equation (40) seems to agree with the solution proposed by Qi and Gao [58] for high values of the Froude number. However, such behavior may be verified by adding new experimental and/or numerical antecedents, which enable us to complement equilibrium scour data in a combined domain of currents and waves acting in opposite directions.

5. Conclusions

A comparison of the results obtained numerically in this research with experimental data from other authors, show that CFD REEF3D adequately models a hydrodynamic field in a combined domain of waves and currents, correctly depicting both the instantaneous water surface and velocities developed by the flow of a freestream and also the flow in the presence of a pile.

The average hydrodynamics determined for each of the simulated scenarios showed that in 6 (E01 to E06) out of 8 cases, currents were the main flow mechanism over the waves, despite the relative velocity of current (U_{cw}) proposed by Sumer and Fredsøe [26], which indicate a combined flow regime. This was clearly reflected in the velocity profiles traced in points P1 to P8, where it could be verified that codirectional and opposite flow cases converge to the same vertical distributions of velocity.

The equilibrium scour estimated by the projection of the numerical data with the equation by Sheppard et al. [54], enabled us to estimate values close to those described in the literature and to use the numerical model effectively to solve a time scale lower than the equilibrium. From this projection, it was verified that the dimensionless scour would be less when waves and currents come from opposite directions.

The U_{cw} parameter is an indicator that adequately measures the interactions between of currents and waves under the condition of codirectional flow. However, it is recommended to modify this parameter for currents and waves from opposite directions, since it does not properly account for the interaction of both forcings. For such purposes, it is recommended to use the modified Froude number (F'_{ra}) proposed in this paper.

Author Contributions: The contributions from the author M.Q. corresponded to the numerical modeling, analysis of the results, discussion and conclusions of the document. A.T. and Y.N. contributed to the analysis of results, discussion and conclusions of the study. All authors contributed to the general development of the document through meetings and research sessions.

Funding: This publication has been partially-funded by CONICYT doctoral fellowship (21140091), CONICYT Project AFB180004, and the Department of Civil Engineering, Faculty of Physical and Mathematical Science, University of Chile.

Acknowledgments: Matías Quezada thanks the support received by Humberto Díaz and Ecotecnos S.A. Company for providing access to the company's computational resources.

Conflicts of Interest: The authors declare no conflict of interest

References

1. Umeyama, M. Reynolds stresses and velocity distributions in a wave-current coexisting environment. *J. Waterw. Port Coast. Ocean Eng.* **2005**, *131*, 203–212. [CrossRef]
2. Umeyama, M. Changes in turbulent flow structure under combined wave-current motions. *J. Waterw. Port Coast. Ocean Eng.* **2009**, *135*, 213–227. [CrossRef]
3. Umeyama, M. Coupled PIV and PTV measurements of particle velocities and trajectories for surface waves following a steady current. *J. Waterw. Port Coast. Ocean Eng.* **2010**, *137*, 85–94. [CrossRef]
4. Faraci, C.; Foti, E.; Marini, A.; Scandura, P. Waves plus currents crossing at a right angle: Sandpit case. *J. Waterw. Port Coast. Ocean Eng.* **2011**, *138*, 339–361. [CrossRef]
5. Lim, K.Y.; Madsen, O.S. An experimental study on near-orthogonal wave–current interaction over smooth and uniform fixed roughness beds. *Coast. Eng.* **2016**, *116*, 258–274. [CrossRef]
6. Faraci, C.; Scandura, P.; Musumeci, R.E.; Foti, E. Waves plus currents crossing at a right angle: Near-bed velocity statistics. *J. Hydraul. Res.* **2018**, *56*, 464–481. [CrossRef]
7. Hjorth, P. *Studies on the Nature of Local Scour*; Lund University: Lund, Sweden, 1975.
8. Melville, B.W. Local Scour at Bridge Sites. Ph.D. Thesis, University of Auckland, Auckland, New Zealand, 1975.
9. Ettema, R. Scour at Bridge Piers. Ph.D. Thesis, University of Auckland, Auckland, New Zealand, 1980.
10. Chiew, Y.M.; Melville, B.W. Local scour around bridge piers. *J. Hydraul. Res.* **1987**, *25*, 15–26. [CrossRef]
11. Melville, B.W.; Chiew, Y.-M. Time scale for local scour at bridge piers. *J. Hydraul. Eng.* **1999**, *125*, 59–65. [CrossRef]
12. Oliveto, G.; Hager, W.H. Temporal evolution of clear-water pier and abutment scour. *J. Hydraul. Eng.* **2002**, *128*, 811–820. [CrossRef]
13. Link, O.; Pfleger, F.; Zanke, U. Characteristics of developing scour-holes at a sand-embedded cylinder. *Int. J. Sediment Res.* **2008**, *23*, 258–266. [CrossRef]

14. Diab, R.; Link, O.; Zanke, U. Geometry of developing and equilibrium scour holes at bridge piers in gravel. *Can. J. Civil Eng.* **2010**, *37*, 544–552. [CrossRef]
15. Link, O.; Henríquez, S.; Ettmer, B. Physical scale modelling of scour around bridge piers. *J. Hydraul. Res.* **2019**, *57*, 227–237. [CrossRef]
16. Sumer, B.M.; Fredsøe, J.; Christiansen, N. Scour around vertical pile in waves. *J. Waterw. Port Coast. Ocean Eng.* **1992**, *118*, 15–31. [CrossRef]
17. Sumer, B.M.; Christiansen, N.; Fredsøe, J. The horseshoe vortex and vortex shedding around a vertical wall-mounted cylinder exposed to waves. *J. Fluid Mech.* **1997**, *332*, 41–70. [CrossRef]
18. Sumer, B.M.; Fredsøe, J. Wave scour around a large vertical circular cylinder. *J. Waterw. Port Coast. Ocean Eng.* **2001**, *127*, 125–134. [CrossRef]
19. Eadie Robert, W.; Herbich John, B. Scour about a Single, Cylindrical Pile Due to Combined Random Waves and a Current. In Proceedings of the 20th International Conference on Coastal Engineering, Taipei, Taiwan, 9–14 November 1986; pp. 1858–1870.
20. Kawata, Y.; Tsuchiya, Y. Local Scour around Cylindrical Piles Due to Waves and Currents Combined. In Proceedings of the 21st International Conference on Coastal Engineering, Malaga, Spain, 20–25 June 1988; pp. 1310–1322.
21. Raaijmakers, T.; Rudolph, D. Time-dependent scour development under combined current and waves conditions-laboratory experiments with online monitoring technique. In Proceedings of the 4th International Conference on Scour and Erosion, ICSE, Tokyo, Japan, 5–7 November 2008; pp. 152–161.
22. Rudolph, D.; Raaijmakers, T.; Stam, C.-J. Time-dependent scour development under combined current and wave conditions–Hindcast of field measurements. In Proceedings of the 4th International Conference on Scour and Erosion, ICSE, Tokyo, Japan, 5–7 November 2008; pp. 340–347.
23. Zanke, U.C.; Hsu, T.-W.; Roland, A.; Link, O.; Diab, R. Equilibrium scour depths around piles in noncohesive sediments under currents and waves. *Coast. Eng.* **2011**, *58*, 986–991. [CrossRef]
24. Ong, M.C.; Myrhaug, D.; Hesten, P. Scour around vertical piles due to long-crested and short-crested nonlinear random waves plus a current. *Coast. Eng.* **2013**, *73*, 106–114. [CrossRef]
25. Sumer, B.M.; Fredsøe, J. Scour around pile in combined waves and current. *J. Hydraul. Eng.* **2001**, *127*, 403–411. [CrossRef]
26. Qi, W.-G.; Gao, F.-P. Physical modeling of local scour development around a large-diameter monopile in combined waves and current. *Coast. Eng.* **2014**, *83*, 72–81. [CrossRef]
27. Quezada, M.; Tamburrino, A.; Niño, Y. Numerical simulation of scour around circular piles due to unsteady currents and oscillatory flows. *Eng. Appl. Comput. Fluid Mech.* **2018**, *12*, 354–374. [CrossRef]
28. Teles, M.J.; Pires-Silva, A.A.; Benoit, M. Numerical modelling of wave current interactions at a local scale. *Ocean Model.* **2013**, *68*, 72–87. [CrossRef]
29. Markus, D.; Hojjat, M.; Wüchner, R.; Bletzinger, K.-U. A CFD approach to modeling wave-current interaction. *Int. J. Offshore Polar Eng.* **2013**, *23*, 29–32.
30. Zhang, J.-S.; Zhang, Y.; Jeng, D.-S.; Liu, P.-F.; Zhang, C. Numerical simulation of wave–current interaction using a RANS solver. *Ocean Eng.* **2014**, *75*, 157–164. [CrossRef]
31. Ahmad, N.; Bihs, H.; Myrhaug, D.; Kamath, A.; Arntsen, Ø.A. Numerical modelling of pipeline scour under the combined action of waves and current with free-surface capturing. *Coast. Eng.* **2019**, *148*, 19–35. [CrossRef]
32. Bihs, H. Three-Dimensional Numerical Modeling of Local Scouring in Open Channel Flow. Ph.D. Thesis, Norwegian University of Science and Technology, Trondheim, Norway, 2011.
33. Afzal, M.S. 3D Numerical Modelling of Sediment Transport under Current and Waves. Master's Thesis, Norwegian University of Science and Technology, Trondheim, Norway, 2013.
34. Wilcox, D.C. *Turbulence Modeling for CFD*; DCW Industries: La Canada, CA, USA, 1994.
35. Berthelsen, P.A. A decomposed immersed interface method for variable coefficient elliptic equations with non-smooth and discontinuous solutions. *J. Comput. Phys.* **2004**, *197*, 364–386. [CrossRef]
36. Tseng, Y.-H.; Ferziger, J.H. A ghost-cell immersed boundary method for flow in complex geometry. *J. Comput. Phys.* **2003**, *192*, 593–623. [CrossRef]
37. Leveque, R.J.; Li, Z. The immersed interface method for elliptic equations with discontinuous coefficients and singular sources. *SIAM J. Numer. Anal.* **1994**, *31*, 1019–1044. [CrossRef]

38. Jiang, G.-S.; Shu, C.-W. Efficient implementation of weighted ENO schemes. *J. Comput. Phys.* **1996**, *126*, 202–228. [CrossRef]
39. Patankar, S.V.; Spalding, D.B. A calculation procedure for heat, mass and momentum transfer in three-dimensional parabolic flows. In *Numerical Prediction of Flow, Heat Transfer, Turbulence and Combustion*; Elsevier: Amsterdam, The Netherlands, 1983; pp. 54–73.
40. Dean, R.G.; Dalrymple, R.A. *Water Wave Mechanics for Engineers and Scientists*; World Scientific Publishing Company: Singapore, 1991; Volume 2.
41. Schäffer, H.A.; Klopman, G. Review of Multidirectional Active Wave Absorption Methods. *J. Waterw. Port Coast. Ocean Eng.* **2000**, *126*, 88–97. [CrossRef]
42. Exner, F.M. Uber die Wechselwirkung Zwischen Wasser und Geschiebe in Flussen. *Akad. Wiss. Wien Math. Naturwiss. Klasse* **1925**, *134*, 165–204.
43. Paola, C.; Voller, V.R. A generalized Exner equation for sediment mass balance. *J. Geophys. Res. Earth Surface* **2005**, *110*. [CrossRef]
44. Meyer-Peter, E.; Müller, R. Formulas for Bed-Load Transport. In Proceedings of the International Association for Hydraulic Structures Research (IAHSR) Second meeting, Stockholm, Sweden, 7 June 1948.
45. Van Rijn, L.C. Sediment transport, part II: Suspended load transport. *J. Hydraul. Eng.* **1984**, *110*, 1613–1641. [CrossRef]
46. Rouse, H. Modern conceptions of the mechanics of fluid turbulence. *Trans ASCE* **1937**, *102*, 463–505.
47. Shields, A. Application of Similarity Principles and Turbulence Research to Bed-Load Movement. Ph.D. Thesis, Technical University Berlin, Prussian Research Institute for Hydraulic Engineering, Berlin, Germany, 1936.
48. Yalin, M.S. *Mechanics of Sediment Transport*; Pergamon Press: New York, NY, USA, 1972.
49. Dey, S. Experimental study on incipient motion of sediment particles on generalized sloping fluvial beds. *Int. J. Sediment Res.* **2001**, *16*, 391–398.
50. Wu, W.; Rodi, W.; Wenka, T. 3D numerical modeling of flow and sediment transport in open channels. *J. Hydraul. Eng.* **2000**, *126*, 4–15. [CrossRef]
51. Miles, J.; Martin, T.; Goddard, L. Current and wave effects around windfarm monopile foundations. *Coast. Eng.* **2017**, *121*, 167–178. [CrossRef]
52. Soulsby, R. *Dynamics of Marine Sands: A Manual for Practical Applications*; Thomas Telford: London, UK, 1997; ISBN 0-7277-2584-X.
53. Fredsøe, J.; Deigaard, R. *Mechanics of Coastal Sediment Transport*; World Scientific Publishing Company: Singapore, 1992; ISBN 981-4365-68-8.
54. Sheppard, D.M.; Odeh, M.; Glasser, T. Large scale clear-water local pier scour experiments. *J. Hydraul. Eng.* **2004**, *130*, 957–963. [CrossRef]
55. Sumer, B.M.; Petersen, T.U.; Locatelli, L.; Fredsøe, J.; Musumeci, R.E.; Foti, E. Backfilling of a scour hole around a pile in waves and current. *J. Waterw. Port Coast. Ocean Eng.* **2013**, *139*, 9–23. [CrossRef]
56. Mostafa, Y.E.; Agamy, A.F. Scour around single pile and pile groups subjected to waves and currents. *Int. J. Eng. Sci. Technol. IJEST* **2011**, *3*, 8160–8178.
57. Van Rijn, L.C. *Principles of Sediment Transport in Rivers, Estuaries and Coastal Seas*; Aqua Publications: Amsterdam, The Netherlands, 1993; Volume 1006.
58. Qi, W.; Gao, F. Equilibrium scour depth at offshore monopile foundation in combined waves and current. *Sci. China Technol. Sci.* **2014**, *57*, 1030–1039. [CrossRef]

 *water*

Article

Temporal Evolution of Seabed Scour Induced by Darrieus-Type Tidal Current Turbine

Chong Sun [1], Wei Haur Lam [1,2,*], Su Shiung Lam [3], Ming Dai [4] and Gerard Hamill [5]

[1] State Key Laboratory of Hydraulic Engineering Simulation and Safety, Tianjin University, Tianjin 300350, China; chong@tju.edu.cn
[2] First R&D Services, A-08-16 M Suites, 283 Jalan Ampang, Kuala Lumpur 50450, Malaysia
[3] Pyrolysis Technology Research Group, Eastern Corridor Renewable Energy Group (ECRE), School of Ocean Engineering, Universiti Malaysia Terengganu, Kuala Nerus 21030, Malaysia; lam@umt.edu.my
[4] School of Engineering, University of Plymouth, Plymouth PL4 8AA, UK; Y.Dai@plymouth.ac.uk
[5] School of Natural and Built Environment, Architecture, Civil & Structural Engineering and Planning, Queen's University Belfast, David Keir Building, Stranmillis Road, Belfast BT9 5AG, UK; g.a.hamill@qub.ac.uk
* Correspondence: joshuawhlam@hotmail.com or wlam@tju.edu.cn

Received: 26 March 2019; Accepted: 25 April 2019; Published: 28 April 2019

Abstract: The temporal evolution of seabed scour was investigated to prevent damage around a monopile foundation for Darrieus-type tidal current turbine. Temporal scour depths and profiles at various turbine radius and tip clearances were studied by using the experimental measurements. Experiments were carried out in a purpose-built recirculating water flume associated with 3D printed turbines. The scour hole was developed rapidly in the initial process and grew gradually. The ultimate equilibrium of scour hole was reached after 180 min. The scour speed increased with the existence of a rotating turbine on top of the monopile. The findings suggested that monopile foundation and the rotating turbine are two significant considerations for the temporal evolution of scour. The scour depth is inversely correlated to the tip-bed clearance between the turbine and seabed. Empirical equations were proposed to predict the temporal scour depth around turbine. These equations were in good agreement with the experimental data.

Keywords: seabed scour; tidal current turbine; ocean renewable energy

---

1. Introduction

Tidal current energy is a proven type of ocean renewable energy gaining more attention after the successful installation of the first grid-connected tidal turbine in Strangford Loch, Belfast [1]. The vertical-axis tidal current turbine, commonly known as the Darrieus turbine, is able to capture the tidal current from various flow directions without changing turbine facing as horizontal-axis turbine [2]. Scour induced by turbines is an important engineering problem in tidal power projects. Dey et al. [3] reported the damages of scour around offshore structures. The contraction of flow by the turbine rotor changes the flow patterns around the turbine and accelerates the flow between the turbine and seabed due to blockage effects [4]. The dominant feature of scour around turbines is the horseshoe vortex system when monopile foundation is installed. The strength of the horseshoe vortex system is heavily affected by sediment bed velocity [5]. However, the slipstream below turbine can be accelerated due to the compression effect of turbine rotor. The development of scour hole induced by tidal current turbine amplifies the flow speed compared to the bridge pier scour due to flow suppression in the narrow area between rotor tip and seabed. This phenomenon has been proved in paper [6]. Hence the tidal turbine-induced scour is more complex.

Researchers simplified the tidal turbine-induced scour to propose the bridge pier scour equations without consideration of the influences of rotating turbine [7]. Seabed scour around a tidal current turbine can be investigated based on the known knowledge in bridge pier scours. Pier-induced scour focuses on the maximum scour depth, temporal scour evolution, flow field around scour hole, and parametric study on the scour process [5]. These scour works can be found in Sumer and Fredsøe's book "The Mechanics of Scour in the Marine Environment" [8] and Whitehouse's book "Scour at Marine Structures" [9]. The empirical equations for scour around bridge piers can be used for scour prediction around turbine [6]. However, the influences of the turbine on the evolution of scour hole are still unknown.

The horseshoe vortex system and downflow near front side of turbine foundation are the main factors for seabed scour. The horseshoe vortex system greatly strengthens the seabed shear stress inside the vortex's action area [10]. The initial scour hole occurs at both sides of the foundation where the shear stress of the seabed is greater than the critical shear stress. During the scour process, the sediment particles inside the hole are washed away downstream under the exerted action from horseshoe vortex. Rate of sediment scour decreases with the continuous reduction of vortex induced forces in time and finally reached the equilibrium state. On the other hand, the wake–vortex system can expand the scour hole downstream. Large scour holes may develop downstream by the combined action of horse-shoe vortex system and wake–vortex system. The wake–vortex system acts like a shovel to remove the bed material which is then carried downstream by shedding eddies from the supporting piles [11]. Bianchini et al. [12] analyzed the wake structure of vertical axis tidal current turbine using CFD (computational fluid dynamic) method and suggested the main vertical structures were generated by the rotors. Chen and Lam [4] found that tip clearance between the turbine and seabed is the important parameter to determine its impact to the scour by CFD simulation using OpenFOAM. They found the presence of turbine rotor changed the boundary layer profile and results in the altering of horseshoe vortex formation. The axial velocity of flow increased by ~10% of the initial velocity, which can cause deeper scour hole. Axial flow acceleration occurs between the turbine and seabed, but limited discussion was made to unveil the scour mechanism and these works are focused on the horizontal-axis turbine induced scour. Wake patterns of turbines are important foundations to investigate flow near seabed under turbine rotor. Wang et al. [13] measured the slipstream wash in terms of field velocities in turbine wake and surrounding flow, and suggested the turbine should be installed at least one diameter height to decrease its impact on seabed. Myers and Bahaj [14] demonstrated that increased velocity deficits and turbulence exist along the rotor centerline. Pinon et al. [15] used "vortex method" to simulate the flow field of horizontal-axis tidal current turbine. The "vortex method" was a velocity-vorticity numerical implementation of the Navier–Stokes equations to compute an unsteady evolution of the turbine wake by some three-dimensional software. The results show that flow velocity reduced behind rotor and gradually recovered downstream. Lam et al. [16] proposed a wake equation to illustrate the wake distribution of horizontal-axis turbine based on axial momentum theory and the experimental data. Flow acceleration can be clearly observed close to the rotating blade tips. Ma et al. [17] proposed a wake model for vertical-axis tidal current turbine. In their model, the minimum velocity position in lateral wake distribution deviated from centerline at efflux plane. The turbine's influence on the flow field reduced with the increase of the axial distance in turbine wake.

For the investigation of scour induced by tidal current turbine, Neill et al. [18] found the energy extraction from tidal current turbine reduced the overall bed level change; the asymmetry tidal region had a greater (20%) increase in sediment transport level compared to the tidal symmetry region. Hill et al. [19] carried out the experiments to examine the seabed scour around the foundation of horizontal axis tidal current turbine on an erodible channel. The works stated the scour development was accelerated compared to the scour around piles when rotor being installed upstream. Zhang [6] completed the numerical simulation to predict the horizontal-axis tidal current turbine scour evolution. Giles et al. [20] investigated the effect of foundation on the scour protection. Regarding hydrodynamic characteristics, Antheaume et al. [21] suggested that the efficiency of a single hydraulic Darrieus turbine could be greatly increased in farm arrangements. The results proved that the hydrodynamic performance of Darrieus-type tidal current turbine could affect the flow field significantly. Recently, Sun et al. [22] proposed an empirical model to predict equilibrium scour depth around Darrieus tidal turbines in water. The temporal evolution of scour depth is important for engineering projects of tidal current turbine. Current works provide empirical equations to predict the temporal evolution of scour around foundation of Darrieus-type tidal current turbine.

2. Methodology

2.1. Experimental Set Up

Experiments were conducted in Marine Renewable Energy Laboratory, State Key Laboratory of Hydraulic Engineering Simulation and Safety, Tianjin University. The experiments used recirculating flume to simulate the tidal current flow and a 3D printing facility to create the tidal current turbine. Sandbox was used to simulate the seabed condition with the selected sand size to simulate the tidal turbine-induced seabed scour. A laser distance meter was used to measure the scour depth at designated positions of the scour regions. The overall experimental installations are shown in Figure 1. To make the experimental designs more clear, the overall diagram is shown in Figure 2.

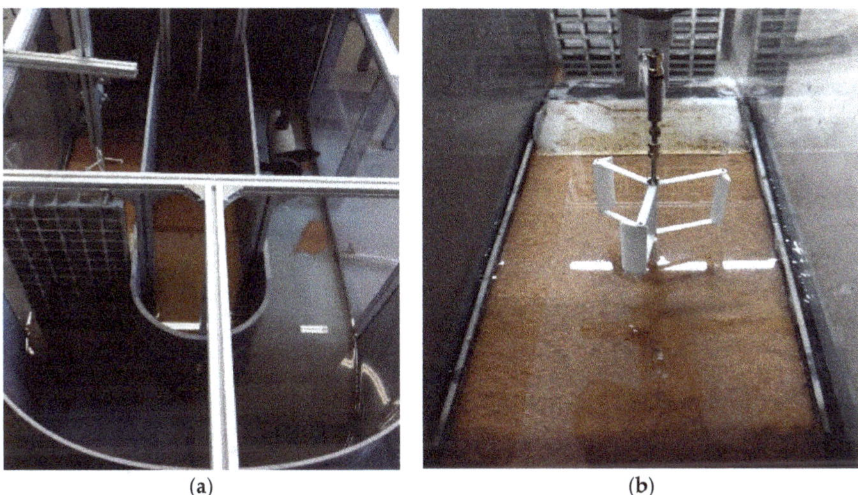

Figure 1. Experimental installation. (**a**) Overall setup of recirculating water flume [22] and (**b**) local setup of scour experimental part.

In the turbine scour experiment, the flume is 0.6 m in height and 0.35 m in width and the central cross-section is 0.8 m in width and 0.6 m in height. The turbine models were created by 3D printing. During the experiment process, flow circulates in a flume. Water is driven by propeller and moves towards the experimental side of flume. Flow moves through the flow-equalizing equipment and becomes more uniform. Then, the flow passes the tidal current turbine and erodes the sediment around supporting piles of turbine. The rotated turbine rotor is driven by Miniature DC rotor to ensure stable speed; the profiles of real-time seabed were measured by laser distance meter with 0.2 mm accuracy. During the experiment, the measurement grid is 5 mm in either the x-axis or y-axis. The seabed profile is measured in real time: the laser distance meter is installed above the water. It can launch laser and measure the distance between target and source by calculating the time difference. During the experiment, the flow moving is stopped by turning off the propeller after certain operational time. The overall sand bed profile is measured by Laser distance meter. Then reopening the propeller and continuing scour experiment until the scour equilibrium. Laser-based instruments provide accurate results when they are utilized to acquire scour pattern. For instance, Dodaro et al. (2014) [23] and Dodaro et al. (2016) [24] employed a 3D laser scanner to measure scour profile downstream of a rigid bed.

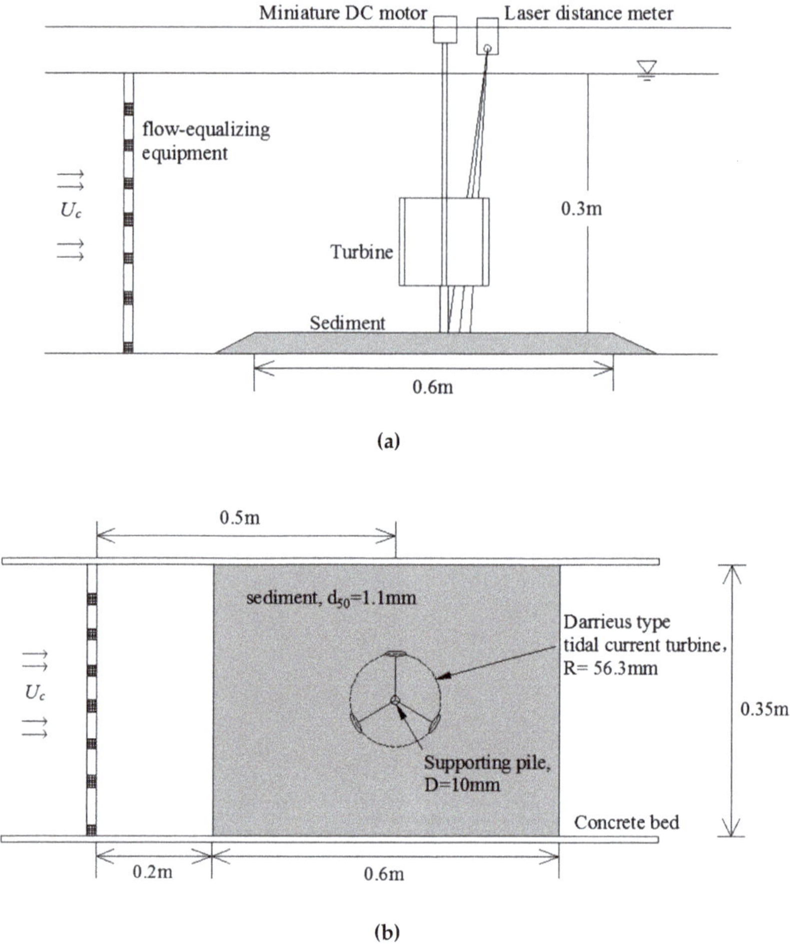

Figure 2. Overall design diagrams of scour experiment: (**a**) front view and (**b**) overhead view.

Various parameters were considered in the series of experiments to investigate the influence factor of turbine scour. The constant parameters are supporting pile diameter, rotor height, blade chord length, water depth, inlet velocity, sediment diameter, and rotational speed. The variable parameters are rotor radius and tip-bed clearance. These two parameters are main variables in our research. The specific value of each parameter is shown in Table 1. It is worth noting that this recirculating flume was used in the previous works to propose empirical model for equilibrium scour depth induced by turbine [22].

Table 1. Experimental parameters.

Parameters	Values
Rotor radius R (mm)	56.3, 45.9, 37.4
Rotor height H (mm)	87.5
Chord length c (mm)	20.4,
Dimensionless tip clearance C/H	0.25, 0.5, 0.75, 1
Monopile diameter D (mm)	10
Rotational speed w (rpm)	110
Depth of water h (m)	0.3
Width of flume b (m)	0.35
Sediment diameter d_{50}(mm)	1.1
Mean current velocity $U_c(m/s)$	0.23

2.2. Scaling Effects

2.2.1. Physical Similarity Relationships

The size of the turbine model is constrained by the dimensions of the recirculating flume. A turbine height of 87.5 mm is chosen so it could occupy ~1/3 of 0.3 m water depth. The size of the turbine is 1:8 of the physical model in paper [25]. Rotor radii of 56.3, 45.9, and 37.4 mm were set-up to ensure the solidity of turbine within the scope of 1.09 to 1.64 to investigate the impact of rotor radius on turbine scour. This solidity has been proved to show great energy extraction efficiency [25]. The turbine model represents a ratio of approximately 1:60 of full scale Darrieus-type tidal current turbine like Kobold project [26]. The ratio of turbine height and water depth was the same as Kobold. The inlet velocity is set as 0.23 m/s ensuring the clear water scours at initial scour stage. In addition, the tip speed ratio (TSR) is an important dimensionless number to influence the hydrodynamic performance of Darrieus-type tidal current turbine. The TSR is designed to be same as paper [25] to make sure the turbine has enough energy extraction efficiency. The TSR can be calculated by Equation (1):

$$\mathrm{TSR} = \frac{\omega R}{U_c} \quad (1)$$

The Reynolds number of the turbine wake induced by the Darrieus-type turbine model used in the experiment was ~287,500, which is calculated using Equation (2).

$$\mathrm{Re} = \frac{\rho U_c D_t}{\mu} \quad (2)$$

where ρ is density of water (kg/m^3), U_c is mean current velocity (m/s), D_t is turbine diameter (m), and μ is the dynamic viscosity of water (*Pa s*). Rajaratnam [27] suggested the effect of viscosity could be neglected as long as the Reynolds number of propeller jet was more than 10,000. Therefore the viscosity effect could be neglected in the analysis of our turbine model.

2.2.2. Scaling Effects of Experimental Setup

The mean sediment diameter was kept at 1.1 mm by filtering the layers of the sieves in the current experiment. The scour depth can be impacted by the critical startup shear stress of sediment with different diameter [10]. However, our research focused on the influence of turbine's tip clearance and rotor radius on temporal scour depth; the impact of sediment diameter is not a main variable. In previous experimental study of scour, the sediment size d_{50} was the same size as naturally occurring sediments—from 0.5 mm to 2.0 mm—such as [18,28]. We choose 1.1 mm sediment size as a normal prevailing condition to ensure that the sand is under clear water scour at the initial stage, in line with reality.

The size of water flume used in our experiment is shown in Figure 2. The fixed wall surfaces on both sides of flume have influence the flow distribution and experimental accuracy. The flume width is

0.35 m, which is ~6 times than turbine radius and 35 times the supporting pile diameter. In Roulund's experiment of scour around pile [29], the flume width was 36 times than pile diameter, which is almost same as our flume. In addition, based on existed experimental research of flow field around tidal current turbine [16], the operational turbine can disturb flow around it, but shows little impact on flow more than 3R distance in radial direction. Hence the flume is wide enough for our experiments.

During the experiments, the incoming velocity is maintained as 0.23 m/s measured by pitot tube in experimental flow region. This flow velocity is less than critical current speed. The formula to calculate the critical current speed can be found in paper [11]. It appears as clear water scour at first, but the initial flow cannot sweep the sediment and the scour phenomenon only occurs around the turbine's supporting pile. It should be pointed out that due to our innovative type of horizontal recirculating flume the flow on the outside of flume is a little quicker than the inside flow at the corner. When the flow moves out of the corner and towards the experimental area, the flow-equalizing equipment can reduce the speed difference and turbulence intensity to make the flow more stable. In addition, the distance between the turbine center and flow-equalizing equipment is ~10*R*. This distance is long enough to minimis the influence of uneven velocity distribution and produce uniform flow in the measurement area. To verify the reliability of flow uniformity in measurement area, the 3D shape of scour hole and the sand dune downstream is shown in Figure 3. In Figure 3, the flow moves from right to left. The white pile is supporting pile of turbine. The outline of scour hole and sand dune is indicated by white line and the centerline of sand dune is sketched by black line. The scour shape shows great symmetry downstream. However, the outside tail of the sand dune is a little longer than the inside tail. This is due to the wake asymmetry of Darrieus-type tidal current turbine which has been proposed in paper [17].

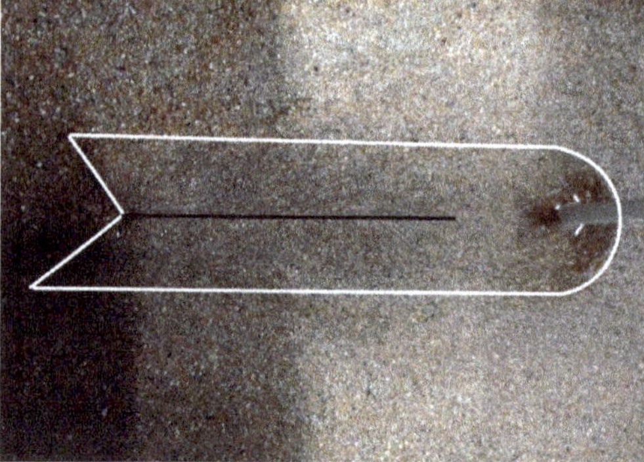

Figure 3. The sediment bed after scour equilibrium in case $C/H = 1.0$, $R/D = 5.63$.

In summary, the turbine model used in our experiments is small scale model compared with real tidal current turbine under the constraints of the experimental setup. However, the hydrodynamic performance of the turbine can be guaranteed based on reasonable TSR and turbine solidity. Meanwhile, the recirculating flume is wide and long enough to ignore the wall boundary's impact on scour process, and ensures the accuracy of experiments. Our primary purpose is to investigate the temporal development of scour depth induced by turbine with different rotor radius and tip clearance, and our experimental conditions are qualified for this work.

3. Temporal Variations of Scour Profiles

Darrieus turbine induced scour has limited data compared to the horizontal turbine induced scour. Previous works simplified the rotating turbine in the scour predictions [20]. Tip-bed clearance and the turbine radius are two main parameters of turbine to influence scour processes. In the current works, the temporal scour evolution are discussed in Section 3 with particular focusses on the influences of tip-bed clearance in Section 3.1 and turbine radius in Section 3.2. The turbine-induced contracted flow leading to scour is considered to occur on top of the monopile scour due to horseshoe effects.

Selected cases with various tip-bed clearances and turbine radii were tested to produce data to investigate the impact of Darrieus turbine to seabed scour. Seabed scour reached equilibrium after ~150 min in the experiments. The temporal variations of scour profiles are presented at 30 min, 90 min, and 150 min. Tip-bed clearance (*C*) was nondimensionalized by using the turbine height (*H*). In the first four cases, the turbine rotor remained 56.25 mm. The equilibrium scour hole depth is 2.2*D*, 2.2*D*, 1.7*D*, and 1.6*D* in cases with different tip clearance (*C/H* = 0.25, 0.5, 0.75, and 1.0 respectively). In other two cases, the tip clearance is remained as 0.5H, and the equilibrium scour hole depth reaches 2*D* when *R* = 45.9 mm, and 2.6*D* when *R*= 37.4 mm.

3.1. Temporal Seabed Scours at Various Tip Clearance

The measured temporal developments of the scour profiles of turbine are presented in Figures 4 and 5 at different tip clearances. Four tip-bed clearances of *C/H* = 1, 0.75, 0.5 and 0.25 respectively were examined. The nondimensional temporal variations of the scour profiles at 30 min, 90 min and 150 min are plotted in the figures. Figure 4 shows scour profiles at centerline of supporting pile (y = 0), and Figure 5 shows scour profiles at y = 0.5*D*. For the x-axis, the x/*D* = 0 is located at monopile central line. For y-axis, *S/D* is the scour depth nondimensionalized by monopile diameter (*D*) in the vertical direction.

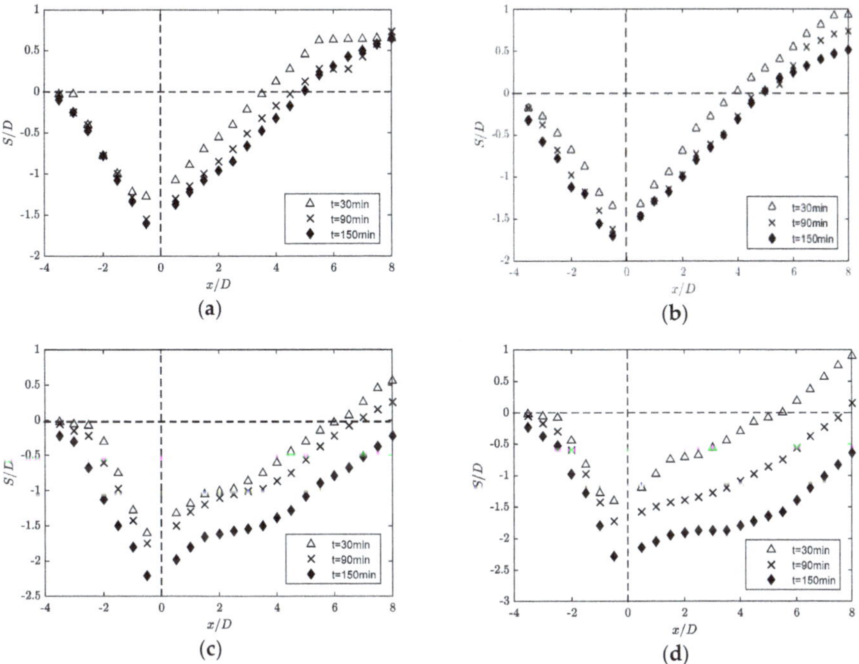

Figure 4. Temporal seabed profiles with different tip clearance at location y/D = 0. (**a**) Tip clearance =1H, (**b**) tip clearance = 0.75H, (**c**) tip clearance = 0.5H, and (**d**) tip clearance = 0.25H.

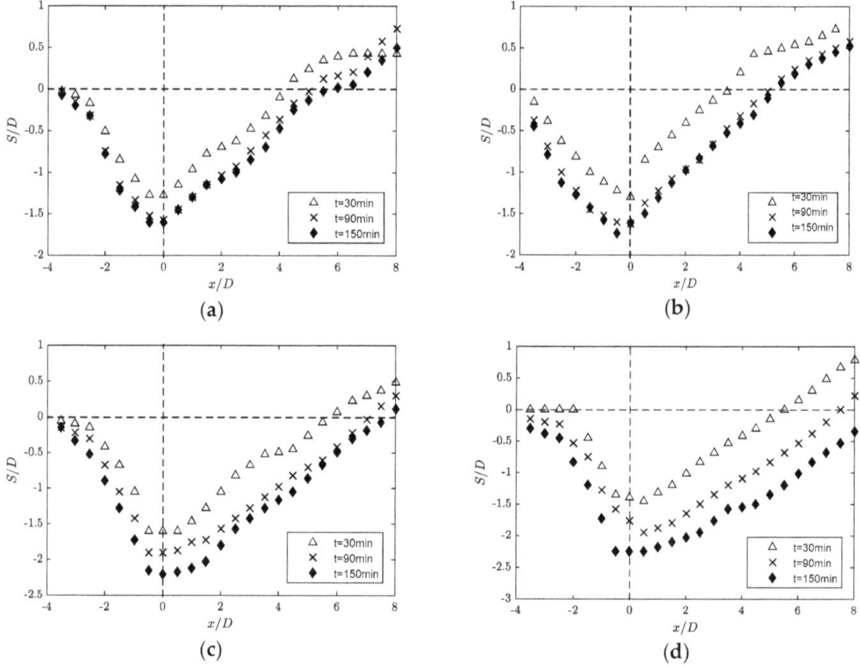

Figure 5. Temporal seabed profiles with different tip clearance at location y/D = 0.5. (**a**) Tip clearance = 1*H*, (**b**) tip clearance = 0.75*H*, (**c**) tip clearance = 0.5*H*, and (**d**) tip clearance = 0.25*H*.

3.1.1. Size of Scour Hole

The size of the scour holes increases with time up to 150 min with limited impact of tip-bed clearances. Tip-bed clearance gives significant impacts to the scour depth (*S/D*), but insignificantly influences the growing trend of the size of scour hole. The size of scour hole increases over time from 30, 90, and 150 min at all tip clearance *C/H* = 1, 0.75, 0.5, and 0.25. In all cases, the scour holes were rapidly developed in the beginning process. Temporal scour depth can reach 75% of the final depth after initial 30 min in the model tests. The increase rate of scour depth slows down with time. After scour equilibrium, the scour depth no longer increases. However, the horizontal size gradually increases with time. For the clearances at *C/H* = 0.25 and 0.5, the deposition moves back quickly, and the deposition mound cannot be observed in the range of −4*D*–8*D* along x-axis after 150 min. The deposition wedge moves backwards during the scour process.

3.1.2. Maximum Scour Depth

Scour depth increases with time for Darrieus tidal turbine at various tip-bed clearances. After equilibrium, the maximum scour depths are 2.2*D*, 2.2*D*, 1.7*D*, and 1.6*D* for *C/H* = 0.25, 0.5, 0.75, and 1.0, respectively. It can be shown that maximum scour depths in cases of *C/H* = 1.0 and *C/H* = 0.75 are much smaller than cases of *C/H* = 0.5 and *C/H* = 0.25. The difference is ~23%. The reason for this is when the turbine is installed higher than 0.5H distance from bed, the scour process is clear water scour. The rotating rotor shows weak influence on scour around the turbine's foundation. The slipstream velocity between seabed and turbine is not great enough to move sediment around scour hole. The scour hole developed mainly due to the action of a horseshoe vortex system. However, when the turbine is installed near the bed (for cases of *C/H* = 0.5 or 0.25 in the experiment), the rotor's effect is notable. The near bed velocity is great enough to move sediment around hole. The scour process is combined with digging by the downflow and slant bed erosion until the dynamic equilibrium is reached.

3.1.3. Position of Maximum Scour Depth

The position of maximum scour occurs at both back sides of the monopile regardless of the clearance distances. The specific location is x/D = −0.5, y/D = 0.5 in each case. This position is also the maximum bed shear stress occurs under the action of horseshoe vortex system. These results also prove that the horseshoe vortex system still the main function for scour around tidal current turbine.

3.1.4. Deposition

The deposition of sediment can be seen in Figure 6: a dune formed behind the foundation. The dune moved slowly downstream with time. When C/H = 0.75, the axial scour hole edge along the centerline at the location of approximately x = 4.5D at 90 min. However the hole edge migrated to a location approximately x = 7.8D downstream after 90min when C/H = 0.25. The migrating speed of dune is inverse proportional with the clearance between seabed and rotor. In each case, the highest sand dune is ~1D. The axial location of the highest sand dune migrates with time. The upstream scour hole slope is much steeper than the downstream slope.

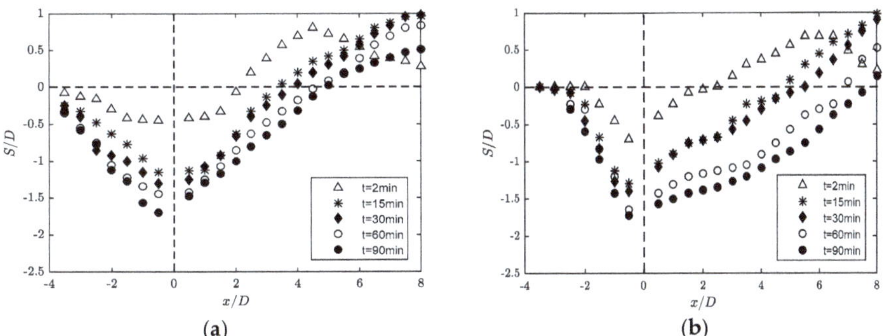

Figure 6. Temporal seabed profiles of different times (t = 2 min, 15 min, 30 min, 60 min, and 90 min) with different tip clearance at location y/D = 0. (**a**) Tip clearance =0.75H and (**b**) tip clearance = 0.25H.

3.1.5. Remark

In general, the maximum scour hole depth occurs at the back side of the foundation. The location is approximately x/D = −0.5, y/D = 0.5. A scour hole is developed around the supporting pile and the deposition mound occurs behind the turbine. The depositional wedge moves further downstream. The hole size shows an inverse correlation with the space between turbine and seabed. The deposited mount moves faster with lower installation height. Furthermore, when the tip clearance is high enough, the turbine has little impact on seabed sour development. The evolution speed of scour shows great increase with a lower installation height. And this influence increases until the appearance of live bed scour. The scour development shows little change with the continuous decrease of tip clearance in live bed scour process. This process shows the same trend with live bed scour at piers [30].

3.2. *Temporal Seabed Scours at Various Rotor Radius*

The influence of rotor radius is investigated at radii of 37.4 mm, 45.9 mm, and 56.3 mm, as shown in Figures 7 and 8. Figure 7 shows temporal scour profiles at y = 0, and Figure 8 shows temporal scour profiles at y = 0.5D.

As discussed in the previous section, tip-bed clearance gives significant temporal impacts to the size of the scour hole, maximum scour depth, and position of the maximum scour. The constant tip-bed clearance C/H = 0.5 was chosen with the changes of turbine radius. A tip-bed clearance of C/H = 0.5 is commonly used distance in the installation.

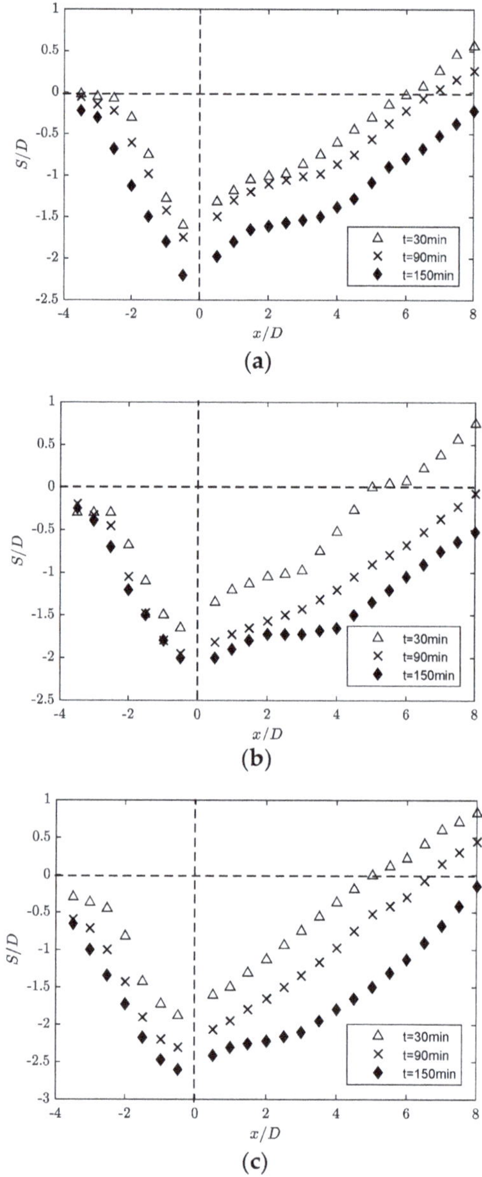

Figure 7. Temporal seabed profiles at different times with different turbine radius at location y/0. (**a**) $R = 56.3$ mm, (**b**) $R = 45.9$ mm, and (**c**) $R = 37.4$ mm.

Rotor radius is an important factor for hydrodynamic performance and wake structure of Darrieus-type tidal current turbine [26]. The energy extraction efficiency is closely related to the tip speed ratio which is a function define by rotor radius, inlet velocity, and rotational speed. The scour process around the turbine foundation is greatly influenced by the turbine energy extraction [18]. Therefore, the rotor radius has significant impact on the temporal evolution of scour depth induced by tidal current turbine.

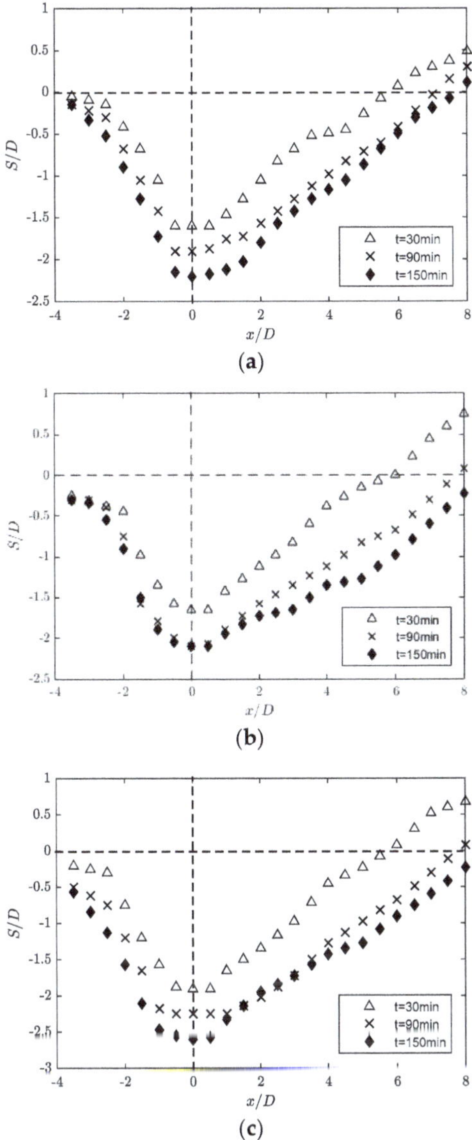

Figure 8. Temporal seabed profiles at different times with different turbine radius at location y/D = 0.5. (**a**) R = 56.3 mm, (**b**) R = 45.9 mm, and (**c**) R = 37.4 mm.

3.2.1. Size of Scour Hole

The scour process is live bed scour in all the tested cases with different rotor radius. The size of scour hole increases with the time until equilibrium state at 150 min. The temporal seabed profiles at 30 min, 90 min, and 150 min can be seen in Figures 7 and 8. The size of scour hole shows a nonlinear relationship with rotor radius. For instance, the maximum size of the scour hole is obtained at R = 37.4 mm, and the minimum size of scour hole is at R = 45.9 mm. In the case of R = 56.3 mm, a medium-sized equilibrium scour hole is obtained.

3.2.2. Maximum Scour Depth

After scour equilibrium, the maximum scour depths were 2.7D, 2D, and 2.2D for R = 37.4 mm, R = 45.9 mm, and R = 56.3 mm, respectively. The maximum scour depth decrease first, and then increases with the increase in rotor radius. The results can be explained that the change of energy extract efficiency of turbine is nonlinear against tip speed ratio. In the experiment, the rotational speed and inlet velocity is remained constant. Hence, the tip speed ratio increases linearly with rotor radius. Accordingly, the equilibrium scour depth shows a complex nonlinear relationship with rotor radius.

3.2.3. Position of Maximum Scour Depth

According to the model tests, the position of the maximum scour occurs at both back sides of the monopile regardless of the rotor radius. This result agrees with previous researches of scour at piers [28] and scour induced by turbine [18]. The specific location is $x/D = -0.5$, $y/D = 0.5$ in each case. These results are same as the cases with different tip clearance. The maximum scour depth position is still decided by horseshoe vortex system. In each case, the equilibrium deposition of sediment cannot be observed within 8D downstream. However the moving process of sand dunes behind the turbine's foundation can be seen at different scour time.

3.3. *Overall Temporal Evolution of Scour Depth*

The temporal evolution of scour depths at different rotor radius and tip clearance are plotted in Figure 9. In each case, the scour hole develops rapidly in the beginning process and then grows gradually until 150 min. At first, at approximately 10 min, the scour depth increases rapidly. The general appearance of scour around piles: turbine scour starts with an increase of flow speed around turbine supporting pile. The sediment is easily washed away by large bed shear stress on both sides of pile. The scour hole expands by the energy of downflow and horseshoe vortex. However, the less bed shear stress and weaker energy of horseshoe vortex exists inside the hole along with the development of scour hole until the equilibrium situation. The maximum final scour depth shows two scour types against different tip clearance when the turbine radius remains unchanged. When $C/H > 0.5$, the equilibrium scour hole depth S/D is less than 1.7. An equilibrium scour depth is reached after ~150 min of run time. However, when $C \leq 0.5$, the equilibrium scour depth is ~2.2D, and the scour equilibrium is reached after ~110 min. The reason is illustrated below.

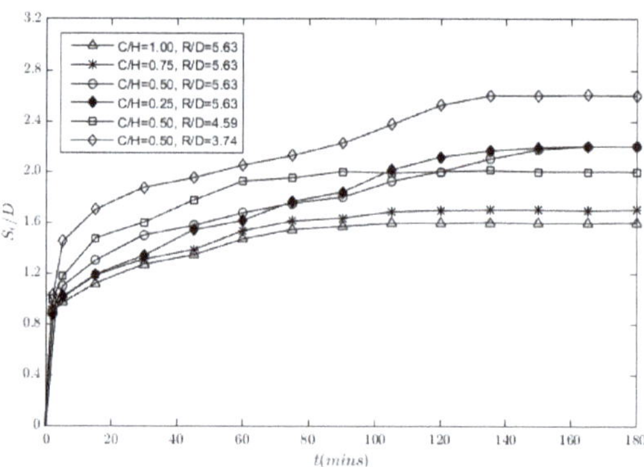

Figure 9. Evolution of maximum scour depth in time with different rotor radius and tip clearance.

For all model tests, even though each test has the same inlet velocity, the bed velocity is different due to different setup of turbine model. The scour depth around pile increases with the increase of flow velocity near sediment bed in clear water scour, but the depth shows little increase when the inlet velocity is faster than critical velocity [11]. In cases 1–4 with same rotor radius but different tip clearance, flow moves faster at lower tip clearance generally. This is due to greater flow acceleration below the turbine at lower tip clearance, thus the velocity of downflow and near-bed flow is amplified. Hence the evolution of the scour hole can be sped up. In cases 3, 5, and 6 with the same tip clearance but different rotor radius, there is also a big difference in scour depth. When dimensionless tip clearance is maintained as $C/H = 0.5$, the equilibrium scour hole depth reaches a maximum value of $2.6D$ when $R/D = 3.74$, and minimum value of $2.0D$ when $R/D = 4.59$. As discussed in Section 3.2, this result reflects the impact of the turbine's hydrodynamic performance on the seabed scour.

To further illustrate the shape of scour hole impacted by turbine's set up, the temporal development of scour width is shown in Figure 10. Final scour hole width is located between $3.5D$ to $4.8D$ with different rotor radius and tip clearance. It can be seen that final hole width shows same trends as scour depth, the max hole width occurs in case $C/H = 0.50$, $R/D = 3.74$. In each case, the final scour hole width is ~2 times that of the scour depth. This can be explained by the development mechanism of the scour hole. The horizontal expansion of the scour hole is a result of a sand slide. With the sustained evolution of scour depth, the slope of hole exceeds sediment repose angle, hence the sand at edge of hole slide into the scour hole. Hence the equilibrium scour hole slope is fixed, which is approximately the sediment repose angle. Thus the horizontal and vertical dimensions of the scour hole show a 2:1 ratio.

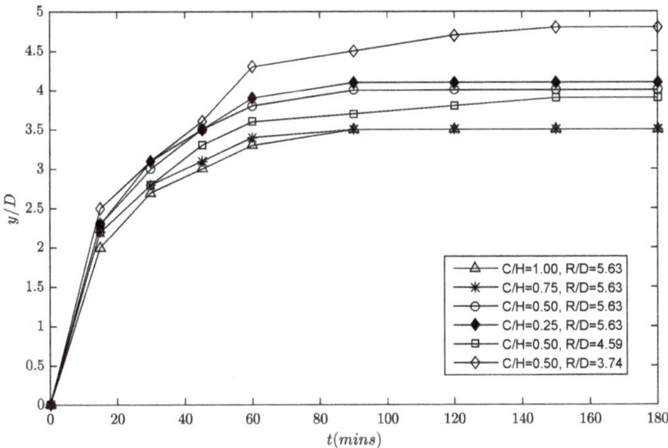

Figure 10. Evolution of scour hole width in time with different rotor radius and tip clearance.

4. Equations to Predict Temporal Evolution of Scour Depth

Tip-bed clearance and turbine radius are used to propose novel empirical equations used to predict Darrieus turbine-induced scour. Temporal scour equations are compared to the established works from bridge pier scour equations, ship propeller scour equations, and relevant tidal turbine scour data.

4.1. Derivation of the Temporal Scour Equations

Previous researchers proposed the importance of the scour and predict the temporal variation of scour depth [31–36]. The temporal scour depth S_d was predicted by a selected percentage of equilibrium scour depth S_e with consideration of the effects of current velocity, pier diameter, and water height. The selected scour equations to compare the proposed temporal scour depth model are listed below. Several numerical equations were found to predict the scour depth induced by propeller jet [37,38]. The flow velocity, propeller diameter, tip clearance, and sediment diameter are the main factors that influence the evolution of the scour hole. During the scour process induced by turbine, the temporal scour hole is deeper and wider compared to the pier-induced scour due to the flow contraction caused by rotating turbine, the increase ratio is ~10–75% with various turbine rotor radius and tip clearance in the model tests. However the scour depth is smaller than the propeller jet scour. Therefore, the pier scour equations underpredict the turbine-induced scour, whereas the propeller scour equations overpredict the temporal scour depth induced by the turbine. Engineers do not have the appropriate empirical equations to predict the temporal development of scour hole around turbine foundation. Accordingly, this paper proposed more applicable empirical equations to predict temporal evolution of scour depth induced by tidal current turbine.

For empirical equations to predict temporal scour depth at piers, Sumer et al. [31] proposed the Equation (3), where S_e is maximum scour depth at equilibrium stage, t_e is the equilibrium time of scour:

$$\frac{S_d}{S_e} = 1 - e^{(-\frac{t}{t_e})} \tag{3}$$

Melville and Chiew [32] proposed the Equation (4), where U_{cr} is critical velocity of sediment, U is approach velocity of flow:

$$\frac{S_d}{S_e} = e^{\left\{-0.03\left|\frac{U_{cr}}{U_c}\ln\left(\frac{t}{t_e}\right)\right|^{1.6}\right\}} \tag{4}$$

Oliveto and Hager [33] proposed the Equation (5), where T is dimensionless time of scour:

$$\frac{S_d}{L_R} = 0.068\sigma^{-1/2}F_d^{1.5}\log T \tag{5}$$

In Equation (5),

$$L_R = D^{2/3}h^{1/3},\ F_d = \frac{U_c}{(g'd_{50})^{1/2}},\ \sigma = (d_{84}/d_{16})^{0.5},\ T = \left(\frac{(g'd_{50})^{0.5}}{L_R}\right)t,\ g' = \left[\frac{\rho_s - \rho}{\rho}\right]g.$$

Lança et al. [34] proposed the Equation (6), where D is the diameter of pile:

$$\frac{S_d}{S_e} = 1 - e^{\left[-a_1\left(\frac{U_c t}{D}\right)^{a_2}\right]} \tag{6}$$

In Equation (6),

$$a_1 = 1.22\left(\frac{D}{d_{50}}\right)^{-0.764},\ a_2 = 0.99\left(\frac{D}{d_{50}}\right)^{0.244}$$

Choi and Choi [35] proposed the Equation (7), where h is flow depth:

$$\frac{S_d}{S_e} = e^{\left\{0.065\left(\frac{U_c}{U_{cr}}\right)^{0.35}\left(\frac{h}{D}\right)^{0.19}\ln\left(\frac{t}{t_e}\right)\right\}} \tag{7}$$

Harris et al. [36] proposed Equations (8) and (9), in these models, Breuser's [34] equation has been used to predict the maximum scour depth:

$$\frac{S_d}{S_e} = 1 - e^{(-\frac{t}{t_e})^n} \tag{8}$$

$$\frac{S_e}{D} = 1.5K_s K_\theta K_b K_d \tanh\left(\frac{h}{D}\right) \tag{9}$$

In our equations, the parameters that influence the temporal scour depth S_t are related in Equation (10) and when dimensionless lead to Equation (11).

$$S_t = f_1(U_c, D, C, d_{50}, R, g, \rho, \rho_s, h, t) \tag{10}$$

$$S_t = f_2\left(\frac{U_c t}{D}, \frac{C}{H}, \frac{d_{50}}{D}, \frac{R}{D}, \frac{U_c}{(gh)^{0.5}}, \frac{\rho_s - \rho}{\rho}\right) \tag{11}$$

Dimensionless tip clearance and turbine radius are important for scour process. ρ_s is the sediment density, which is considered as constant in nature. By use of Vaschy–Buckingham theorem, Equation (12) is obtained.

$$\frac{S_t}{D} = f_3\left(\frac{U_c t}{D}, F_r, \frac{d_{50}}{D}, \frac{C}{H}, \frac{R}{D}\right) \tag{12}$$

$$F_r = U_c/(gh)^{0.5} \tag{13}$$

The Froude number (F_r) is constant when considering constant flow depth and uniform flow velocity, as calculated by Equation (13). $F_r = 0.13$ in the current works. The sediment diameter is remained as 1.1mm at seabed. Equation (12) can be simplified as Equation (14):

$$\frac{S_t}{D} = f_4\left(\frac{U_c t}{D}, \frac{C}{H}, \frac{R}{D}\right) \tag{14}$$

Live bed scour and clear water scour can be found in the experiment against different tip clearance. Clear water scour occurs when $C/H > 0.5$; the temporal scour depth is closely related to current velocity, foundation diameter, rotor radius, and tip clearance. When $C/H \leq 0.5$, live bed scour takes place around monopile foundation, and temporal scour depth has no relevance to tip clearance or rotor radius. Dimensionless analysis produces the empirical equation to predict the temporal evolution of scour hole depth around monopile foundation of Darrieus tidal current turbine in Equations (15)–(19):

a. For clear water scour ($C/H > 0.5$):

$$\frac{S_t}{D} = k_1\left[\log_{10}\left(\frac{U_c t}{D}\right) - k_2\right]^{k_3} \tag{15}$$

where

$$k_1 = 0.269\left(\frac{C}{H}\right)^{1.1}\left|\frac{R}{D} - 5\right|^{-0.09} \tag{16}$$

$$k_2 = -1.088\left(\frac{C}{H}\right)^{0.171}\left|\frac{R}{D} - 5\right|^{-0.3} \tag{17}$$

$$k_3 = 1.233\left(\frac{C}{H}\right)^{-0.7}\left|\frac{R}{D} - 5\right|^{0.15} \tag{18}$$

b. For live bed scour ($C/H \leq 0.5$):

$$\frac{S_t}{D} = 0.131\left[\log_{10}\left(\frac{U_c t}{D}\right) + 1.11\right]^{1.869} \tag{19}$$

Figure 11 shows the comparison between experimental data and data calculated by proposed temporal scour depth equation; R^2 is 0.93. The proposed model well agrees with the experimental data.

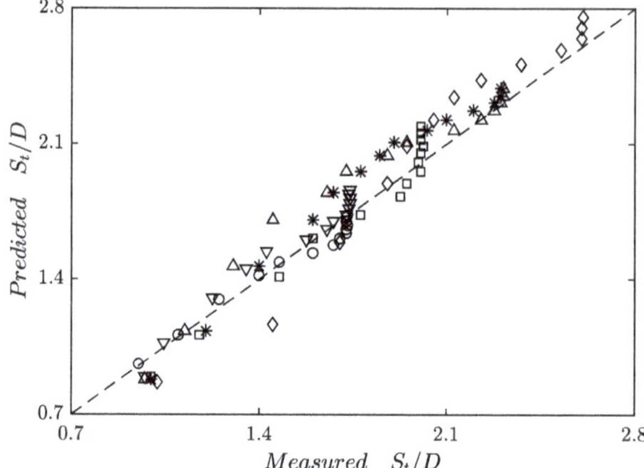

Figure 11. Comparison between observed and computed time-dependent scour depth.

4.2. Comparison of the Proposed Model with the Previous Works

The proposed empirical model was compared to the bridge pier scour model, propeller scour model, and the scour data of tidal turbine.

4.2.1. Comparison with Bridge Pier Scour Equations

Figure 12 shows the comparison between temporal evolution of scour depth around turbine foundation calculated by our proposed model and scour depth at bridge piers predicted by existing equations. In present study, six previously developed time-dependent scour depth equations [31–36] are chosen for checking the accuracy. According to Gaudio et al. (2010) [39] and Gaudio et al. (2013) [40], due to different mathematical structures in scour prediction formulas, the available bridge pier scour equations may give significantly different predictions.

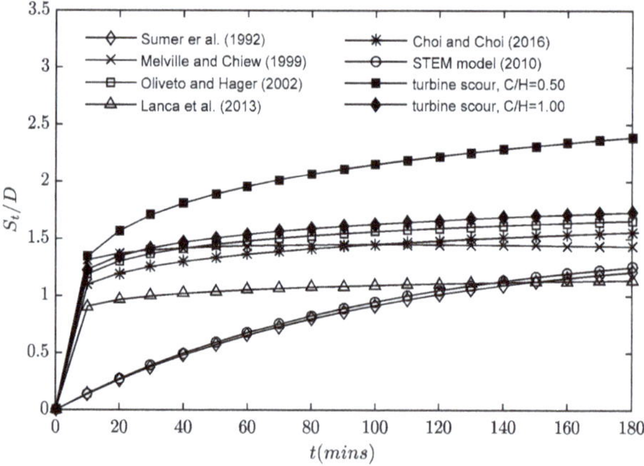

Figure 12. Comparison of time-dependent scour depth around turbine and scour at piers.

For comparison, the equilibrium scour hole depth S_e in Equations (3)–(9) is set to be $1.45D$. This is the equilibrium scour depth around single pile for the same test conditions measured in the current experiment. Two cases (C/H = 0.5 or 1) in the current experiment have been chosen to be references. These two cases are typical turbine scour process as live bed scour or clear water scour. Equations

(4), (5), and (7) show great agreement with the temporal scour depth at $C/H = 1.0$. It shows a great increase in scour depth in the first 10 min and low growth in the later scouring time. When $C/H = 1.0$, the seabed scour around turbine is more like scour at piers. The scour process is not much affected by turbine rotor. When the tip clearance lower than $0.5H$, the evolution of scour hole is greatly affected by turbine rotor. The equilibrium scour hole depth is much more than other cases in Figure 12. The increasing extent is ~50%.

4.2.2. Comparison with the Equations of Ship Propeller Induced Scour

The temporal scour depth induced by turbine and propeller jet has been compared in Figure 13. Two empirical equations for temporal propeller jet induced scour depth prediction proposed by Qurrain [37] and Hong [38] are chosen to compare the scour induced by turbine. These equations are listed below. Temporal scour depth induced by turbine and propeller jet has approximate similar variation trend, as shown in Figure 13. The scour depth increases greatly in the first 10 min. The scour depth increases slowly until equilibrium. The scour mechanisms of these two types of scour are different. In the ship propeller jet-induced scour, the propeller jet causes high shear force which can activate sediment. The maximum scour depth occurs at the centerline of the propeller in a streamwise direction downstream. However, the dominant feature of scour around foundation of turbine is horseshoe vortex. The evidence of turbine rotor disturbs the surrounding flow and accelerates flow under the turbine. Hence the scour hole can be deeper than the scour hole around piers. The position of maximum scour depth is around supporting pile of turbine. Generally, the scour depth induced by propeller jet is much deeper than scour around tidal current turbine at same installation height. In Figure 13, the equilibrium scour hole depth predicted by Hong's empirical equation reaches about 3.5D when C/H = 0.5, which is 58% deeper than equilibrium depth around turbine.

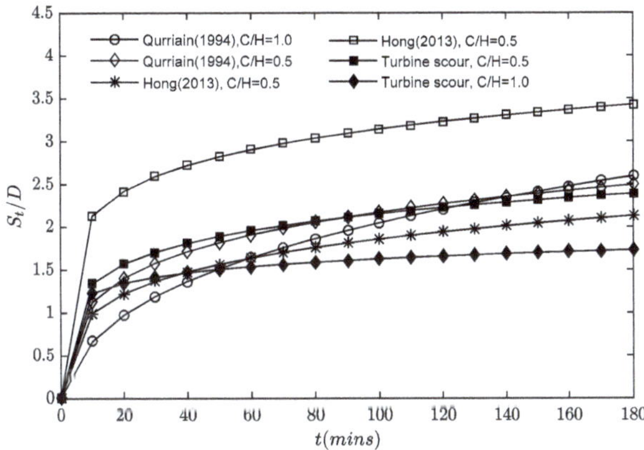

Figure 13. Comparison of time-dependent scour depth around the turbine with scour induced by propeller jet.

For empirical equations to predict temporal scour depth induced by propellers, Qurrain et al. [37] proposed Equation (20), where S_p is maximum scour depth in mm at any time t, in seconds.

$$S_p = \Omega[\ln(t)]^{\Gamma} \tag{20}$$

In Equation (20),

$$\Omega = \left[\frac{C}{d_{50}}\right]^{-4.758}\left[\frac{D_p}{d_{50}}\right]^{2.657}[F_0]^{3.517}, \Gamma = \left[\frac{C}{d_{50}}\right]^{0.758}\left[\frac{D_p}{d_{50}}\right]^{-0.339}[F_0]^{-0.479}.$$

Hong et al. [38] proposed Equation (21), where D_p is propeller diameter, U_0 is efflux velocity.

$$\frac{S_p}{D_p} = k_1\left[\log_{10}\left(\frac{U_0 t}{D_p}\right) - k_2\right]^{k_3} \tag{21}$$

In Equation (21),

$$k_a = 0.014F_0^{1.120}\left(\frac{C}{D_p}\right)^{-1.740}\left(\frac{C}{d_{50}}\right)^{-0.170}$$
$$k_b = 1.882F_0^{-0.009}\left(\frac{C}{D_p}\right)^{2.302}\left(\frac{C}{d_{50}}\right)^{-0.441}$$
$$k_c = 2.477F_0^{-0.073}\left(\frac{C}{D_p}\right)^{0.53}\left(\frac{C}{d_{50}}\right)^{-0.045}$$

4.2.3. Comparison with Published Data of Turbine Scour

In addition, a comparison between the proposed equations in the current study for scour around Darrieus tidal current turbine and experimental data studied by Hill et al. [19] for horizontal axis turbine is presented in Figure 14. All the experiments had approached scour equilibrium after about 180 min. In Hill's experiment, the scour hole develops deeper and faster with upstream installed rotor. For turbine rotors installed upstream, the maximum scour depth was ~2.4D. The temporal scour depth indicated similar pattern with live bed scour type in our proposed equation. For downstream installed turbine rotor, the maximum scour depth was about 1.5D. This data was approximately same as C/H = 1 in our experiment. This was clear water type in our proposed equations. However, the temporal evolution of scour depth before scour equilibrium in downstream rotor condition in Hill's experiment was not match well with the current experiment. This may be caused by the different type of tidal current turbine.

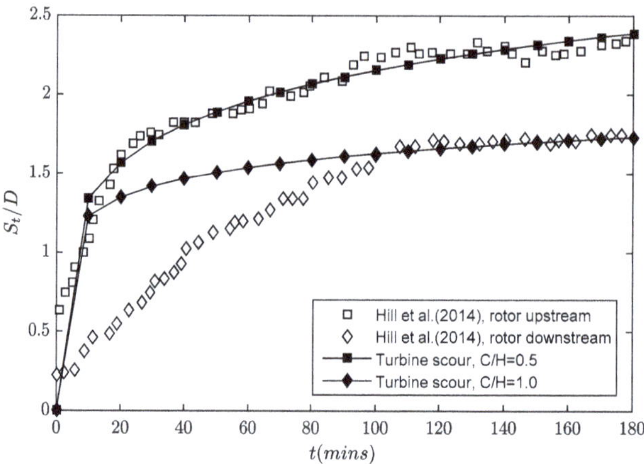

Figure 14. Comparison of time-dependent scour depth around two types of tidal current turbine.

5. Conclusions

The temporal evolution of seabed scour around monopile foundation of a Darrieus-type tidal current turbine is presented in the current study. The temporal scour depth and scour profiles are obtained by experimental measurement using 3D printed models with different rotor radius and tip clearance. Based on experimental data and analysis, empirical equations for temporal evolution of scour depth around turbine is proposed. By comparison analysis with other existing equations for other scour types, the influence of turbine rotor and tip clearance on the development of seabed scour is discussed. The evidence of turbine rotor accelerates the mean velocity and increases the turbulence intensity between turbine and seabed. The scour hole size is amplified. The conclusions are as follows.

1. The empirical equation for predicting the temporal evolution of scour hole depth around the turbine is proposed. In this equation, the turbine radius and tip-bed clearance are the two main influence factors for the turbine as these two factors can influence the hydrodynamic of turbine [7]. With the comparison with experimental data and existing equations, the empirical equation shows great ability to predict the temporal scour depth around tidal current turbine.
2. Rotating turbine gives significant impacts to the turbine induced scour. (a) Tip-bed clearance: The equilibrium scour depth is deeper and evolution speed of scour hole is faster when tip clearance decrease in general. However, the size of scour hole is no longer increasing with smaller tip clearance when C/H is lower than 0.5. This means that live bed scour occurs when tip clearance is lower than $0.5H$. In such conditions, tip clearance achieves the greatest impact on scour. (b) Rotor radius: the size of scour hole shows nonlinear relationship against rotor radius. The equilibrium scour depth decreases first and then increases when turbine rotor increases from 37.4 mm to 56.3 mm. The temporal evolution speed shows the same trend; (c) Compared with other types of scour: the scour hole around turbine foundation is bigger than scour at piers but smaller than scour induced by propeller. Our proposed equations can fit the temporal evolution of scour hole well.
3. The temporal seabed profiles in the streamtube direction around the turbine foundation are plotted. The locations and values of equilibrium scour hole depth at different turbine radius and installation height are investigated. The maximum scour hole depth takes place at location about $x/D = -0.5$, $y/D = 0.5$. Generally, the scour hole size shows an inverse correlation with the space between turbine and seabed. The scour process can be divided as live bed scour or clear water scour due to different tip clearance. The amplification of scour depth is ~10–80% in different cases compared with local scour at piers. Furthermore, the scour hole shows rapid development in the first 10 min of the scour process, and then the scour depth grows slowly until equilibrium. These findings are valid for different tip clearance.

Our empirical model for temporal scour induced by Darrieus-type tidal current turbine has some limits. The model mainly considers the impact of tip clearance and rotor radius. Other influential parameters for scour, like sediment diameter and inlet velocity are considered as conventional conditions in our model. In these general conditions, our model shows great agreement with existed experimental data. However, there is some errors using the equation to calculate temporal scour development in some extreme conditions. In the future, it is suggested to study other parameters affecting the temporal and spatial development of scour induced by tidal current turbine.

Author Contributions: W.H.L., long term research series in tidal turbine and ship propeller jet induced scour with former supervisors M.D. and G.H.; C.S. and W.H.L. contributed to establish temporal scour equation of Darrieus type tidal turbine; C.S. wrote the manuscript with recommendations and validations from W.H.L., S.S.L., M.D. and G.H.

Funding: This research was funded by Natural Science Foundation of Tianjin City: 18JCYBJC21900 and The APC was funded by Natural Science Foundation of Tianjin City: 18JCYBJC21900.

Acknowledgments: The authors wish to extend their gratitude to School of Civil Engineering in Tianjin University for laboratory space; Queen's University Belfast, University of Plymouth, University of Oxford, Dalian University of Technology, University of Malaya (HIR ENG47), Universiti Teknologi Malaysia, and Southern University College for their past support; and professional bodies EI, IEI, IET, BCS, IEM, IEAust, ASCE, ECUK, BEM, FEANI, AFEO, MINDS and academy AAET for membership support and available resources.

Conflicts of Interest: The authors declare no conflict of interest.

References

1. Chong, H.Y.; Lam, W.H. Ocean renewable energy in Malaysia: The potential of the Straits of Malacca. *Renew. Sustain. Energy Rev.* **2013**, 23, 169–178. [CrossRef]

2. Ghasemian, M.; Ashrafi, Z.N.; Sedaghat, A. A review on computational fluid dynamic simulation techniques for Darrieus vertical axis wind turbines. *Energy Convers. Manag.* **2017**, *149*, 87–100. [CrossRef]
3. Dey, S. Local scour at piers, part 1: A review of development of research. *Int. J. Sediment Res.* **1997**, *12*, 23–44.
4. Chen, L.; Lam, W.H. Slipstream between marine current turbine and seabed. *Energy* **2014**, *68*, 801–810. [CrossRef]
5. Sumer, B.M. Mathematical modelling of scour: A review. *J. Hydraul. Res.* **2007**, *45*, 723–735. [CrossRef]
6. Zhang, J.S.; Gao, P.; Zheng, J.H.; Wu, X.G.; Peng, Y.X.; Zhang, T.T. Current-Induced seabed scour around a pile-supported horizontal-axis tidal stream turbine. *J. Mar. Sci. Technol.* **2015**, *23*, 929–936.
7. Chen, L.; Lam, W.H. Methods for predicting seabed scour around marine current turbine. *Renew. Sustain. Energy Rev.* **2014**, *29*, 683–692. [CrossRef]
8. Sumer, B.M. *The Mechanics of Scour in the Marine Environment*; World Scientific Publishing Company: Singapore, 2002.
9. Whitehouse, R. *Scour at Marine Structures: A Manual for Practical Applications*; Thomas Telford Publications: London, UK, 1998.
10. Sumer, B.M.; Fresoe, J.; Christiansen, N.; Hansen, S.B. Bed shear stress and scour around coastal structures. *Coastal Eng.* **1994**, *1995*, 1595–1609.
11. Breusers, H.N.C.; Nicollet, G.; Shen, H.W. Local Scour around Cylindrical Piers. *J. Hydraul. Res.* **1977**, *15*, 211–252. [CrossRef]
12. Bianchini, A.; Balduzzi, F.; Bachant, P.; Ferrara, G.; Ferrari, L. Effectiveness of two-dimensional CFD simulations for Darrieus VAWTs: A combined numerical and experimental assessment. *Energy Convers. Manag.* **2017**, *136*, 318–328. [CrossRef]
13. Wang, D.; Atlar, M.; Sampson, R. An experimental investigation on cavitation, noise, and slipstream characteristics of ocean stream turbines. *Int. J. Power Energy Syst.* **2007**, *221*, 219–231. [CrossRef]
14. Myers, L.E.; Bahaj, A.S. Near wake properties of horizontal axis marine current turbines. In Proceedings of the Eighth European Wave and Tidal Energy Conference, Uppsala, Sweden, 7–10 September 2009.
15. Pinon, G.; Mycek, P.; Germain, G.; Rivoalen, E. Numerical simulation of the wake of marine current turbines with a particle method. *Renewable Energy* **2012**, *46*, 111–126. [CrossRef]
16. Lam, W.H.; Long, C.; Hashim, R. Analytical wake model of tidal current turbine. *Energy* **2015**, *79*, 512–521. [CrossRef]
17. Ma, Y.; Lam, W.H.; Cui, Y.; Zhang, T.; Jiang, J.; Sun, C.; Guo, J.; Wang, S.; Lam, S.; Hamill, G. Theoretical vertical-axis tidal-current-turbine wake model using axial momentum theory with CFD corrections. *Appl. Ocean Res.* **2018**, *79*, 113–122. [CrossRef]
18. Neill, S.P.; Litt, E.J.; Couch, S.J.; Davies, A.G. The impact of tidal stream turbines on large-scale sediment dynamics. *Renewable Energy* **2009**, *34*, 2803–2812. [CrossRef]
19. Hill, C.; Musa, M.; Chamorro, L.P.; Ellis, C.; Guala, M. Local Scour around a Model Hydrokinetic Turbine in an Erodible Channel. *J. Hydraul. Eng.* **2014**, *140*. [CrossRef]
20. Giles, J.; Myers, L.; Bahaj, A.; O'Nians, J.; Shelmerdine, B. Foundation-based flow acceleration structures for marine current energy converters. *IET Renew. Power Gener.* **2011**, *5*, 287–298. [CrossRef]
21. Antheaume, S.; Maître, T.; Achard, J.L. Hydraulic Darrieus turbines efficiency for free fluid flow conditions versus power farms conditions. *Renewable Energy* **2008**, *33*, 2186–2198. [CrossRef]
22. Sun, C.; Lam, W.H.; Cui, Y.; Zhang, T.; Jiang, J.; Guo, J.; Ma, Y.; Wang, S.; Tan, T.; Chuah, J.; et al. Empirical model for Darrieus-type tidal current turbine induced seabed scour. *Energy Convers. Manag.* **2018**, *71*, 478–490. [CrossRef]
23. Dodaro, G.; Tafarojnoruz, A.; Stefanucci, F.; Adduce, C.; Calomino, F.; Gaudio, R.; Sciortino, G. An experimental and numerical study on the spatial and temporal evolution of a scour hole downstream of a rigid bed. In Proceedings of the International Conference on Fluvial Hydraulics, River Flow, Lausanne, Switzerland, 3–5 September 2014; pp. 3–5.
24. Dodaro, G.; Tafarojnoruz, A.; Sciortino, G.; Adduce, C.; Calomino, F.; Gaudio, R. Modified Einstein sediment transport method to simulate the local scour evolution downstream of a rigid bed. *J. Hydraul. Eng.* **2016**, *142*. [CrossRef]
25. Dai, Y.M.; Lam, W. Numerical study of straight-bladed Darrieus-type tidal turbine. *Inst. Civ. Eng. Energy* **2009**, *162*, 67–76. [CrossRef]

26. Brinck, D.; Jeremejeff, N. The Development of a Vertical Axis Tidal Current Turbine. Master's Thesis, KTH Royal Institute of Technology, Stockholm, Sverige, 1 November 2013.
27. Rajaratnam, N. Erosion by plane turbulent jets. *J. Hydraul. Res.* **1981**, *19*, 339–358. [CrossRef]
28. Melville, B.W. Local Scour at Bridge Sites. Ph.D. Thesis, University of Auckland, Auckland, New Zealand, 1975.
29. Roulund, A.; Sumer, B.M.; Fredsøe, J.; Michelsen, J. Numerical and experimental investigation of flow and scour around a circular pile. *Int. J. Fluid Mech.* **2005**, *534*, 351–401. [CrossRef]
30. Melville, B.W. Live-bed Scour at Bridge Piers. *J. Hydraul. Eng.* **1984**, *110*, 1234–1247. [CrossRef]
31. Sumer, B.M.; Christiansen, N.; Fredsøe, J. *Time Scale of Scour around a Vertical Pile;* International Society of Offshore and Polar Engineers: Mountain View, CA, USA, 1992.
32. Melville, B.W.; Chiew, Y.M. Time Scale for Local Scour at Bridge Piers. *J. Hydraul. Eng.* **1999**, *126*, 59–65. [CrossRef]
33. Oliveto, G.; Hager, W.H. Temporal Evolution of Clear Water Pier and Abutment Scour. *J. Hydraul. Eng.* **2002**, *128*, 811–820. [CrossRef]
34. Lança, R.M.; Fael, C.S.; Maia, R.J.; Pêgo, J.P.; Cardoso, A.H. Clear-Water Scour at Comparatively Large Cylindrical Piers. *J. Hydraul. Eng.* **2013**, *139*, 1117–1125. [CrossRef]
35. Choi, S.; Choi, B. Prediction of time dependent local scour around bridge piers. *Water Environ. J.* **2016**, *30*, 14–21. [CrossRef]
36. Harris, J.M.; Whitehouse, R.J.S.; Benson, T. The time evolution of scour around offshore structures. *Marit. Eng.* **2015**, *163*, 3–17. [CrossRef]
37. Qurrain, M.R.M. Influence of the Sea Bed and Berth Geometry on the Hydrodynamics of the Wash from a Ship's Propeller. Ph.D. Thesis, The Queen's University of Belfast, Belfast, Northern Ireland, 1994.
38. Hong, J.H.; Chiew, Y.M.; Cheng, N.S. Scour Caused by a Propeller Jet. *J. Hydraul. Eng.* **2013**, *139*, 1003–1012. [CrossRef]
39. Gaudio, R.; Grimaldi, C.; Tafarojnoruz, A.; Calomino, F. Comparison of formulae for the prediction of scour depth at piers. In Proceedings of the 1st IAHR European Division Congress, Heriot-Watt University, Edinburgh, UK, 4–6 May 2010.
40. Gaudio, R.; Tafarojnoruz, A.; De Bartolo, S. Sensitivity analysis of bridge pier scour depth predictive formulae. *J. Hydroinf.* **2013**, *15*, 939–951. [CrossRef]

Article

Visible Light Communication System for Offshore Wind Turbine Foundation Scour Early Warning Monitoring

Yung-Bin Lin [1,*], Tzu-Kang Lin [2], Cheng-Chun Chang [3], Chang-Wei Huang [4], Ben-Ting Chen [1], Jihn-Sung Lai [5] and Kuo-Chun Chang [1]

1 National Center for Research on Earthquake Engineering, 200, Sec. 3, Xinhai Rd., Taipei 106, Taiwan
2 Department of Civil Engineering, National Chiao Tung University, 1001 University Rd., Hsinchu 300, Taiwan
3 Department of Electrical Engineering, National Taipei University of Technology, 1, Sec. 3, Zhongxiao E. Rd., Taipei 106, Taiwan
4 Department of Civil Engineering, Chung Yuan Christian University, 200, Chung Pei Rd., Taoyuan 320, Taiwan
5 Hydrotech Research Institute, National Taiwan University, 1, Sec. 4, Roosevelt Rd., Taipei 106, Taiwan
* Correspondence: yblin@narlabs.org.tw

Received: 4 June 2019; Accepted: 11 July 2019; Published: 17 July 2019

Abstract: Offshore wind farms have a superior wind source to terrestrial wind farms, but they also face more severe environmental conditions such as severe storms, typhoons, and sea waves. Scour leads to the excavation of sediments around the foundations of structures, reducing the safe capacity of the structures. The phenomenon of pier scour is extremely complex because of the combined effects of the vortex system involving time-dependent flow patterns and sediment transport mechanisms. A real-time scour monitoring system can improve the safety of structures and afford cost-effective operations by preventing premature or unnecessary maintenance. This paper proposes an on-site scour monitoring system using visible light communication (VLC) modules for offshore wind turbine installations. A flume experiment revealed that the system was highly sensitive and accurate in monitoring seabed scour processes. This arrayed-VLC sensory system, proposed in this paper, has considerable potential for safety monitoring and also can contribute to improving the accuracy of empirical scour formulas for sustainable maintenance in the life cycle of offshore structures.

Keywords: visible light communication system; offshore wind turbine foundation; scour; early warning monitoring; life cycle

1. Introduction

Wind energy, in particular, offshore wind power, has been recognized as one of the highest growing and the most important future renewable energy source [1,2]. Due to severe environmental conditions—such as severe storms, typhoons, ocean currents, and waves—offshore wind farms face more challenges in the issues of structure safety. Scour around the foundation of structure leads to the excavation of sediment deposits, reducing the safe capacity of the structure [3,4].

The scour mechanism of offshore wind turbine foundation caused by waves and currents is quite similar to bridge scour. From the studies of the past 30 years, flooding and foundation scouring was the primary cause of 600 bridges failed or collapse in the United States [5–14]. Average damage repair on highways cost of flooding in the United States is estimated to be $50 million per year [11]. A scour monitoring and early warning system must be developed for evaluating the safety of structures. Moreover, conducting timely reinforcement and maintenance processes in response to seabed topographical changes induced by current erosion and scour processes around these structures is also needed. Information obtained from scour monitoring systems can help engineers to design relatively safe and cost-effective offshore wind farms.

The phenomenon of pier scour is extremely complex because of the combined effects of the vortex system involving time-dependent flow patterns and sediment transport mechanisms. Scour processes around structures have received considerable attention over the past decades. Numerous studies have explored the mechanisms of hydraulic scour around foundations and have presented several formulas for scour depth estimation around piers. Most studies on scour have applied experimental flumes and mainly focused on the application of empirical regression equations for estimating the maximum scour depth. However, field data are limited because of the difficulties associated with long-term measurement processes. Without sufficient field measurement data, such empirical equations may not be sufficiently accurate for field applications. In general, when a steady current encounters a cylindrical vertical foundation, the flow rate increases around the periphery of the foundation, producing a complex hydraulic flow such as a bow wave, a downflow in front of the pier, a horseshoe vortex, and a highly turbulent wake in the downstream region of the foundation. These combined effects of hydraulic scour lead to the erosion of sediments from the foundation in all directions and reduce the loading capacity of the foundation, thereby compromising the safety of the supported structure [5–17]. Uncertainties regarding the maximum scour depth around offshore wind turbines lead to complications in their design and risks in their operation. Several methods have been proposed for estimating or monitoring the maximum scour depth around structures. A real-time scour monitoring system can improve the safety of structures and afford cost-effective operations by preventing premature or unnecessary maintenance [17].

In general, the difficulty associated with developing measuring instruments with data acquisition systems is ensuring their durability in monitoring large-scale hydraulic and transportation structures under severe conditions. The Keulegan–Carpenter number (KC) was applied to realize the foundation uncertainties of marine wind farm structures scour [18]. There were many works focused on the vibration-based approaches to monitoring the structural health status of the wind turbine [19–21]. For example, a distributed-spring foundation model to estimate the variation of natural frequencies and provide a strategy for addressing scour-induced damage around monopile foundations has used [22]. Full-scale offshore wind turbines with tripod structure were analyzed using real structural features and three-dimensional (3D) finite element models; the results show that scouring has a slight effect on natural frequency data [23]. Another study employed nonlinear springs to simulate the interaction between the foundation of a wind turbine and soil subjected to different wind, wave, and current loads—reflecting operational conditions—to determine the effects of scour on stiffness properties. The results revealed that scour considerably altered the eigen frequency of the structure compared with that of an offshore monopile wind turbine with scour protection [24,25]. Furthermore, a study applied 3D computational fluid dynamics (CFD) to develop a numerical model to examine the seabed boundary-layer flow around monopile and hexagonal gravity-based foundations of offshore wind turbines; the flow was examined to determine the formation of horseshoe vortices and flow structures to estimate potential scour for engineering designs. The results showed that the horseshoe vortex size for the hexagonal gravity-based foundation was smaller than that for the monopile foundation [26–30]. Another study also proposed a wireless network monitoring system connected to an array of small capacitive sensor probes installed around a foundation for observing scour and sediment deposition processes; this system is similar to a sonar scanning approach [31].

The foundation of an offshore wind turbine constitutes approximately 35% of the installation cost of such a turbine [32]. Construction sites for offshore wind farms are typically surveyed using different investigation approaches or hydraulic models such as bathymetry, seismometry, and side-scan sonar techniques before and after the main construction phase. However, as mentioned, the seabed topography changes constantly because of sea currents. Because uncertainties regarding seabed erosion and scour constitute a major risk for offshore wind farm development, the design and operation of offshore wind turbines should mainly focus on addressing the uncertainty regarding the maximum scour depth around the foundations of such turbines. In order to protect against the erosion of the offshore wind turbine foundation, rock dump is usually laid to prevent removal of the sediment base. However, edge scour still continues to occur despite the foundation protective devices installed [33]. With the advancement of artificial intelligence (AI) technology, machine learning (ML) and deep learning (DL) will have a better contribution to offshore wind turbine condition monitoring [34–49]. Artificial intelligence (AI) is basically an algorithm for regression analysis of existing big data rules which include machine learning, deep learning, genetic algorithm, neural network, and fuzzy. Generally, the multilayer perceptron (MLP) neural network is commonly used as an AI model prediction. Feature extraction from the multiple linear regression (MLR) and multivariate nonlinear regression (MNLR) properties of supervised and unsupervised learning need to compare with existing empirical equations. To accurately predict the scouring process by means of inductive modeling, the AI modeling process still requires a large amount of data as training, test, and vilified dataset to analyze the sensitivity of the model. Once the scour depth can be measured, empirical formulas for measuring scour processes can be developed. Most of the current formulas are based on laboratory-based research models, engineering design assessments, and measurement experience after in situ scouring. However, due to the lack of reliable and durable instrumentation technology, scour data from real-time monitoring systems is still insufficient.

An on-site scour monitoring system using visible light communication (VLC) for offshore wind turbine is proposed in this paper. Specifically, the monitoring system consists of arrays of small VLC modules attached directly to a pile structure and use the topology of the underwater optical wireless sensor network to enable remote data acquisition. Experiments conducted in flumes have revealed that the system was highly sensitive and accurate to monitor seabed scour processes. The proposed robust sensory monitoring system has considered for further on-site applications and as an indicator to improve the empirical scour formulas for sustainable maintenance in the life cycle of offshore structures.

2. Underwater VLC Turbidity and Velocity Characteristics

VLC, a novel free-space optical wireless communication technology, entails the combination of white and colored light-emitting diodes (LEDs) to utilize visible light (375–780 nm) as a transmission medium.

VLC is becoming an alternative choice for wireless technology because of its low operating cost, low maintenance cost, long-term service stability, broad bandwidth, and ubiquitous infrastructure support. Numerous studies have been conducted in both industry and academia to develop and commercialize VLC systems. Particularly, underwater wireless communication is of great interest to the marine industry and scientific society [50]. With the rapid development of solid-state lighting and semiconductor technology, VLC modules equipped with LEDs as light sources are expected to be mass produced at low cost. This technology has potential for use in a wide range of both indoor and outdoor applications for free communication services. Indoor VLC for an optical wireless communication system using LED lights was firstly proposed in 2004 with its high brightness, reliability, lower power consumption, and long lifetime advantages [50].

Measuring water turbidity has been widely developed over the past few decades. Theoretically, water turbidity was measured based on absorption, attenuation, and scattering effects by using spectrometers or photometric devices. For example, the acoustic Doppler velocimeter (ADV) measured the flowing velocity [51] while the optical laser Doppler velocimeter (LDV) [52] estimated the Doppler frequency shift (DFS) of coherent sound or light caused by the particle concentration in water. However, because of its size and inconvenient implementation, ADV and LDV are not suitable for measuring seabed turbidity and scouring. For offshore wind turbine foundation scour monitoring, the attenuation and absorption characters of the VLC measurement system in seawater, particularly the turbidity of the scouring suspension particles, need to be studied first.

Typically, Beer's law (also known as Beer–Lambert law) is a well-known optical law and commonly applied to derive the relationship in between absorption coefficient, optical path length, and the media concentration in spectroscopy from a continuous wave [53,54].

$$\mathrm{h(D)} = h_c\, e^{-c(\lambda)D} \tag{1}$$

where $\mathrm{c}\ (\lambda) = \mathrm{a}(\lambda) + \mathrm{b}(\lambda)$ is the cumulative attenuation coefficient of the medium, $\mathrm{a}(\lambda)$ and $\mathrm{b}(\lambda)$ denote the absorption coefficient and scattering coefficient, respectively. Typically, λ stands for the light wavelength, D is the communication distance, h(D) is the output or detected intensity, h_c is the input intensity.

VLC system implemented for underwater turbidity and scour laboratory demonstration in this paper, a nearby 2 cm distance of transmitter and receiver are arranged for less multiple scattering effects and avoided long distance channel attenuation loss. However, there are higher power optical lasers and higher intensity LEDs for long distance optical wireless communications (OWCs) which have less length intensity dispersion and improved the channel scattering effects [55]. A VLC system in the order of 100–200 m, and up to 300 m has been used to transmit data in the water environment [56,57]. However, due to the properties of oceanic turbulence such as the suspended particles, salinity, and temperature, the line-of-sight path attenuation loss is estimated by the radiative transfer equation (RTE). The vector RTE, implies energy conservation of a light wave traversing a scattering medium, is calculated by

$$\left[\frac{1}{c}\frac{\partial}{\partial t} + \alpha\cdot\nabla\right]\psi(t,\rho,\alpha) = \int_{4\pi} \xi(\rho,\alpha,\alpha')\psi(\rho,\alpha,\alpha')d\alpha' - \kappa(\lambda)\psi(t,\rho,\alpha) + \Phi(t,\rho,\alpha) \tag{2}$$

Herein, α is the direction vector while ρ is the position vector. ∇ presents the divergence operator with respect to ρ, ψ is the irradiance, Φ is the internal source radiance, and ξ is the volume scattering function.

Experimental results obtained in our previous study demonstrated the feasibility of the VLC modules for executing both water turbidity and water flow velocity measurements [58]. To prevent the effects of ambient light, the proposed system applies a sinusoidal signal to modulate the VLC module. Figure 1 describes both the sinusoidal signal from the VLC transmitter and the interference signals from ambient light sources, obtained on the receiver side. Normally, VLC modules operate at a frequency of a few megahertz. The frequency of ambient light intensity changes is slower than that of the designed sinusoidal signal, bandpass filters implemented on the VLC receiver. Therefore, the VLC can receive the in-band sinusoidal signal and eliminate the interference signals of out-of-band frequencies; thus, the problems associated with ambient light interference are prevented in the proposed system.

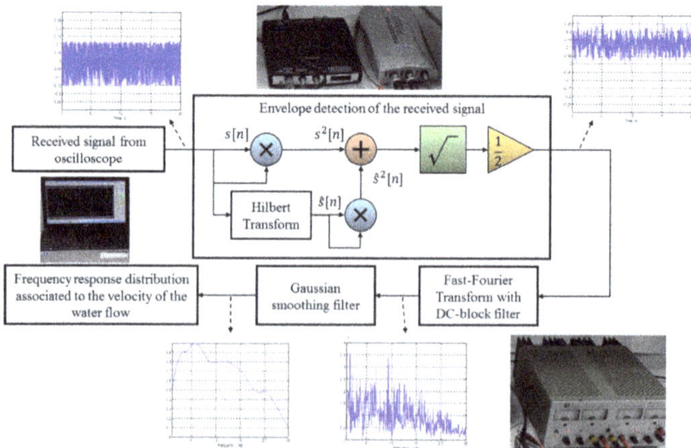

Figure 1. Flowchart of signal processing for water flow velocity/turbidity measurement.

2.1. Water Turbidity Measurement

For water turbidity measurement, assume that $\mathbf{y}_t$ is the sampled version of the received signal $y(t)$ at a sampling rate of f_t and window size of N_t-point. The root mean square (RMS) value of $\mathbf{y}_t$ can be computed as

$$y_{RMS} = \sqrt{\frac{\mathbf{y}_t\mathbf{y}_t^{\mathrm{T}}}{N_t}} \tag{3}$$

Assume that y_0 is the reference RMS value of transparent water. The difference between y_{RMS} and y_0 can then be calculated as

$$\Delta_{RMS} = y_0 - y_{RMS} \tag{4}$$

Notably, Equation (4) is the attenuated power energy caused by the attenuated light path through the turbid water. The proposed system applies Δ_{RMS} to measure water turbidity.

2.2. Water Flow Velocity Measurement

Figure 1 presents the signal processing setup for water flow velocity measurement, where $\mathbf{y}_v = [y_{v,1}, \ldots, y_{v,N_v}]$ is the sampled version of the received signal $y(t)$ at a sampling rate of f_v and window size of N_v-point. The envelope of the received signal $\mathbf{y}_v$ is computed to outline the characteristics of the signal. Let the Hilbert transform of $\mathbf{y}_v$ be $\tilde{\mathbf{y}}_v = [\tilde{y}_{v,1}, \ldots, \tilde{y}_{v,N_v}]$. The envelope of the received signal $\mathbf{y}_v$ can be expressed as

$$\hat{\mathbf{y}}_v = [\hat{y}_{v,1}, \ldots, \hat{y}_{v,N_v}] \tag{5}$$

in which $\hat{y}_i = \sqrt{y_{v,i}^2 + \tilde{y}_{v,i}^2}$, $i = 1, \ldots, N_v$. The frequency of the envelope from the received signal can be obtained as

$$\mathbf{y}_F = [y_{F,1}, \ldots, y_{F,N_v}] = \Im(\hat{\mathbf{y}}_v) \tag{6}$$

where $\Im(\bullet)$ describes the Fourier transform operation. Herein, only half of $\mathbf{y}_F$ is considered due to the symmetric property of the frequency response (i.e., $\left[y_{F,1}, \ldots, y_{F,N_v/2}\right]$). To remove the direct current (DC), the system applies a frequency-domain DC-block filter with the coefficients

$$\mathbf{h}_{DC} = \left[h_1, \ldots, h_{N_v/2}\right] \tag{7}$$

where $h_i = \begin{cases} 1, \; i \geq p_{cut} \\ 0, \; i < p_{cut} \end{cases}, \; i = 1, \ldots, N_v/2.$

In Equation (7), $p_{cut} = [(f_v/2 + f_{cut})(N_v - 1)/f_v + 1] - N_v/2$, and f_{cut} represents the DC-block filter of the desired cutoff frequency. The DC-block filter provides

$$\mathbf{y}_{DC} = \left[y_{DC,1}, \ldots, y_{DC,N_v/2}\right] \tag{8}$$

in which $y_{DC,i} = y_{F,i}h_i$, $i = 1, \ldots, N_v/2$.

The Gaussian smoothing filter is applied to smooth the frequency response $\mathbf{y}_{DC}$

$$\mathbf{y}_G = \mathbf{y}_{DC} \otimes \mathbf{g} \tag{9}$$

where $\mathbf{g} = \left[g_1, \ldots, g_L\right]$ shows the L-point filter coefficients and $g_i = e^{\left(-\frac{1}{2\sigma^2}\right)[(2/L-1)(i-1)-1]^2}$, $i = 1, 2 \ldots L$, herein, the σ contains the variance of the Gaussian coefficients. In the proposed system, the flowing velocity is measured from the frequency response in Equation (9).

Figure 2 depicts the experimental system for the water turbidity and water flow velocity measurement processes. The water flow was generated by a pump driven by a 1/6-hp motor in a 0.35 m wide and 2 m long flume. Four release holes were established at the end of the water channel to control the amount of water released. The sluice has been applied to the water flow channel to steadily control the water level of the flowing water. The turbidity of the flowing water was slight adding fine sand to the water. Sieving sediments of uniform sand with a diameter of 0.88 mm were used in this experiment. A 0.2 × 0.2 m pier made by a transparent acrylic column with VLC sensors was placed in the middle of the flume.

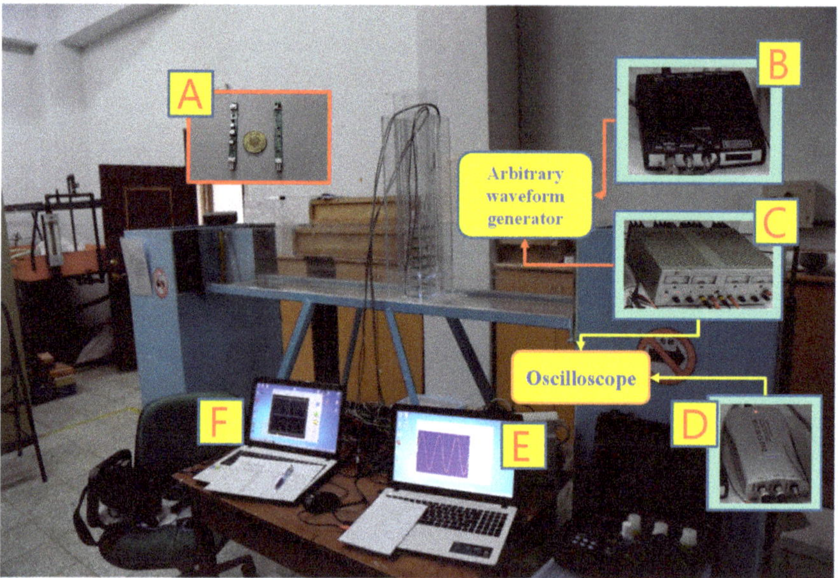

Figure 2. Experimental setup consisting of **A**: VLC sensor (Tx & Rx); **B**, **C**, **D**: waveform signal generator; **E**: input waveform signal monitoring; **F**: sensor signal response monitoring.

2.3. Experimental Setup for Water Turbidity Measurement

A sinusoidal signal with the frequency of 1 MHz was generated from the arbitrary waveform generator (AWG) in this turbidity measurement and stayed an amplitude of 0.2 V. The received signal was then obtained from the oscilloscope at a sampling rate of $f_t = 1.024\,\text{Ghz}$ with a window size of $N_t = 10,240$. From Equations (3) and (4), Δ_{RMS} values were computed. In this experiment, signal measurements were conducted at water turbidity levels of 0, 200, 400, 600, 800, 1000, and 1200 ppm and at two water flow velocities of 83.14q and 136.40q, where $q = (\text{Liter})/(\text{second} \times \text{meter}^2)$.

2.4. Experimental Setup for Water Flow Velocity Measurement

As mentioned, the sinusoidal signal was generated from the AWG. Signals were captured on the oscilloscope using a sampling rate of $f_v = 50\,\text{hz}$ with a window size of $N_v = 500$ to estimate the flowing velocity. The envelope of the received signal $\hat{\mathbf{y}}_v$ was then calculated using Equation (5) and the DC-blocked frequency of the received signal $\mathbf{y}_{DC}$ has computed from Equations (6)–(8). Finally, the Gaussian smoothing filter with a variance of $\sigma = 1.8$ and a window size of $L = 45$ was substituted into the smoothed frequency $\mathbf{y}_G$ in Equation (9). The flowing velocities were set to be 25.98q, 83.14q, and 136.40q in the experiments.

Figure 3a shows the received signals at turbidity levels of 0, 200, 600, and 1000 ppm and at a water flow velocity of 83.14q. The attenuation of the amplitude of the received signals increased in accordance with the water turbidity levels. Figure 3b illustrates the relationship between Δ_{RMS} values and water turbidity levels at the water flow velocities of 83.14q and 136.40q. It seems that a linear relationship was observed between the Δ_{RMS} values and trifling water turbidity levels. Furthermore, Δ_{RMS} values computed at different flowing velocities were approximately the same. It shows that Δ_{RMS} is independent of the flowing velocity in the experiment.

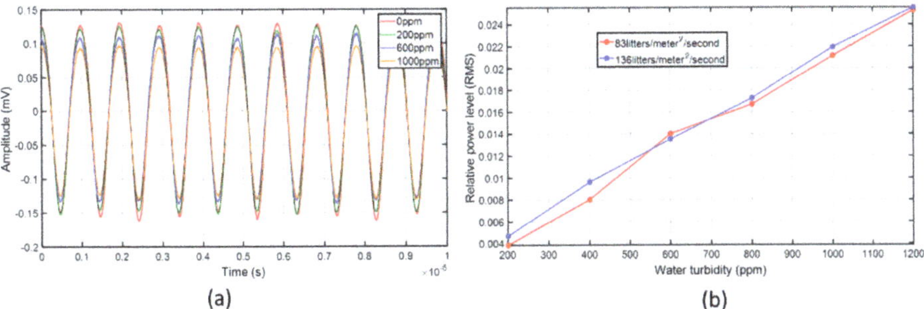

Figure 3. (**a**) Received signals at turbidity levels of 0, 200, 600, and 1000 ppm and at water a flow velocity of 83.14*q*; (**b**) water turbidity levels versus Δ_{RMS} values.

Figure 4 shows the turbidity effect in nephelometric units (NTU) of the output voltage value for VLC, blue LED, and infrared LED (IR LED). VLC turbidity data was tested in the flowing flume with suspension particles distribution while the blue LED and IR LED turbidity data is obtained from the standard specimen. As shown in Figure 4, the VLC data have a slightly variated than the blue LED and IR LED, this variation is because the flowing suspension particles of the scattering, attenuation, and absorption effects have a significant influence on the measurement in the water. It is well-known that the on-line resolution of the experiment progress is highly dependent on the measurement angle of the sensor between the transmitter and the receiver. In addition, the ambient indoor light would be a noise resource which affects the performance of VLC during the experiment progress. Despite the influence of these factors, the nonlinear nature of turbidity which actually responds in exponential form to the light intensity was obviously obtained from the VLC monitoring system.

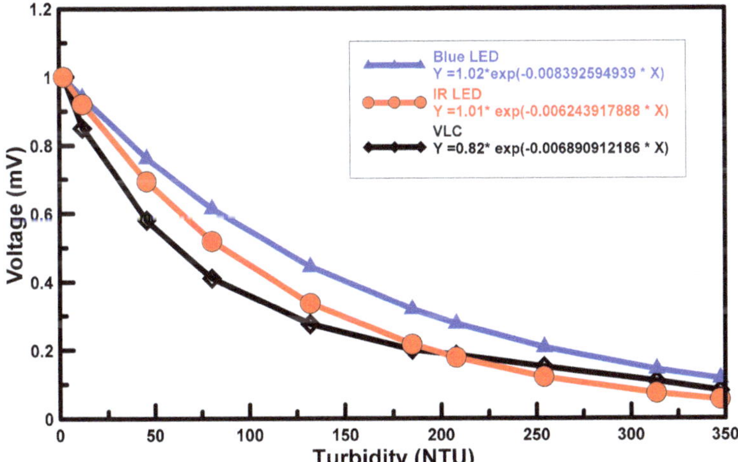

Figure 4. Turbidity effect in nephelometric units (NTU).

According to the results of the underwater turbidity and water flow velocity experiments in Section 2, we see that the communication light path of the VLC modules can be sensitively affected by turbulent movement of particles in water. In the following sections, the study further examines the effects of this notable phenomenon on real-time scour measurement by conducting real-time scour measurement and Hilbert transform analysis.

3. VLC Scour Experiment

An experiment was conducted in a 6 m wide and 30 m long flume, as depicted in Figure 5. Eight VLC sensors including the transmitter (Tx) and the receiver (Rx) were installed in the pier. Sensor 1 was located 5-cm below the bed surface. These sensors were arranged in two rows of eight sensors separated by 5-cm intervals in the vertical direction. The sampling rate was set to 100 Hz to record the time history throughout the scour experiment. Uniform sand with a diameter of 0.88 mm was paved as the bed material in the flume. The approach velocity of the steady current was set to be 0.5 m/s. All dynamic data were monitored through the experimental setup shown in Figure 5. Figure 5 illustrates the scour responses recorded by individual sensors. These responses were further analyzed using the proposed Hilbert–Huang transform (HHT) process.

Figure 5. Hydraulic flume test and experimental setup.

Initially, all the sensors are embedded in the soil and the transmit signal cannot be detected by the receiver sensors. At the early stage, the scour depth increases significantly while inflow runs through the pier. As the embedded VLC sensor scoured from the soil due to scour, the significant signal of the waveform can be easily obtained. An obvious example of the sensors 3 and 4 is shown in Figure 6. The variation of waveform magnitude is due to the turbidity and velocity of the turbulent flow that contains time-depended suspension particles. As seen in Figure 6, the scour depth increases gradually after 1 h. Around 2.5 h, a total scour depth of 30 cm is measured in this test, which may be close to the equilibrium state. As experiment finished, the scour hole can be observed as shown in Figure 5. A small camera recorded the scouring process as a comparison to valid the scour depth evolution measured by the VLC. An empirical scour formula with the nonlinear nature of flowing water which responds in exponential form to the light intensity for the VLC monitoring system is also obtained as shown in Figure 6.

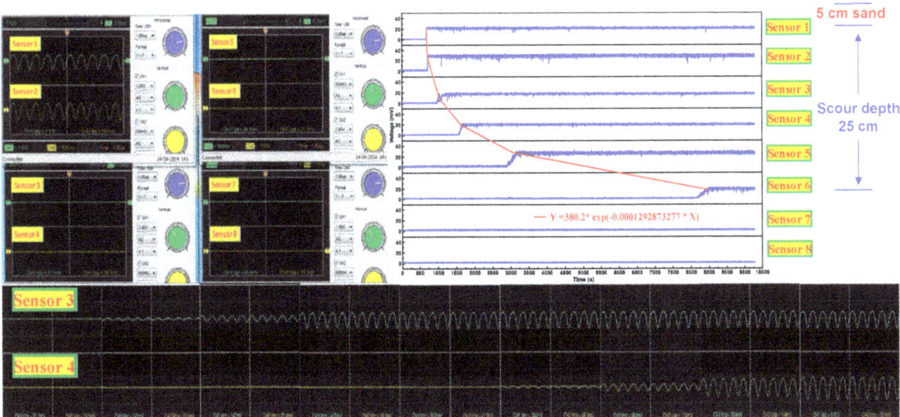

Figure 6. Monitoring interface and recorded scour responses.

4. HHT and Data Analysis

4.1. Instantaneous Frequency

Traditionally, a Fourier spectrum is evaluated using sine and cosine basis functions with a fixed amplitude. However, signals vary with time limiting the applicability of the fast Fourier transform; moreover, obtaining the instantaneous frequency (IF) at any specific time is impractical. For a structure affected by an earthquake, understanding the frequency variation is imperative. The Hilbert transform is widely used for nonlinear and nonstationary cases, facilitating the analysis of a time-varying signal. A measured signal can be expressed in the form of a complex number to determine the instantaneous amplitude $a(t)$ and instantaneous phase $\theta(t)$, and the IF $\omega(t)$ can then be determined [59,60]. The Hilbert transform can be defined as the convolution between $X(t)$ and $1/t$. For any time series $X(\tau)$, the Hilbert transform $Y(t)$ can be expressed as

$$Y(t) = \frac{1}{\pi} P \int_{-\infty}^{\infty} \frac{X(\tau)}{t-\tau} d\tau \tag{10}$$

where P represents the Cauchy principal value.

Combining $X(t)$ and $Y(t)$ into a conjugate complex number yields an analytic signal $Z(t)$

$$Z(t) = X(t) + iY(t) = a(t)e^{i\theta(t)} \tag{11}$$

For example,

$$a(t) = \sqrt{X^2(t) + Y^2(t)} \tag{12}$$

$$\theta(t) = \tan^{-1}(Y(t)/X(t)) \tag{13}$$

$$\omega(t) = (d\theta(t)/dt) \tag{14}$$

According to the analysis, the time–frequency–amplitude distribution of the time series can be obtained.

4.2. Empirical Mode Decomposition

In contrast to other decomposition methods, empirical mode decomposition (EMD) does not entail predetermining a basis function. In EMD, such a function is directly obtained from the signal data; therefore, this method has considerable flexibility. EMD entails decomposing an original signal into a finite number of intrinsic mode function (IMF) components; specifically, a signal is approximated as a sum of zero-mean amplitude modulation and frequency modulation components. The finite number of IMF components can be divided into high- and low-frequency partitions until a monotonic function (trend) remains. The original data can be regarded as the sum of all IMF components and trends. During analysis, if the time difference between the extreme values represents the time scalar of the intrawave, an optimal vibration modal resolution can be achieved and can be applied to nonzero mean values as well as non-zero-crossing data. Thus, the original signal can be re-presented as

$$X(t) = \sum_{i=1}^{n} c_i + r_n \quad (15)$$

where C_i is the ith intrinsic mode functions; r_n is the residual.

4.3. Ensemble EMD

EMD is often used as a signal disassembly method. EMD is used to decompose mixed signals of different scales into different IMF components. However, a limitation of this approach is the phenomenon called 'mode mixing'. Specifically, during the EMD process, a low-amplitude oscillation or an intermittent signal may exist in several IMF components; for example, a single modal component may be decomposed into different IMF components, or a single IMF component may contain two different modal signals, resulting in mode mixing within the IMF component. To solve this issue, Ensemble EMD (EEMD) is proposed [61]. When EEMD is performed, a white noise $w_i(t)$ signal with a limited amplitude is added to the original signal $X(t)$; thus, the original signal is transformed as

$$X_i(t) = X(t) + w_i(t) \quad (16)$$

4.4. Hilbert Spectrum

As indicated in Equation (17), an IMF component can be converted from a time-domain to a frequency-domain component; this process is called Hilbert spectral analysis (Figure 7)

$$X(t) = \sum_{j=1}^{n} a_j(t)\, exp\left(i \int \omega_j(t)dt\right) \quad (17)$$

Although the Hilbert transform can process a monotonous trend and consider it a part of a longer amplitude, the remaining energy may be excessively strong, considering the uncertainties of longer-term trends and other low-energy elements and information contained in a high-frequency component. The preceding formula provides a time function for each amplitude and frequency component, and this formula can be expanded using a Fourier expression as

$$X(t) = \sum_{j=1}^{\infty} a_j e^{i\omega_j t} \quad (18)$$

where a_j and w_j are constants. Comparing Equations (17) and (18) reveals that the IMF can be represented by a generalized Fourier expansion. Variables within the amplitude and the IF cannot only improve the expansion but also render it applicable to unsteady signals. In terms of the expansion of the IMF, the amplitude and the frequency modulation are clearly separated. The amplitude of the time function and the IF can be combined as the time–frequency–amplitude spectrum; this spectrum is referred to as the Hilbert amplitude spectrum.

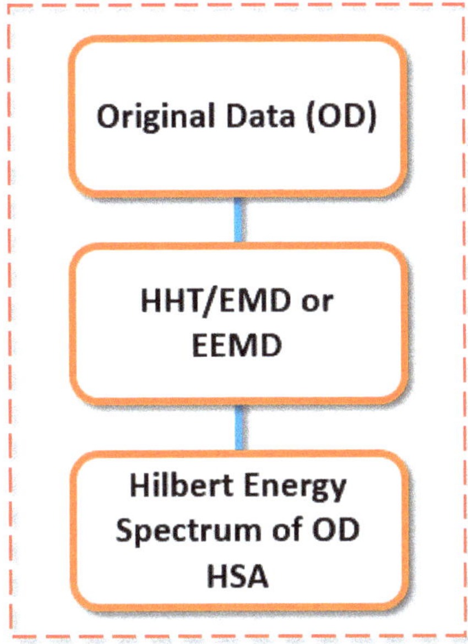

Figure 7. HHT calculation process.

4.5. Analysis Results

From recorded data, the time history data and HHT diagrams obtained for the first six sensors are plotted in Figure 8. The recorded time history is plotted on the top of each subplot and the corresponding Hilbert amplitude spectrum is shown in the bottom. The time-varying IF demonstrated nonlinear characteristics of the signals. The x-axis represents the time history measured in seconds, and the energy density is illustrated at frequencies between 0 and 5 Hz, which contains the most energy of the vibration.

The HHT diagram for channel 1 shown in Figure 8a reveals a large energy distribution indicated by the significant yellow bar at approximately 90 s. This distribution can be attributed to the rapid jump in the time domain. Similarly, Figure 8b also reveals a yellow bar, indicating a sudden increase in the beginning of the time history. According to the color distribution in Figure 8b, more energy was covered in channel 2 than in channel 1, which fits well with the measured time histories.

The HHT diagrams for channels 3 and 4 shown in Figure 8c,d, respectively, reveal a gradual shift of the yellow bar, which indicates the large variation section of energy in the time domain. Based on the result, the yellow bars in HHT diagrams can be used to estimate the occurrence time of possible scour processes. Figure 8d indicates the scour phenomenon to occur between 150 and 200 s and then it continues till the end of the experiment; this figure also reveals the energy distribution in channel 4 to be greater than that in channel 3.

Finally, as observed from the time history, significant vibration can be reflected from channel 5. Furthermore, the HHT diagram for channel 5 shown in Figure 8e reveals that the yellow energy bar is shifted to 300 s. It can be treated as a proper indicator to evaluate the scour depth, which follows the trend observed from the experiment. Additionally, the HHT diagram for channel 6 shown in Figure 8f reveals two peaks. In contrast to the other five sensors, the first yellow energy bar located at 300 s with the fundamental frequency between 0 and 2 Hz is not considered as scour phenomenon in the HHT diagram. As only slight energy is shown in the diagram, it can be neglected as a small vibration in the time domain. By combining Figures 7 and 8, a possible backbone scour curve can be established to provide a rapid estimation and early warning for scour processes around a bridge.

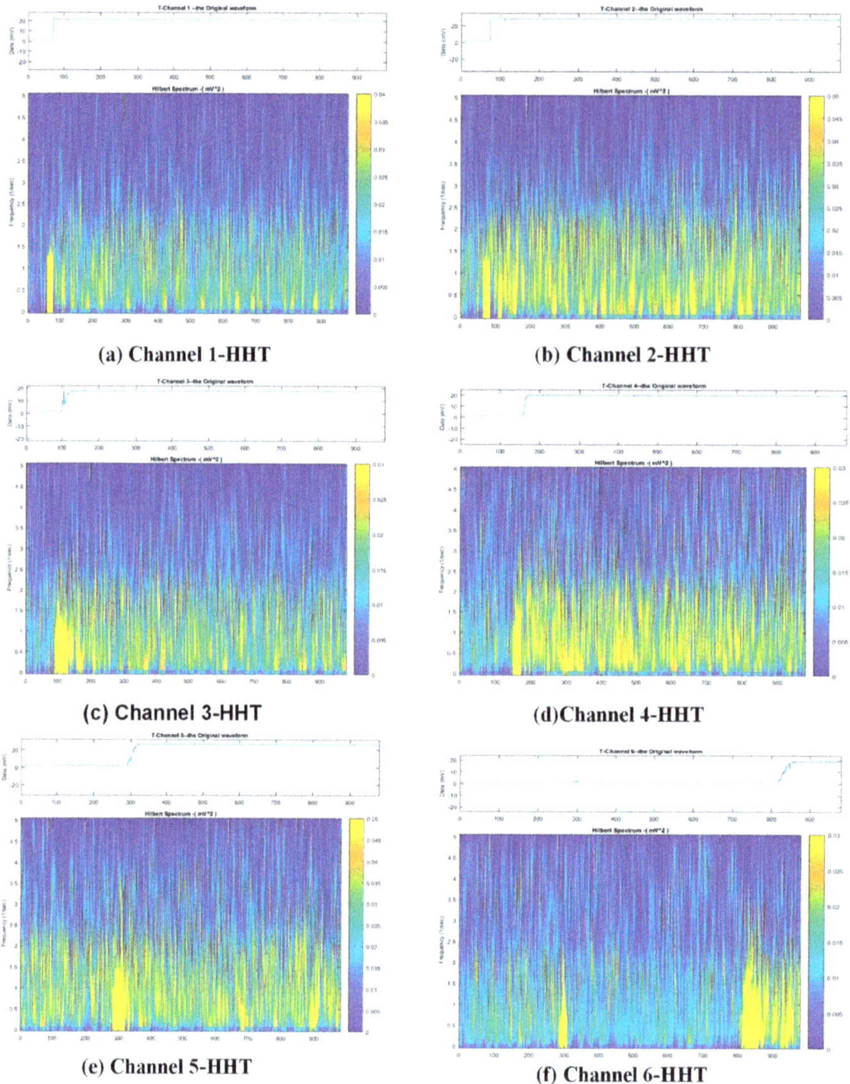

Figure 8. HHT diagrams for the first six sensors.

5. Summary

Offshore wind farms face more severe environmental conditions such as severe storms, typhoons, ocean currents, and waves. Flow induced scour around the foundation of structure is extremely complex. Scour leads to the excavation of sediments, reducing the safe capacity of the structures. The phenomenon of pier scour combines the effects of the vortex system involving time-dependent flow pattern and sediment transport mechanism. A real-time scour monitoring system can improve the safety of structures and afford cost-effective operations by preventing premature or unnecessary maintenance. Numerous studies have explored the mechanisms of hydraulic scour around foundations and have presented several formulas for scour depth estimation around piers. However, scour data from real-time monitoring systems are still inadequate due to the lack of reliable and durable instrumentation techniques.

Normally, water turbidity was measured based on absorption, attenuation, and scattering effects by using spectrometers or photometric devices. However, conventional estimation instruments are usually bulky and costly. Experimental results obtained in this study demonstrated the feasibility of the VLC modules for executing both water turbidity and water flow velocity measurements.

According to the results of the underwater turbidity and water flow velocity experiments, the communication light path of the VLC modules can be sensitively affected by turbulent movement of particles in water. These notable phenomenon effects could be further implemented as an early warning structural health monitoring system by conducting real-time scour measurement and Hilbert transform analysis. In this present study, an on-site scour monitoring system for offshore wind turbines has been proposed; specifically, the monitoring system consists of arrays of small VLC modules attached directly to a pile foundation structure and linked to a wireless network to enable remote data acquisition has demonstrated. From the flume experiment results, it revealed that the system was highly sensitive to seabed scour processes. The proposed robust sensory monitoring system has considered for further on-site applications and as an indicator to improve the empirical scour formulas for sustainable maintenance in the life cycle of offshore structures.

From the result analysis, the VLC ray is easily affected by the suspended particles in the water and the turbidity, especially in the scouring process. Hence, the proposed arrayed-transmission measurement method will be limited by the turbidity effect. In the future work, it may be necessary to cooperate with the reflection approach, as a supplement comparison, for simultaneous measuring the back-scattering characters. Combining these two transmission and back-scattering implementations would provide a better real-time approach to measure and discriminate turbidity, flow velocity, and scour depth. These data can be useful to establish a local scouring formula to evaluate structural safety.

Author Contributions: Y.-B.L., developed the methodology and took the lead in writing the manuscript. C.-C.C., and T.-K.L., provide data analyzed and results. C.-W.H., and B.-T.C., responsible for image measurement and data collection. J.-S.L., and K.-C.C., revised the manuscript and suggestion to the manuscript.

Funding: This study is funded by Ministry of Science and Technology, Taiwan, under Grants MOST: 107-2625-M-492 -003, 106-2625-M-492-011, and 105-2625-M-492-016.

Acknowledgments: The authors would like to gratefully acknowledge the experimental sites from the National Center for Research on Earthquake Engineering in Taiwan. In addition, the authors appreciate the Hydrotech Research Institute of National Taiwan University for facilities and support.

Conflicts of Interest: The authors declare no conflict of interest.

References

1. Igwemeziea, V.; Mehmanparasta, A.; Koliosa, A. Current trend in offshore wind energy sector and material requirements for fatigue resistance improvement in large wind turbine support structures—A review. *Renew. Sustain. Energy Rev.* **2019**, *101*, 181–196. [CrossRef]
2. Willis, D.J.; Niezrecki, C.; Kuchma, D.; Hines, E.; Arwade, S.R.; Barthelmie, R.J.; DiPaola, M.; Drane, P.J.; Hansen, C.J.; Inalpolat, M.; et al. Wind energy research: State-of-the-art and future research directions. *Renew. Energy* **2018**, *125*, 133–154. [CrossRef]
3. Ma, H.; Yang, J.; Chen, L. Effect of scour on the structural response of an offshore wind turbine supported on tripod foundation. *Appl. Ocean Res.* **2018**, *73*, 179–189. [CrossRef]
4. Li, H.; Ong, M.C.; Leira, B.J.; Myrhaug, D. Effects of soil profile variation and scour on structural response of an offshore monopile wind turbine. *J. Offshore Mech. Arct. Eng.* **2018**, *140*, 042001. [CrossRef]
5. Chiew, Y.M.; Melville, B.W. Local scour around bridge piers. *J. Hydraul. Eng. ASCE* **1987**, *25*, 15–26. [CrossRef]
6. Melville, B.W.; Chiew, Y.M. Time scale for local scour at bridge piers. *J. Hydraul. Eng. ASCE* **1999**, *125*, 59–65. [CrossRef]
7. Melville, B.W. Pier and abutment scour: Integrated approach. *J. Hydraul. Eng. ASCE* **1997**, *123*, 125–136. [CrossRef]

8. Briaud, J.L.; Ting, F.C.K.; Chen, H.C. Erosion function apparatus for scour rate predictions. *J. Geotech. Geoenviron. Eng.* **2001**, *127*, 105–113. [CrossRef]
9. Briaud, J.L.; Ting, F.C.K.; Chen, H.C. SRICOS: Prediction of scour rate in cohesive soils at bridge piers. *J. Geotech. Geoenviron. Eng.* **1999**, *125*, 237–246. [CrossRef]
10. Shirole, A.M. Bridge management to the Year 2020 and beyond. *Transp. Res. Rec.* **2010**, *2202*, 159–164. [CrossRef]
11. Lagasse, P.F.; Richardson, E.V. ASCE compendium of stream stability and bridge scour papers. *J. Hydraul. Eng. ASCE* **2001**, *127*, 531–533. [CrossRef]
12. Wardhana, K.; Hadipriono, F.C. Analysis of recent bridge failures in the United States. *J. Perform. Constr. Facil.* **2003**, *17*, 144–150. [CrossRef]
13. Roulund, A.; Sumer, B.M.; Fredsoe, J. Numerical and experimental investigation of flow and scour around a circular pile. *J. Fluid Mech.* **2005**, *534*, 351–401. [CrossRef]
14. Lin, Y.B.; Chen, J.C.; Chang, K.C. Real-time monitoring of local scour by using fiber Bragg grating sensors. *Smart Mater. Struct.* **2005**, *14*, 664–670. [CrossRef]
15. Lin, Y.B.; Lai, J.S.; Chang, K.C. Flood scour monitoring system using fiber Bragg grating sensors. *Smart Mater. Struct.* **2006**, *15*, 1950–1959. [CrossRef]
16. Lin, Y.B.; Lai, J.S.; Chang, K.C.; Chang, W.Y.; Lee, F.Z.; Tan, Y.C. Using MEMS sensors in the bridge scour monitoring system. *J. Chin. Inst. Eng.* **2010**, *33*, 25–35. [CrossRef]
17. Prendergast, L.J.; Gavin, K. A review of bridge scour monitoring techniques. *J. Rock Mech. Geotech. Eng.* **2014**, *6*, 138–149. [CrossRef]
18. Luengo, J.; Negro, V.; Garcia-Barba, J. New detected uncertainties in the design of foundations for offshore Wind Turbines. *Renew. Energy* **2019**, *131*, 667–677. [CrossRef]
19. Oliveira, G.; Magalhaes, F.; Cunha, A. Vibration-based damage detection in a wind turbine using 1 year of data. *Struct. Control Health Monit.* **2018**, *25*, 1–22. [CrossRef]
20. Prendergast, L.J.; Gavin, K.; Doherty, P. An investigation into the effect of scour on the natural frequency of an offshore wind turbine. *Ocean Eng.* **2015**, *101*, 1–11. [CrossRef]
21. Tseng, W.C.; Kuo, Y.S.; Lu, K.C.; Chen, J.W.; Chung, C.F.; Chen, R.C. Effect of scour on the natural frequency responses of the meteorological mast in the Taiwan Strait. *Energies* **2018**, *11*, 823. [CrossRef]
22. Tseng, W.C.; Kuo, Y.S.; Chen, J.W. An investigation into the effect of scour on the loading and deformation responses of monopile foundations. *Energies* **2017**, *10*, 1190. [CrossRef]
23. Chen, W.I.; Wong, B.L.; Lin, Y.H.; Chau, S.W.; Huang, H.H. Design and analysis of jacket substructures for offshore wind turbines. *Energies* **2016**, *9*, 264. [CrossRef]
24. Yang, W.; Tian, W. Concept research of a countermeasure device for preventing scour around the monopole foundations of offshore wind turbines. *Energies* **2018**, *11*, 2593. [CrossRef]
25. Yu, T.; Lian, J.; Shi, Z. Experimental investigation of current-induced local scour around composite bucket foundation in silty sand. *Ocean Eng.* **2016**, *117*, 311–320. [CrossRef]
26. Esteban, M.D.; Counago, B.; Lopez-Gutierrez, J.S. Gravity based support structures for offshore wind turbine generators: Review of the installation process. *Ocean Eng.* **2015**, *110*, 281–291. [CrossRef]
27. McGovern, D.J.; Ilic, S.; Folkard, A.M. Time development of scour around a cylinder in simulated tidal currents. *J. Hydraul. Eng.* **2014**, *140*, 04014014. [CrossRef]
28. Michalis, P.; Saafi, M.; Judd, M. Capacitive sensors for offshore scour monitoring. *Proc. Inst. Civ. Eng. Energy* **2013**, *166*, 189–196. [CrossRef]
29. Harris, J.M.; Whitehouse, R.J.S.; Benson, T. The time evolution of scour around offshore structures. *Proc. Inst. Civ. Eng. Energy Marit. Eng.* **2010**, *163*, 3–17. [CrossRef]
30. Ong, M.C.; Trygsland, E.; Myrhaug, D. Numerical study of seabed boundary layer flow around monopile and gravity-based wind turbine foundations. *J. Offshore Mech. Arct. Eng.* **2017**, *139*, 042001.
31. Oh, K.Y.; Nam, W.; Ryu, M.S.; Kim, J.Y.; Epureanu, B.I. A review of foundations of offshore wind energy convertors: Current status and future perspectives. *Renew. Sustain. Energy Rev.* **2018**, *88*, 16–36. [CrossRef]
32. Guan, D.W.; Chiew, Y.M.; Melville, B.W.; Zheng, J.H. Current-induced scour at monopile foundations subjected to lateral vibrations. *Coast. Eng.* **2019**, *144*, 15–21. [CrossRef]
33. Petersen, T.U. Scour around Offshore Wind Turbine Foundations. Ph.D. Thesis, Technical University of Denmark, Lyngby, Denmark, 2014.

34. Prendergast, L.J.; Reale, C.; Gavin, K. Probabilistic examination of the change in eigen frequencies of an offshore wind turbine under progressive scour incorporating soil spatial variability. *Mar. Struct.* **2018**, *57*, 87–104. [CrossRef]
35. Rivier, A.; Bennis, A.C.; Pinon, G.; Magar, V.; Gross, M. Parameterization of wind turbine impacts on hydrodynamics and sediment transport. *Ocean Dyn.* **2016**, *66*, 1285–1299. [CrossRef]
36. Nielsen, A.W.; Liu, X.F.; Sumer, B.M.; Fredsoe, J. Flow and bed shear stresses in scour protections around a pile in a current. *Coast. Eng.* **2013**, *72*, 20–38. [CrossRef]
37. Riahi-Madvar, H.; Dehghani, M.; Seifi, A.; Salwana, E.; Shamshirband, S.; Mosavi, A.; Chau, K.W. Comparative analysis of soft computing techniques RBF, MLP, and ANFIS with MLR and MNLR for predicting grade-control scour hole geometry. *Eng. Appl. Comput. Fluid Mech.* **2019**, *13*, 529–550. [CrossRef]
38. Khan, M.; Tufail, M.; Azamathulla, H.M.; Ahmad, I.; Muhammad, N. Genetic functions-based modelling for pier scour depth prediction in coarse bed streams. *Proc. Inst. Civ. Eng. Water Manag.* **2018**, *171*, 225–240. [CrossRef]
39. Ebtehaj, I.; Bonakdari, H.; Moradi, F.; Gharabaghi, B.; Khozani, Z.S. An integrated framework of Extreme Learning Machines for predicting scour at pile groups in clear water condition. *Coast. Eng.* **2018**, *135*, 1–15. [CrossRef]
40. Eghbalzadeh, A.; Hayati, M.; Rezaei, A.; Javan, M. Prediction of equilibrium scour depth in uniform non-cohesive sediments downstream of an apron using computational intelligence. *Eur. J. Environ. Civ. Eng.* **2018**, *22*, 28–41. [CrossRef]
41. Chou, J.S.; Pham, A.D. Nature-inspired metaheuristic optimization in least squares support vector regression for obtaining bridge scour information. *Inf. Sci.* **2017**, *399*, 64–80. [CrossRef]
42. Ebtehaj, I.; Sattar, A.M.A.; Bonakdari, H.; Zaji, A.H. Prediction of scour depth around bridge piers using self-adaptive extreme learning machine. *J. Hydroinform.* **2017**, *19*, 207–224. [CrossRef]
43. Lashkar-Ara, B.; Ghotbi, S.M.H.; Najafi, L. Prediction of scour in plunge pools below outlet bucket using artificial intelligence. *KSCE J. Civ. Eng.* **2016**, *20*, 2981–2990. [CrossRef]
44. Mesbahi, M.; Talebbeydokhti, N.; Hosseini, S.A.; Afzali, S.H. Gene-expression programming to predict the local scour depth at downstream of stilling basins. *Sci. Iran.* **2016**, *23*, 102–113. [CrossRef]
45. Choi, S.U.; Choi, B.; Choi, S. Improving predictions made by ANN model using data quality assessment: An application to local scour around bridge piers. *J. Hydroinform.* **2015**, *17*, 977–989. [CrossRef]
46. Hosseini, R.; Amini, A. Scour depth estimation methods around pile groups. *KSCE J. Civ. Eng.* **2015**, *19*, 2144–2156. [CrossRef]
47. Cheng, M.Y.; Cao, M.T. Hybrid intelligent inference model for enhancing prediction accuracy of scour depth around bridge piers. *Struct. Infrastruct. Eng.* **2015**, *11*, 1178–1189. [CrossRef]
48. Turan, K.H.; Yanmaz, A.M. Reliability-based optimization of river bridges using artificial intelligence techniques. *Can. J. Civ. Eng.* **2011**, *38*, 1103–1111. [CrossRef]
49. Zounemat-Kermani, M.; Teshnehlab, M. Using adaptive neuro-fuzzy inference system for hydrological time series prediction. *Appl. Soft Comput.* **2008**, *8*, 928–936. [CrossRef]
50. Komine, T.; Nakagawa, M. Fundamental analysis for visible-light communication system using LED lights. *IEEE Trans. Consum. Electron.* **2004**, *50*, 100–107. [CrossRef]
51. Zhang, Y.; Stritlien, K.; Bellingham, J.G.; Baggeroer, A.B. Acoustic doppler velocimeter flow measurement from an autonomous underwater vehicle with applications to deep ocean convection. *J. Atmos. Ocean. Technol.* **2001**, *18*, 2038–2051. [CrossRef]
52. Nezu, I.; Rodi, M.A.W. Open-channel flow measurements with a laser doppler anemomter. *J. Hydraul. Eng.* **1986**, *112*, 335–355. [CrossRef]
53. Kocsis, L.; Herman, P.; Eke, A. The modified Beer-Lambert law revisited. *Phys. Med. Biol.* **2006**, *51*, 91–98. [CrossRef]
54. Zeng, Z.; Fu, S.; Zhang, H.; Dong, Y.; Cheng, J. A Survey of Underwater Wireless Optical Communication. *IEEE Commun. Surv. Tutor.* **2017**, *19*, 204–238. [CrossRef]
55. Wang, C.; Yu, H.Y.; Zhu, Y.J. A long distance underwater visible light communication system with single photon avalanche diode. *IEEE Photonics J.* **2016**, *8*, 7906311. [CrossRef]
56. Al-Kinani, A.; Wang, C.X.; Zhou, L.; Zhang, W. Optical wireless communication channel measurements and models. *IEEE Commun. Surv. Tutor.* **2018**, *20*, 1939–1962. [CrossRef]

57. Kaushal, H.; Kaddoum, G. Underwater Optical Wireless Communication. *IEEE Access* **2016**, *4*, 1518–1547. [CrossRef]
58. Chang, C.C.; Wu, C.T.; Lin, Y.B.; Gu, M.H. Water velocimeter and turbidity-meter using visible light communication modules. In Proceedings of the Sensors, 2013 IEEE, Baltimore, MD, USA, 3–6 November 2013.
59. Huang, N.E.; Shen, Z.; Long, S.R.; Wu, M.C.; Shih, S.H.; Zheng, Q.; Tung, C.C.; Liu, H.H. The empirical mode decomposition method and the Hilbert spectrum for non-stationary time series analysis. *Proc. R. Soc. Lond. A* **1998**, *454*, 903–995. [CrossRef]
60. Su, S.C.; Wen, K.L.; Huang, N.E. A New Dynamic Building Health Monitoring Method Based on the Hilbert-Huang Transform. *Terr. Atmos. Ocean. Sci.* **2014**, *25*, 289–318. [CrossRef]
61. Wu, Z.H.; Huang, N.E. Ensemble Empirical Mode Decomposition: A Noise-Assisted Data Analysis Method. *Adv. Adapt. Data Anal.* **2009**, *1*, 1–41. [CrossRef]

Article

Scour Evolution Downstream of Submerged Weirs in Clear Water Scour Conditions

Dawei Guan [1], Jingang Liu [2], Yee-Meng Chiew [3,*] and Yingzheng Zhou [2]

1 Key Laboratory of Coastal Disaster and Defense (Hohai University), Ministry of Education, Nanjing 210024, China
2 College of Harbour, Coastal and Offshore Engineering, Hohai University, Nanjing 210024, China
3 School of Civil and Environmental Engineering, Nanyang Technological University, Singapore 639798, Singapore
* Correspondence: cymchiew@ntu.edu.sg; Tel.: +65-6790-5256

Received: 15 July 2019; Accepted: 16 August 2019; Published: 22 August 2019

Abstract: Although weirs or dikes in the riverine and coastal environments are frequently overtopped, few studies have hitherto examined the evolution of the scour process downstream of these structures under the submerged condition. This paper presents an experimental investigation on time evolution of the scour process downstream of submerged weirs with a uniform coarse sand. The clear-water scour experiments were carried out in a tilting recirculation flume. Different flow intensities and overtopping ratios (approach flow depth/weir height) were adopted in the experiments. Experimental observations show that the scour hole downstream of submerged weirs develops very fast in the initial stage, before progressing at a decreasing rate and eventually reaching the equilibrium stage. The results show that an increase of the overtopping ratio or flowrate can generate larger scour depth and volume downstream of the weir. Moreover, geometrical similarity of the scour hole that formed downstream of the weir was observed in the tests. Finally, empirical equations for predicting scour hole geometrical evolutions downstream of the submerged weirs were presented. The results of this study are useful in the development of numerical/analytical models capable of estimating the scour depth downstream of weirs in the river or coastal areas, for which the overtopping conditions are present.

Keywords: sediment transport; clear-water scour; submerged weir; temporal development

1. Introduction

Weirs and dikes are common hydraulic structures used in river and coastal engineering for the purposes of bank protection, bed stabilization, channel regulation, and scour protection [1–4]. Local scour downstream of weirs is an important subject in the field of hydraulic and coastal engineering. Since these weir-like structures are frequently overtopped during high flood events [5], it is essential to study local scouring under the submerged condition. When water flow approaches a weir, the flow patterns are modified by the presence of the structure and local scour holes often develop around the weir. The scour hole dimensions downstream of weirs are governed by the flow field and turbulence characteristics immediately downstream of the weir [3]. For practical purposes, the equilibrium scour depth downstream of the weir-like structures is the most important parameter to determine the depth of foundation. Therefore, many studies [6–21] have been carried out to investigate the effects of many variables, providing the publication of a series of empirical scour equations. However, limited research on the temporal development of scour hole downstream of submerged weirs [22,23] can be found in published literature. Moreover, the peak flood flow in the field generally lasts only a few hours or a number of days, which is insufficient to generate the equilibrium scour depth. Therefore, the final scour depth may be overly conservative if a design is based on the clear-water (i.e., no sediment inflow to the scour hole) equilibrium scour equation [24]. The time scale of local scour at submerged weirs under clear-water scour conditions provides the knowledge of short-term scour dimensions, and could be used for the estimation of scour dimensions during short floods [25]. Besides scour dimensions, geometrical similarity of the scour profiles is also an important issue for the investigation of scour at submerged weirs. Previous studies have shown that the geometrical similarity of the scour holes has important implications on the prediction of the eroded volume downstream of the weirs, which is useful for the design of scour countermeasures, such as the volume and extent of the riprap stones needed [11]. This paper presents an experimental study on time development of the scour hole downstream of a submerged weir. The flow intensity and overtopping ratio effects on the scour process are discussed. The temporal evolution of the scour dimension, scour profile and the relationship between the scour dimension (geometrical similarity) are analyzed based on the experimental results. All the experiments are confined to clear-water scour conditions.

2. Experimental Set-Up and Methods

The experiments were carried out in a 12 m long, 0.3 m deep, and 0.44 m wide glass-sided, tilting flume in the Hydraulic Laboratory at the University of Auckland. The schematic of the flume is shown in Figure 1a and the detailed description of the flume can be found in [3]. The sediment used in the experiments was a uniform coarse sand with median diameter, d_{50} = 0.85 mm, and relative submerged particle density, Δ = 1.65. The weir used in the experiments was a 10 mm thick rectangular plastic plate, with the same width as the flume. A constant flow depth, h = 150 mm, was maintained for all the experiments, and three weir heights, z = 30 mm, 40 mm, and 50 mm, were used, which resulted in three different overtopping ratios z/h = 0.20, 0.27, and 0.33, respectively. In total, five clear-water scour tests were conducted. The detailed experimental conditions are shown in Table 1. For all the tests, the velocity profiles of the approach flow upstream of the weir were measured by using a three-component, downward-looking Nortek Vectrino+ acoustic velocimeter on the centerline of the flume (see Figure 1b). For each point measurement, two minute samples were collected with a sampling rate of 200 Hz. After the flow measurements, all the raw velocity data from the velocimeter were filtered by using WinADV [26], so that low quality data that had spikes, SNR (signal-noise ratio) < 15 or COR (correlation) < 70, were removed [27]. The average approach velocities, U_0, were estimated from the processed velocity profiles. The critical average approach velocities for bed sediment entrainment, U_c, were calculated from the logarithmic mean velocity equation for a rough bed [28]. Since the flow intensities used, which is defined as U_0/U_c, are quite small (0.73–0.92), the local water level rise at the weir for all the tests is negligible.

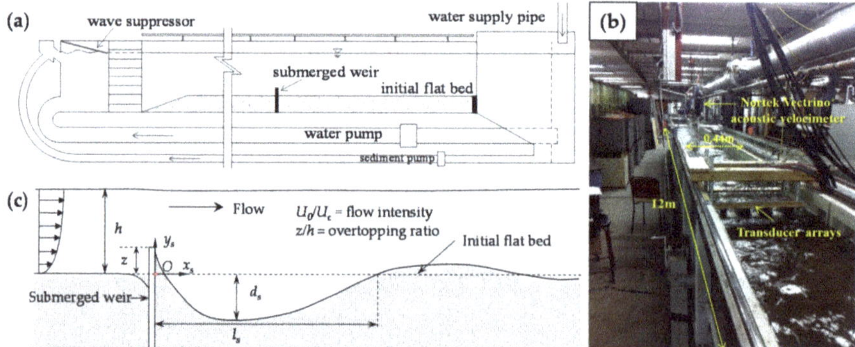

Figure 1. Experimental set-up and definition of experiential parameters: (a) schematic of experimental flume, reproduced from [3]; (b) experimental flume and measurement devices; (c) definition of experimental parameters in 2D.

All the experiments were started with a flat bed. The three-dimensional scour geometry downstream of the weirs was measured by using a Seatek Multiple Transducer Array (MTA) system as a function of time for each clear-water scour test. The MTA system comprises 32 transducers, which can be used to detect the distance from the transducers' surface to a reflective object (bed surface) under the water. The measuring accuracy of the MTA is ±1 mm. In this study, a total of 25 transducers were used to measure the scour profiles. As shown in Figure 1, the transducers were mounted in a rectangular grid on a moving carriage on the top rail of the flume. For each scour profile measurement, the MTA was operated at 5 Hz and could scan the whole scour region within one minute. The detailed arrangement of the transducers and bed profile contouring method can be found in [3]. After the 3D bed profile contouring, the maximum scour depth d_s, scour length l_s, and scour volume V_s were obtained from the processed 3D scour geometry by using a compiled MATLAB code [3].

Table 1. Summary of experimental conditions.

Tests	z (mm)	U_0 (m/s)	z/h	U_0/U_c	Runtime (min)
CS1	40	0.266	0.27	0.73	6240
CS2	40	0.296	0.27	0.82	6060
CS3	40	0.313	0.27	0.86	6150
CS4	30	0.333	0.20	0.92	5640
CS5	50	0.314	0.33	0.87	5850

For all the runs: h = 150 mm, d_{50} = 0.85 mm, U_c = 0.362 m/s.

3. Results and Discussion

3.1. Temporal Evolution of Scour Depth and Scour Length

The 3D bed geometry at three different stages of the scour hole development for Test CS2 is shown in Figure 2.

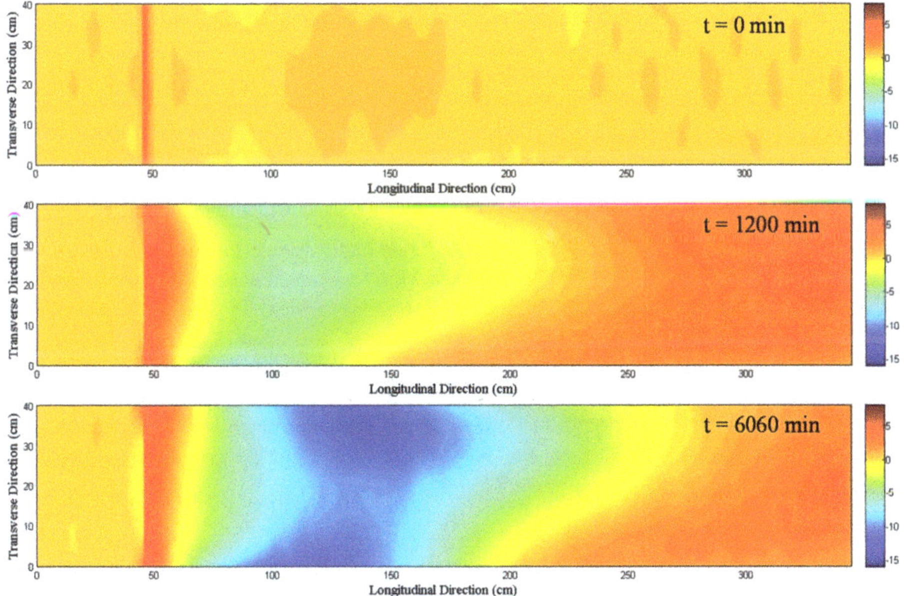

Figure 2. Temporal evolution of bed topography contours (Test CS2) in cm (flow from left to right).

The maximum scour depth downstream of the weir, which is located at longitudinal distance = 50 cm in the figure, occurs close to the weir at the beginning of each test before moving downstream as the scour hole develops. Although the tests were conducted in the straight flume with a 2D weir, the scour hole that forms downstream of the weir exhibits 3D characteristics. The maximum scour depth in the developing scour hole downstream of the weir was found to be close to the flume sidewalls rather than along the centerline of the flume. This is due to the secondary flows in the transverse sections of the flume. According to previous studies [3], the secondary flows are characterized by paired circular flow cells, which are quasi-symmetrically located on both sides of the centerline sand ridge.

The temporal evolution of scour depth and scour length for all the tests is shown in Figure 3. The results show that the maximum scour depth and scour length develop very quickly in the first 20 h, before progressing at a decreasing rate. This trend is consistent with local scour at other structures, such as monopile foundations, submarine pipelines, etc. [29–32].

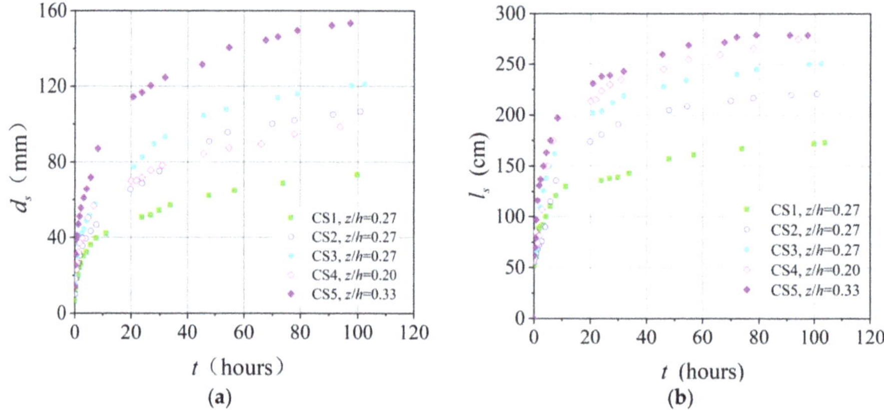

Figure 3. Temporal evolution of scour depth (**a**) and scour length (**b**).

Due to the insufficient scour time, none of the tests, arguably, has reached the final equilibrium stage (see Figure 3a). The equilibrium clear-water scour depth, which is reached asymptotically with time, may take a very long time to form, perhaps infinite time [24]. Therefore, the final equilibrium scour time and depth will not be discussed in this paper. As shown in Figure 3a, the development of scour depth, d_s, is influenced by the flow intensity, U_0/U_c, and overtopping ratio, z/h. The increase of U_0/U_c or z/h accelerates the scour rate and increases the magnitude of scour depth. Figure 3b shows the temporal development of scour length l_s. The trends are similar to those in Figure 3a. The development rate and magnitude of l_s are affected by U_0/U_c and z/h. However, the influence of the overtopping ratio z/h on ls appears to be less significant (see data trends of Tests CS4 and CS5). It should be noted that the values of U_0/U_c for Tests CS3, CS4, and CS5 are not exactly the same in this study, thus more work is needed to determine clearer effects of overtopping ratios on the scour dimensions.

Over the past decades, extensive studies [24,30,31,33–35] have shown that exponential functions could be used to describe the temporal evolution of clear-water scour process around a variety of hydraulic structures, such as bridge piers, bed sills, abutments, etc. The typical form of the exponential function is as follows:

$$\frac{Y_s}{Y_{se}} = 1 - \exp\left[-C\left(\frac{t}{T}\right)^n\right] \tag{1}$$

where Y_s = scour dimension (depth or length) at any time t; Y_{se} = scour dimension (depth or length) at the equilibrium stage; T = time scale (represents the equilibrium scour time in most cases); and C, n = coefficient and exponent to be determined experimentally. Considering the universality of

Equation (1), it is reasonable to adopt the same form for describing the time evolution of clear-water scour at submerged weirs. However, unlike previous studies, the scour processes in this study may not have not reached equilibrium, i.e., the parameters Y_{se} and T are not available to fit Equation (1). As mentioned in the previous paragraph, the temporal evolution of scour dimensions was found to be significantly influenced by the magnitude of weir height z, thus the length scale z may be an appropriate parameter to substitute for the parameter Y_{se} in Equation (1). Therefore, in this study, Equation (1) is modified as follows:

$$\frac{Y_s}{Y_{sz}} = C_1\left\{1 - \exp\left[-C_2\left(\frac{t}{T_z}\right)^n\right]\right\} \tag{2}$$

where T_z = characteristic time at $d_s = z$; Y_{sz} = scour dimension (depth or length) at T_z (estimated by interpolating the observed data in Figure 3a); C_1, C_2, and n = coefficients and exponent to be determined experimentally. Based on the experimental data in Figure 3 and the form of Equation (2), the temporal evolution of scour depth and scour length can be derived as follows:

$$\frac{d_s}{d_{sz}} = 4.22\left\{1 - \exp\left[-0.27\left(\frac{t}{T_z}\right)^{0.40}\right]\right\} \tag{3a}$$

$$\frac{l_s}{l_{sz}} = 2.93\left\{1 - \exp\left[-0.50\left(\frac{t}{T_z}\right)^{0.38}\right]\right\} \tag{3b}$$

where d_{sz} and l_{sz} = scour depth and scour length at T_z, respectively. Figure 4 shows d_s/d_{sz} and l_s/l_{sz} versus t/T_z with an excellent fit. The regression analysis yielded coefficients of determination of $R^2 = 0.994$ and 0.976 for Equations (3a) and (3b), respectively. This indicates that Equation (3) has a high adaptability for the prediction of the scour dimensions downstream of submerged weirs in clear-water scour conditions.

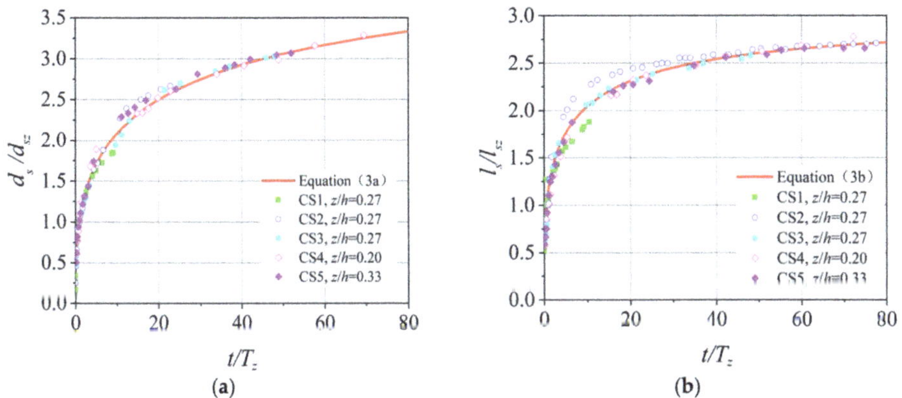

Figure 4. Temporal evolution of dimensionless scour depth (**a**) and scour length (**b**).

It may also be inferred from Equation (3) that the final equilibrium scour depth $d_{se} = 4.22d_{sz} = 4.22z$ and scour length $l_{se} = 2.93l_{sz}$ when the scour time t approaches infinity. However, this simply is an inference at this stage; its verification by more data with sufficiently longer experimental time is needed in future studies.

3.2. Temporal Evolution of Scour Hole Profiles

It is observed that when the scour process downstream of the submerged weir moves into the slow development stage (t > 20 h in this study, see Figure 3), the scour patterns exhibits a high similarity in spite of the different experimental conditions. This is similar to the scour process downstream of bed sills [11,22]. To further explore this similarity, an empirical model is proposed in this section to predict the temporal evolution of the scour hole profile along the centerline of the flume. Based on this model, hydraulic engineers may design the scour countermeasures downstream of submerged weirs more economically. In this study, y_s and x_s are, respectively, defined as the vertical and longitudinal coordinates for describing the scour hole measured with respect to the origin O, as shown in Figure 1c. Based on the experimental data analysis, the temporal evolution of the dimensionless scour hole profile can be described as:

$$\frac{y_s(t)}{y_{s,m}(t)} - \frac{z}{h} = m_1 \exp\left(m_2 \frac{x_s(t)}{x_{s,m}(t)}\right)\left[1 - \exp\left(m_3 \frac{x_s(t)}{x_{s,m}(t)}\right)\right] \quad (4)$$

where y_s (t) and x_s (t) = vertical and longitudinal coordinates for describing the scour hole measured with respect to the origin O at any time t (> 20 h); $y_{s,m}$ (t) and $x_{s,m}$ (t) = maximum vertical and longitudinal coordinates of the scour profiles at any time t (>20 h), respectively, i.e., $y_{s,m}$ $(t) = -d_s$ (t) and $x_{s,m}$ $(t) = l_s$ (t); m_1, m_2, m_3 = coefficients. By fitting the experimental data to Equation (4), the following prediction equation for describing the temporal evolution of the scour hole profile is obtained:

$$\frac{y_s(t)}{y_{s,m}(t)} - \frac{z}{h} = 181.98 \exp\left(-3.96 \frac{x_s(t)}{x_{s,m}(t)}\right)\left[1 - \exp\left(0.06 \frac{x_s(t)}{x_{s,m}(t)}\right)\right] \quad (5)$$

The coefficients of determination, R^2 = 0.903. Comparisons of the results obtained from the proposed Equation (5) with the experimental data in this study are shown in Figure 5. It can be seen from the figure that the proposed equation may be used to predict the temporal evolution of scour hole profiles well.

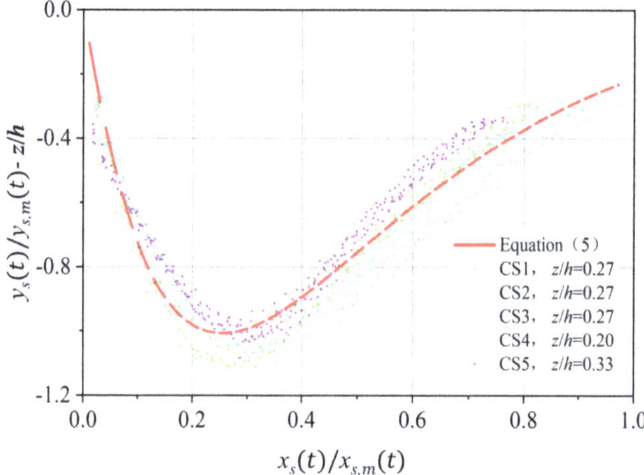

Figure 5. Temporal evolution of dimensionless scour profiles for t > 20 h.

3.3. Relationship between Scour Dimensions

The temporal development of the scour volumes for all the tests was extracted from the scour contour profiles. Although the secondary flows in the cross sections of the flume influence the shape of the scour hole in the transverse direction, the scour hole that forms downstream of submerged weirs can still be approximately regarded as two-dimensional. The mean longitudinal scour area, A_s, of the scour hole downstream of the weir can be approximately considered to be a triangle. Therefore, the scour area A_s can be expressed as:

$$A_s = \frac{V_s}{b} = B \cdot d_s \cdot l_s \quad (6)$$

where b = weir width, B = coefficient to be determined.

Figure 6 plots the relationship between the scour area, A_s, and the product of d_s and l_s. It can be seen that the data trends are linear for all the tests, which may be used to infer that the assumed form of Equation (6) is valid and that the scour hole profiles exhibit geometric similarity.

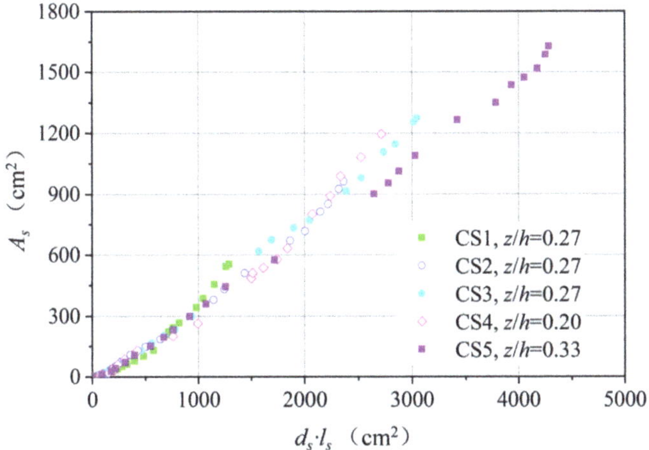

Figure 6. Relationship between averaged longitudinal scour area and scour hole dimensions.

For all the tests, the flow regimes over the submerged weir were surface flow regime. Under this flow regime, the flow remains as a surface jet downstream of the weir; and the scour hole downstream of the submerged weir is caused by the increasing jet thickness and turbulence mixing with the tailwater [17,19]. As indicated in [11], the geometric similarity of the scour hole profiles can be affected by the jet thickness and drop ratio. Therefore, the coefficient B could be expressed as a function of the overtopping ratio:

$$B = f\left(\frac{z}{h}\right) \quad (7)$$

Substituting Equation (7) into Equation (6) yields:

$$A_s = e_1 \cdot \left(\frac{z}{h}\right)^{e_2} \cdot d_s \cdot l_s \quad (8)$$

where e_1 and e_2 are coefficients to be determined. A regression analysis is carried out by using the data in Figure 6, which yields $e_1 = 0.308$ and $e_2 = -0.156$ with a coefficient of determination of $R^2 = 0.991$. The final form of Equation (8) is as follows:

$$A_s = 0.308 \cdot \left(\frac{z}{h}\right)^{-0.156} \cdot d_s \cdot l_s \quad (9)$$

A comparison of the measured and calculated averaged longitudinal scour area is shown in Figure 7, which implies that the fitted Equation (9) can provide a good estimate of the scour area downstream of a 2D submerged weir. The area of the computed final scour hole may be used as the area for rock placement as a type of armoring countermeasure against scour.

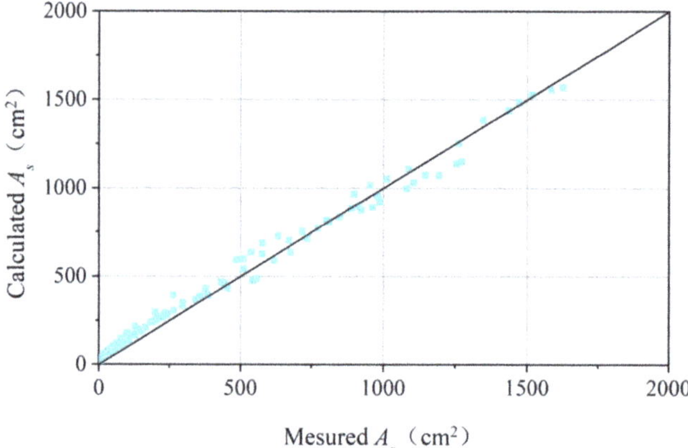

Figure 7. Comparison of measured and calculated averaged longitudinal scour area.

4. Conclusions

This paper presents an experimental study on the temporal development of scour downstream of submerged weirs. Observations show that the dimensions of scour holes downstream of a submerged weir develop very quickly in the initial stage, before progressing at a decreasing rate until the equilibrium is reached. The increase of the flow intensity or overtopping ratio accelerates the scour rate and increases the scour magnitude. Exponential equations were developed to estimate the temporal variation of the scour depth and scour length downstream of the submerged weir. The geometrical similarity of scour holes downstream of the weir was observed in the tests, and an empirical model for describing the temporal evolution of the scour hole profile was developed. A relation between the scour areas and scour dimensions was also proposed. The present study is useful in development of models capable of estimating scour depth downstream of weirs in rivers or coastal areas, for which the overtopping conditions are very common. The proposed equations for scour hole dimensions, profiles, and areal sizes will be good tools for hydraulic engineers in the design of scour countermeasures.

Author Contributions: D.G. conceived, designed and performed the experiments; all authors contributed to analysis of the data, writing and discussion of the paper.

Funding: This work was funded by the Young Scientists Fund of the National Natural Science Foundation of China (grant number 51709082) and the Fundamental Research Funds for the Central Universities (grant number 2018B13014).

Acknowledgments: The first author would like to thank Bruce W. Melville and Jian-Hao Hong for their helpful advice and technical support. The authors would also like to thank The University of Auckland for providing the experimental facilities for this study.

Conflicts of Interest: The authors declare no conflicts of interest.

References

1. Breusers, H.N.C.; Raudkivi, A.J. *Scouring*; Balkema: Rotterdam, The Netherlands, 1991.
2. Guan, D.; Melville, B.; Friedrich, H. Bed load influence on scour at submerged weirs. In Proceedings of the 35th World Congress of IAHR, Chengdu, China, 8 September 2013.
3. Guan, D.; Melville, B.W.; Friedrich, H. Flow Patterns and Turbulence Structures in a Scour Hole Downstream of a Submerged Weir. *J. Hydraul. Eng.* **2014**, *140*, 68–76. [CrossRef]
4. Wang, L.; Melville, B.W.; Whittaker, C.N.; Guan, D. Effects of a downstream submerged weir on local scour at bridge piers. *Hydro Res.* **2018**, *20*, 101–109. [CrossRef]
5. Tregnaghi, M.; Marion, A.; Coleman, S.; Tait, S. Effect of Flood Recession on Scouring at Bed Sills. *J. Hydraul. Eng.* **2010**, *136*, 204–213. [CrossRef]

6. Bormann, N.E.; Julien, P.Y. Scour Downstream of Grade-Control Structures. *J. Hydraul. Eng.* **1991**, *117*, 579–594. [CrossRef]
7. Lenzi, M.A.; Marion, A.; Comiti, F.; Gaudio, R. Local scouring in low and high gradient streams at bed sills. *J. Hydraul. Res.* **2002**, *40*, 731–739. [CrossRef]
8. Marion, A.; Lenzi, M.A.; Comiti, F. Local scouring at grade-control structures in alluvial mountain rivers. *Water Resour. Res.* **2003**, *39*, 39. [CrossRef]
9. Marion, A.; Lenzi, M.A.; Comiti, F. Effect of sill spacing and sediment size grading on scouring at grade-control structures. *Earth Surf. Process. Landforms* **2004**, *29*, 983–993. [CrossRef]
10. Ben Meftah, M.; Mossa, M. Scour holes downstream of bed sills in low-gradient channels. *J. Hydraul. Res.* **2006**, *44*, 497–509. [CrossRef]
11. Marion, A.; Gaudio, R.; Tregnaghi, M. Affinity and similarity of local scour holes at bed sills. *Water Resour. Res.* **2007**, *43*, 11417. [CrossRef]
12. Tregnaghi, M. Local Scouring at Bed Sills under Steady and Unsteady Conditions. Ph.D. Thesis, Università degli studi di Padova, Padova, Italy, 2008.
13. Chinnarasri, C.; Kositgittiwong, D. Laboratory study of maximum scour depth downstream of sills. *Proc. ICE Water Manag.* **2008**, *161*, 267–275. [CrossRef]
14. Lin, B.S.; Yeh, C.H.; Lien, H.P. The experimental study for the allocation of ground-sills downstream of check dams. *Int. J. Sediment Res.* **2008**, *23*, 28–43. [CrossRef]
15. Tregnaghi, M.; Marion, A.; Bottacin-Busolin, A.; Tait, S.J.; Bottacin-Busolin, A. Modelling time varying scouring at bed sills. *Earth Surf. Process. Landforms* **2011**, *36*, 1761–1769. [CrossRef]
16. Guan, D.; Melville, B.; Friedrich, H. A preliminary study on scour at submerged weirs in live bed conditions. In Proceedings of the 7th International Conference on Fluvial Hydraulics (River Flow), École polytechnique fédérale de Lausanne, Lausanne, Switzerland, 13 September 2014.
17. Guan, D.; Melville, B.; Friedrich, H. Live-Bed Scour at Submerged Weirs. *J. Hydraul. Eng.* **2015**, *141*, 04014071. [CrossRef]
18. Guan, D. Scour at Submerged Weirs. Ph.D. Thesis, The University of Auckland, Auckland, New Zealand, 2015.
19. Guan, D.; Melville, B.; Friedrich, H. Local scour at submerged weirs in sand-bed channels. *J. Hydraul. Res.* **2016**, *54*, 1–13. [CrossRef]
20. Tregnaghi, M.; Marion, A.; Coleman, S. Scouring at Bed Sills as a Response to Flash Floods. *J. Hydraul. Eng.* **2009**, *135*, 466–475. [CrossRef]
21. Wang, L.; Melville, B.W.; Guan, D.; Whittaker, C.N. Local Scour at Downstream Sloped Submerged Weirs. *J. Hydraul. Eng.* **2018**, *144*, 04018044. [CrossRef]
22. Gaudio, R.; Marion, A. Time evolution of scouring downstream of bed sills. *J. Hydraul. Res.* **2003**, *41*, 271–284. [CrossRef]
23. Wang, L.; Melville, B.W.; Guan, D. Effects of Upstream Weir Slope on Local Scour at Submerged Weirs. *J. Hydraul. Eng.* **2018**, *144*, 4018002. [CrossRef]
24. Melville, B.W.; Chiew, Y.-M. Time Scale for Local Scour at Bridge Piers. *J. Hydraul. Eng.* **1999**, *125*, 59–65. [CrossRef]
25. Lu, J.-Y.; Hong, J.-H.; Su, C.-C.; Wang, C.-Y.; Lai, J.-S. Field Measurements and Simulation of Bridge Scour Depth Variations during Floods. *J. Hydraul. Eng.* **2008**, *134*, 810–821. [CrossRef]
26. Wahl, T.L. Analyzing ADV Data Using WinADV. In Proceedings of the Joint Conference on Water Resources Engineering and Water Resources Planning and Management, Minneapolis, MI, USA, 30 July–2 August 2000.
27. Goring, D.G.; Nikora, V.I. Despiking Acoustic Doppler Velocimeter Data. *J. Hydraul. Eng.* **2002**, *128*, 117–126. [CrossRef]
28. Melville, B.W. Pier and Abutment Scour: Integrated Approach. *J. Hydraul. Eng.* **1997**, *123*, 125–136. [CrossRef]
29. Guan, D.; Hsieh, S.-C.; Chiew, Y.-M.; Low, Y.M. Experimental study of scour around a forced vibrating pipeline in quiescent water. *Coast. Eng.* **2019**, *143*, 1–11. [CrossRef]
30. Guan, D.; Chiew, Y.-M.; Melville, B.W.; Zheng, J. Current-induced scour at monopile foundations subjected to lateral vibrations. *Coast. Eng.* **2019**, *144*, 15–21. [CrossRef]
31. Wei, M.; Chiew, Y.-M.; Guan, D. Temporal Development of Propeller Scour around a Sloping Bank. *J. Waterw. Port Coastal Ocean Eng.* **2018**, *144*, 06018005. [CrossRef]

32. Link, O.; González, C.; Maldonado, M.; Escauriaza, C. Coherent structure dynamics and sediment particle motion around a cylindrical pier in developing scour holes. *Acta Geophys.* **2012**, *60*, 1689–1719. [CrossRef]
33. Hoffmans, G.J.C.M. Jet Scour in Equilibrium Phase. *J. Hydraul. Eng.* **1998**, *124*, 430–437. [CrossRef]
34. Cheng, N.S.; Chiew, Y.M.; Chen, X. Scaling Analysis of Pier-Scouring Processes. *J. Eng. Mech.* **2016**, *142*, 6016005. [CrossRef]
35. Guan, D.; Chiew, Y.M.; Wei, M.; Hsieh, S.C. Characterization of horseshoe vortex in a developing scour hole at a cylindrical bridge pier. *Int. J. Sediment Res.* **2019**, *34*, 118–124. [CrossRef]

Article

Scouring of Replenished Sediment through Reservoir Flood Discharge Affects Suspended Sediment Concentrations at Downstream River Water Intake

Fong-Zuo Lee [1], Jihn-Sung Lai [1,*], Wen-Dar Guo [2] and Tetsuya Sumi [3]

1 Hydrotech Research Institute, National Taiwan University, Taipei 10617, Taiwan; windleft@gmail.com
2 National Science and Technology Center for Disaster Reduction, New Taipei City 23143, Taiwan; wdguo@ncdr.nat.gov.tw
3 Water Resources Research Center, Disaster Prevention Research Institute, Kyoto University, Goka-sho, Uji-shi 611-0011, Japan; sumi.tetsuya.2s@kyoto-u.ac.jp
* Correspondence: jslai525@ntu.edu.tw

Received: 5 August 2019; Accepted: 20 September 2019; Published: 25 September 2019

Abstract: Dredging is a commonly used sedimentation management strategy to remove mechanically deposited sediment from reservoirs. However, dredged sediment disposal is costly. Dredged sediment can be considered a beneficial resource and used for riverbed replenishment to prevent downstream riverbed degradation and improve aquatic habitats. This study investigated the feasibility of using dredged deposits with cohesive sediment for replenishment at the Shihmen Reservoir. Using the criterion of critical scour velocity, we conducted hydraulic assessments and identified the feasible replenishment area as the experimental domain. A physical model was developed to mimic the scouring process in the replenishment area. By applying dynamic similarity for scouring fine replenished sediment, we derived the regression relationship between flow-critical velocity and sediment-dry density, and used it for model ratio scaling of the grain size, dry density, and concentration in the physical model. Scoured sediment concentrations were measured to study the scour ratio at various flood discharges. Experimental results indicated that the scour ratio was related to factors such as flood discharge, flood duration, and water content of the replenished sediment. The reduction ratio of the concentration of sediment scoured from the replenishment area to the concentration of sediment at the downstream water intake was approximately 90% in the present study.

Keywords: dredging; reservoir sedimentation; replenishment; cohesive sediment; scour

1. Introduction

For reservoirs worldwide, storage preservation through sustainable sediment management operations is a critical issue due to the severe problem of reservoir sedimentation. The available total storage capacity of global reservoirs has been decreasing in recent years because of the lack of effective sediment management practices [1]. Sediment trapping in a reservoir or dam interferes with the course of sediment transport through a river system; such interference exerts a considerable influence on the sustainability of future water supplies from the reservoir, engenders channel erosion, and negatively affects the downstream ecosystem. Morphological effects on downstream river channels include riverbed incision, riverbank instability, damage to embankments and levees, and channel width variation [2–5].

Studies have investigated many management strategies and efficient countermeasures for reducing the influence of sedimentation [3,6–8]. Dredging is a commonly used measure to remove mechanically deposited sediment from reservoirs. The disposal of dredged sediment is costly due to the involved engineering, permission, and placement site requirements. However, dredged sediment can be considered as a resource that provides effective environmental benefits and may be added to the downstream river of a dam to compensate for the lack of sediment supply. Accordingly, an approach called sediment replenishment (sediment augmentation) can be adopted for the effective replenishment due to dam interruption. [3,9,10]. Since the 1980s, sediment has been added into rivers to replenish the downstream river reaches of dams. The sediment replenishment method has been implemented to prevent downstream riverbed degradation and to improve the suitable living habitat of aquatic animals. Field investigations and laboratory experiments have been conducted for rivers in America, Europe, Japan, and Taiwan to improve knowledge about the transport processes during sediment replenishment [11–13].

In field practice, the sediment replenishment method can be planned and practiced in specific hydrological and geographic areas. Okano et al. (2004) [14] investigated reservoir sedimentation management by depositing coarse sediment (mainly sand and gravel sizes) with a volume of 0.3–25 × 10^3 m^3 for replenishing downstream areas. They summarized sediment replenishment projects conducted at eight dams in Japan from 1999 to 2003. Moreover, researchers have investigated the influences of sediment replenishment conducted using appropriate grain sizes on the downstream river of Yahagi dam and the upper basin of Kizu river [15,16]. Researchers also analyzed the interaction between the relative flow field and morphological evolution during field experiments of sediment replenishment [13]. The conceptual idea of the replenishment method is to place sediment in the downstream floodplain before the arrival of floods. If the replenishment areas experience floods, the replenished sediment can be scoured and transported further downstream. As mentioned, field tests in Japan have focused on the deposition of coarse materials in reservoirs to subsequently replenish downstream rivers of the dams in order to create a suitable environment for aquatic organisms. The applied replenishment method requires a sufficient amount of discharge in order to scour replenished sediment; replenished sediment is thus always placed on floodplains before the discharge of water from a reservoir during the wet season. Sediment deposited in a reservoir can be periodically dredged or excavated and then temporarily deposited on the floodplain for transportation to the downstream river of a dam. Moreover, the replenished volume is based on the sediment transport capacity of the channel and environmental conditions. According to reports in Japan regarding the replenished volume, up to 10% of the annual deposition volume in a reservoir could be successfully executed [6,7]. Ock et al. (2013) [17] reviewed methods in the context of sediment replenishment and compared implementation activities undertaken in the Nunome river of Japan and Trinity river of California, USA. According to sediment placement or injection types, sediment replenishment methods were implemented with mechanical rehabilitation for re-creating gravel or sand bar features through fluvial processes. This comparative study provided useful information to adopt proper methods corresponding to river specific high-flow and sediment regimes. In 2009, a field test for coarse sediment replenishment was conducted at the downstream floodplain of Shihkang dam across the Dachia river in Central Taiwan [12]. The total replenished volume of the coarse sediment excavated (median grain size: 80 mm) from the Shigang Reservoir was approximately 50 × 10^3 m^3. The field experiment revealed that the replenished sediment was entirely scoured further downstream in a typhoon flood event with a peak discharge of 5400 m^3/s in 2009.

On the basis of the aforementioned studies, field and laboratory experiments for sediment replenishment have mainly focused on coarse materials, with few fine sediment examples being provided. In fact, fine replenished sediment dredged from reservoirs is valuable for coastline or estuary restoration and for the respiration of aquatic organisms. However, if a high concentration of fine replenished sediment is scoured by a reservoir flood discharge, the water quality of water treatment plants connected to the relevant reservoir may deteriorate.

In the present study, the Shihmen Reservoir was used as the study site to investigate the feasibility of fine-sediment replenishment. First, possible areas for replenishment were analyzed through a numerical simulation conducted using a two-dimensional (2D) numerical model [18]. On the basis of the framework of the finite volume method, the 2D numerical model was used to solve the shallow water equations coupled with the advection–diffusion equation in order to simulate the suspended sediment transport phenomenon in the river. A physical model was built to mimic the scouring process of replenished sediment in the downstream river reach of Shihmen dam. Experiments involving different water content levels of dredged fine sediment were conducted in the physical model to investigate the incipient motion of the replenished sediment with cohesiveness. Through the experiments, the scour ratio was analyzed using relevant factors such as flood discharge, flood duration, and water content of the replenished sediment. The effect of the scouring of fine sediment from the replenishment area on the downstream water intake was also investigated.

2. Description of Study Site

The Shihmen Reservoir was built in 1964 in northern Taiwan (Figure 1). Shihmen dam is a 133.1-m-high embankment dam and is equipped with spillways, flood diversion tunnels, a bottom outlet, a power plant inlet, a sediment-sluicing tunnel, and an irrigation inlet. The design capacity of the three spillways is 11,400 m^3/s, that of the flood diversion tunnels is 2400 m^3/s, that of the bottom outlet is 34 m^3/s, that of the power plant inlet is 56 m^3/s, that of the sediment-sluicing tunnel is 300 m^3/s, and that of the irrigation inlet is 18.4 m^3/s. At a normal water level of EL. 245 m, the reservoir pool is approximately 16.5 km in length and forms a water surface area of 8 km^2. The initial storage capacity is 30.9×10^6 m^3. The reservoir is multifunctional and is used for water supply, irrigation, electric power generation, and flood mitigation; it is also a tourist attraction. In particular, the reservoir supplies water to 2.1 million people in Taoyuan City and New Taipei City for daily use.

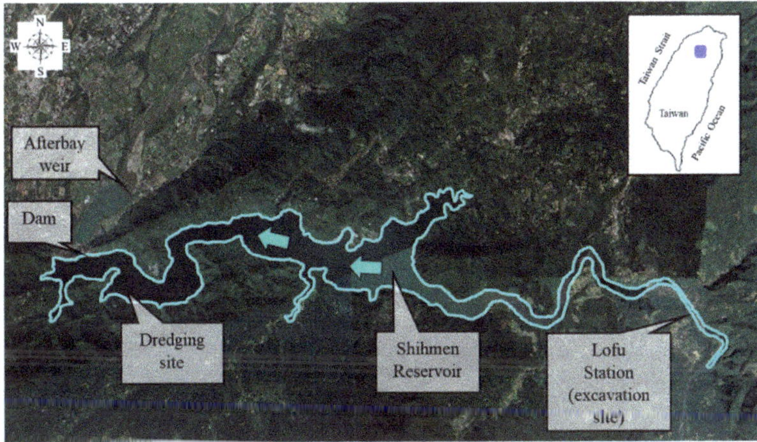

Figure 1. Location map and geometry of Shihmen Reservoir.

According to the data of a survey conducted in 2018, the reservoir's storage capacity was estimated to be 65.72% of its initial capacity. The average annual deposited sediment volume from 1963 to 2018 was approximately 1.93×10^6 m^3/year. Except for hydraulic desilting measures, mechanical excavation constitutes the strategies adopted for maintaining the Shihmen Reservoir. The mechanical excavation includes the dredging and dry excavation. The dredging is mainly performed under the water surface, and the main purpose of dry excavation is to remove exposed sediment. The dry excavation site in this study is located at the Lofu gauge station, and the averaged exaction volume between 2009 and 2018 was approximately 0.47×10^6 m^3/year. The dredging site is near the dam, and the dredged sediment volume between 2009 and 2018 was nearly 0.36×10^6 m^3/year. Fine sediment deposited in front of Shihmen Dam is typically removed through a pipeline by a hydraulic dredging system that applies siphon dredging; before executing the dredging process, the system calculates the head difference between the reservoir water surface and the pipeline outlet. However, finding a disposal location for the dredged sediment is a problem. Therefore, one of the implemented reservoir-desilting strategies entails conducting sediment replenishment by placing the reservoir deposits on the downstream floodplain, which may be used in the Shihmen Reservoir.

Referring to the particle size distribution presented in Figure 2, the dredged sediment near the dam site is almost entirely composed of silt and clay; thus, it is classified as a cohesive material. The sediment excavated at the Lofu gauge station is coarse sand. In this study, sediment obtained from both dredging and dry excavation was used for sediment replenishment investigation on the downstream floodplain. According to the regulation of the water pollution control action [19], the replenished material should be placed at the floorplan and confined by a filter structure that serves as a permeable barrier to avoid the replenished fine sediment from polluting the water in nearby main channels during low-flow periods.

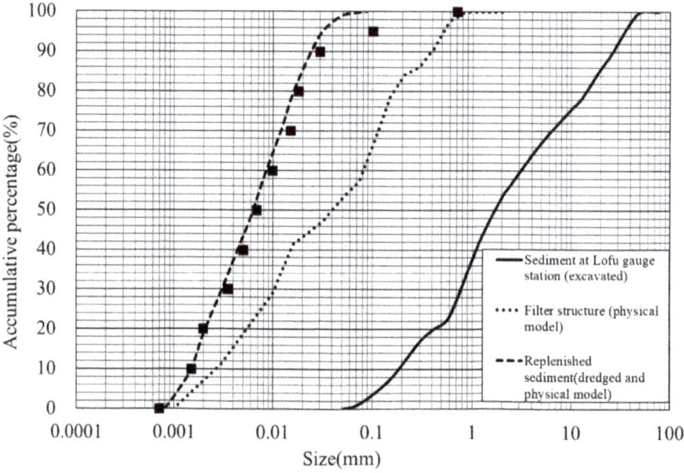

Figure 2. Sediment grain size distribution sampled in the field and used in the physical model.

The available flood discharges from the reservoir and the flood duration are key factors for implementing sediment replenishment. Based on historical records from 1911 to 2017, three typhoons hit Taiwan each year on average. According to a frequency analysis, the 2-, 10-, and 20-year return period floods have peak discharges of 1700, 3500, and 6100 m^3/s, respectively [20].

3. Hydraulic Assessment of Sediment Replenishment

If the flow intensity exceeds a specific value and sediment particles begin to move, the flow condition that corresponds to a critical value is called incipience. The main parameters of the incipience condition are usually related to the flow velocity and water depth. Therefore, in this study, the flow velocity fields and water depths were simulated using the 2D numerical model developed by Guo et al. (2011) [18]. The advection–diffusion equation is commonly used to describe the suspended sediment transport phenomenon. For determining the process of suspended sediment transport using the developed numerical model, the 2D shallow water equations coupled with the advection–diffusion equation served as the governing equations. Flow velocity fields and water depths in the candidate replenishment areas were compared, as described in this section. Subsequently, the feasible area for the physical model test was identified and selected.

3.1. 2D Numerical Model

The conservation form of the governing equations is expressed as follows:

$$\frac{\partial \mathrm{Q_v}}{\partial t}+\frac{\partial \mathrm{F_{adv}}}{\partial x}+\frac{\partial \mathrm{G_{adv}}}{\partial y}=\frac{\partial \mathrm{F_{diff}}}{\partial x}+\frac{\partial \mathrm{G_{diff}}}{\partial y}+\mathrm{S_s} \tag{1}$$

Here,

$$\mathrm{Q_v}=\begin{bmatrix} h \\ hu \\ hv \\ hC \end{bmatrix},\ \mathrm{F_{adv}}=\begin{bmatrix} hu \\ hu^2+gh^2/2 \\ huv \\ huC \end{bmatrix},\ \mathrm{G_{adv}}=\begin{bmatrix} hv \\ huv \\ hv^2+gh^2/2 \\ hvC \end{bmatrix},$$
$$\mathrm{F_{diff}}=\begin{bmatrix} 0 \\ hT_{xx}/\rho \\ hT_{xy}/\rho \\ D_x\partial(hC)/\partial x \end{bmatrix},\ \mathrm{G_{diff}}=\begin{bmatrix} 0 \\ hT_{xy}/\rho \\ hT_{yy}/\rho \\ D_y\partial(hC)/\partial y \end{bmatrix},\ \mathrm{S_s}=\begin{bmatrix} 0 \\ gh(s_{0x}-s_{fx}) \\ gh(s_{0y}-s_{fy}) \\ q_{se}-q_{sd} \end{bmatrix} \tag{2}$$

where $\mathrm{Q_v}$ is the conserved physical vector; $\mathrm{F_{adv}}$ and $\mathrm{G_{adv}}$ are the advection flux vectors in the x and y directions, respectively; $\mathrm{F_{diff}}$ and $\mathrm{G_{diff}}$ are the diffusion flux vectors in the x and y directions, respectively; $\mathrm{S_s}$ is the source term; h is the water depth; u and v are the depth-averaged velocity components in the x and y directions, respectively; ρ is the density of water; T_{xx}, T_{xy}, and T_{yy} are the depth-averaged turbulent stresses; C is the depth-averaged suspended sediment concentration; g is the gravitational acceleration; D_x and D_y are the diffusion coefficients. s_{0x} and s_{0y} are the bed slopes in the x and y directions, respectively, and they are expressed as:

$$s_{0x}=-\partial z_b/\partial x \text{ and } s_{0y}=-\partial z_b/\partial y \tag{3}$$

z_b is the bed elevation; s_{fx} and s_{fy} are the friction slopes in the x and y directions, respectively; and T_{xx}, T_{xy}, and T_{yy} are the depth-averaged turbulent stresses. On the basis of Manning formulas, the friction slopes can be expressed as:

$$s_{fx}=\frac{un_{\mathrm{M}}^2\sqrt{u^2+v^2}}{h^{\frac{4}{3}}},\ s_{fy}=\frac{vn_{\mathrm{M}}^2\sqrt{u^2+v^2}}{h^{\frac{4}{3}}} \tag{4}$$

where n_{M} is the Manning roughness coefficient.

According to the particle size ranges, suspended sediment particles can be divided into cohesive and non-cohesive particles [21]. For cohesive sediment, the deposition flux (depositional rate) can be determined using the following relation [22].

$$q_{sd}=Pw_sC,\ P=1-\sqrt{\tau_{bx}^2+\tau_{by}^2}/\tau_{cd} \tag{5}$$

where P is the probability of deposition and τ_{cd} is the critical shear stress of deposition. Moreover, w_s is the settling velocity and defined as:

$$w_s=250\,d_m^{0.2} \tag{6}$$

where d_m is the median particle size (μm). The entrainment flux (erosion rate) for cohesive materials was given by Liu et al. (2002) [23] and can be computed from:

$$q_{se}=E(\sqrt{\tau_{bx}^2+\tau_{by}^2}/\tau_{ce}-1) \tag{7}$$

where E is the erosion parameter and τ_{ce} is the critical shear stress of erosion. In general, the critical shear stresses of deposition and erosion differ in various study areas. Based on the experimental data and model calibration, Liu et al. (2002) [23] provided the suggested values of τ_{cd}(= 0.05 N/m^2) and τ_{ce}(= 0.1 N/m^2) for modeling cohesive sediment transport in Equations (5) and (7), respectively.

For the numerical discretization in the framework of the finite volume method, the finite-volume multi-stage (FMUSTA) scheme was adopted to estimate the numerical flux through each cell interface in order to solve Equation (1) [24]. In the treatment of source terms, the hydrostatic reconstruction method was employed to prevent the imbalance between flux gradients and source terms. The presented 2D numerical model is practically suitable for complex flow conditions (mixed regimes) and is appropriate for special problems related to wetting and drying. A detailed description of the numerical formulation was provided by Guo et al. (2011) [18].

3.2. Numerical Model Validation

For the validation of the numerical model, the simulation domain was selected as the downstream river reach. The reach is approximately 16.5 km long from the Shihmen afterbay weir to the Yuanshan weir, as shown in Figure 3. The water intake is located on the right-hand-side bank of the Yuanshan weir, which delivers withdrawal water to the Bansin water treatment plant and supplies water. Field data collected at the Dasi bridge gauge station during two historical typhoon flood events—Typhoon Changmi and Typhoon Krosa—were used for model validation. By comparing the measured water level and suspended sediment concentration hydrographs, this study determined that the simulated results obtained using the adopted 2D numerical model were highly consistent with the measured data (Figure 4), despite the existence of some discrepancy; for example, a discrepancy was observed in the peak sediment concentration, for which the average relative error was 2.2% for Typhoon Changmi. Accordingly, the adopted 2D numerical model can model velocity and concentration fields in natural rivers with irregular topography.

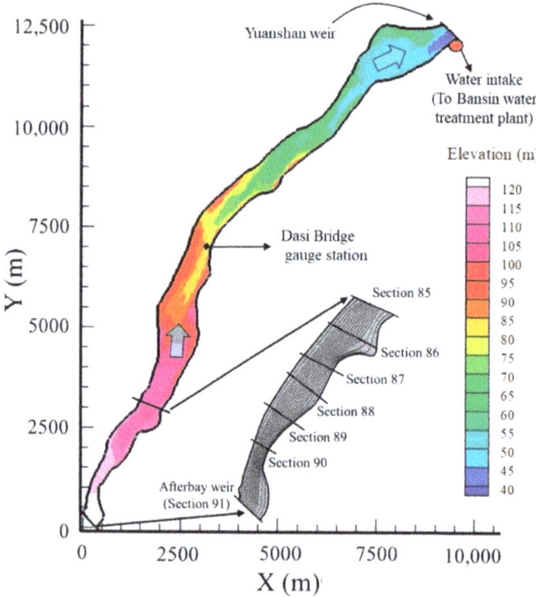

Figure 3. Simulation domain along the downstream river from Shihmen afterbay weir to Yuanshan weir; some of the grid meshes also displayed.

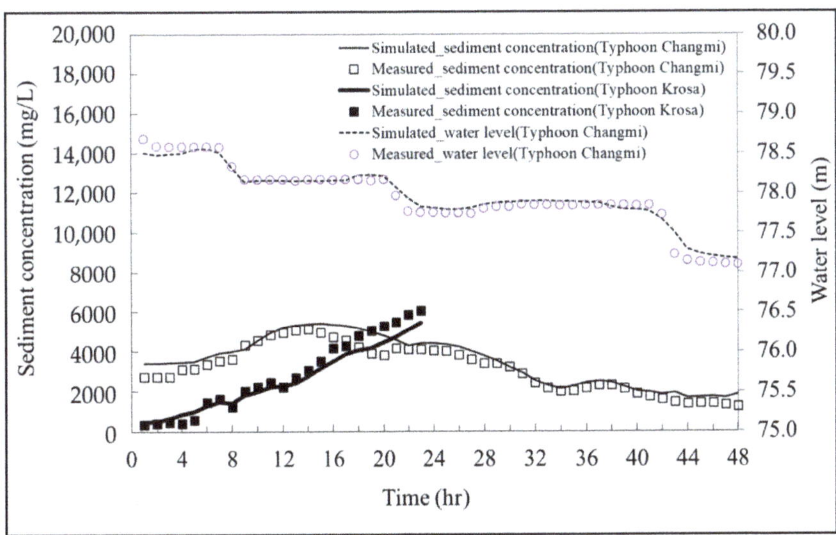

Figure 4. Model validation using data of two typhoon flood events at the Dasi bridge gauge station.

After the validation of the numerical model, this study evaluated the most feasible replenishment area for placing the dredged sediment. As presented in Figure 3, the approximately 3.6-km river reach located between the Shihmen afterbay weir (Section 91) and Section 85 may be a suitable area for sediment replenishment due to the cost constraints of dredging pipelines. In part of the simulation domain, the simulated results pertaining to the flow velocity and water depth in the river reach between Section 91 and Section 85 were analyzed, as discussed in the following section.

3.3. *Hydraulic Assessment of the Feasible Replenishment Area*

During the selection of suitable replenishment areas, locations near a dam should be considered to reduce the implementation cost of the dredging pipeline system and maintenance work. Moreover, the replenishment areas should be designed for construction on a floodplain. Accordingly, in this study, the candidate replenishment areas were tentatively selected and marked as Zone 1, Zone 2, and Zone 3 (Figure 5).

To resolve the problem associated with the unavailability of suitable sites for the disposal of materials dredged from reservoirs, the replenishment method is typically considered for disposing fine sediment dredged from reservoirs. Hjulstrom (1935) [25] found that the critical scour (or erosion) velocity is proportional to the particle size, and resistance to scour increases for sediment finer than 50 μm; these findings indicate that factors other than particle size and weight influence scour phenomena [26,27]. Sediment dredged from around the dam site exhibited a finer level of cohesiveness than the median particle size (d_{50} = 0.006 mm; Figure 2); therefore, we used d_{25} (0.0023 mm) to represent the replenished sediment particle size, signifying that a critical velocity of approximately 1.5 m/s would be required to initiate the scouring process [28]. Moreover, according to the regulation of water pollution control action [19], the replenishment area should be confined by a filter structure that serves as a permeable barrier to avoid the replenished fine sediment from polluting the water in nearby main channels during low-flow periods. Therefore, sediment that was excavated from the Lofu site and had a median particle size d_{50} of 1.76 mm (classified as coarse sand; Figure 2) was used as the filter material. We used d_{75} (9.5 mm, classified as medium gravel) to represent the replenished sediment particle size; hence, a critical velocity of approximately 1.05 m/s would be required to initiate scouring for the filter structure [28]. On the basis of the preceding analysis, we may adopt a higher critical scour velocity (e.g., 1.5 m/s) with sufficient flow depth to evaluate the incipient motion of both fine sediment and coarse sediment. Although the concept of critical scour velocity is used herein for illustrative purposes, it indeed provided the preliminary criterion to assess the feasible replenishment area.

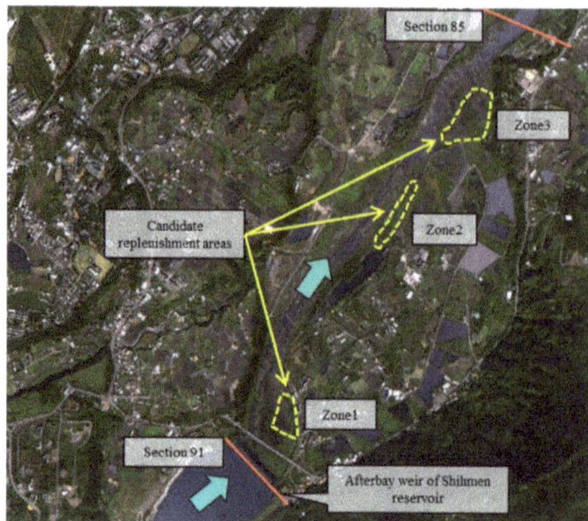

Figure 5. Candidate sediment replenishment areas.

The validated 2D numerical model was used to model the flow velocity distribution field in the candidate replenishment areas. The simulated results for water depth (Figure 6) revealed that the flow current was confined within the main channel at a discharge of 200 m^3/s. Accordingly, the 2-, 10-, and 20-year return period floods were determined to have peak discharges of 1700, 3500, and 6100 m^3/s, respectively [20]. We adopted the flood discharges of 1700 and 6100 m^3/s in two test cases separately to execute the 2D numerical model for the preliminary selection of the feasible replenishment areas.

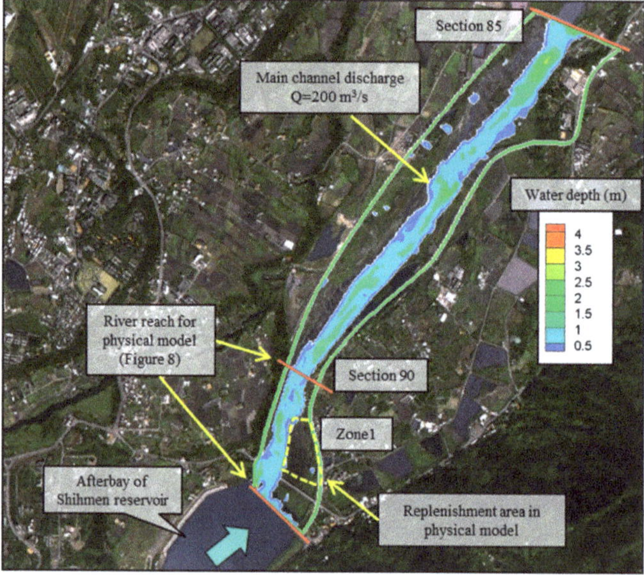

Figure 6. Downstream river reach for evaluation with a discharge of 200 m^3/s.

According to the simulated results obtained from the test case that involved the flood discharge of 1700 m^3/s, the water depth was lower than 1 m in both Zone 2 and Zone 3 (Figure 7a). In Zone 1, the water depth was higher than 1 m in most portions of the replenishment area. In addition, the velocity was lower than 1 m/s in both Zone 2 and Zone 3. However, most portions of Zone 1 had the velocity of more than 1 m/s, whereas the velocity near the main channel was higher than 2 m/s. In the test case that involved the discharge of 6100 m^3/s, the water depth was still lower than 1 m in most portions of Zone 2 and Zone 3, whereas the depth in Zone 1 increased to more than 3 m in most portions of the replenishment area (Figure 7b). The velocity in most portions of Zone 1 was more than 1.5 m/s. However, the velocities in most portions of Zone 2 and Zone 3 were lower than 1 m/s, except for the portion near the main channel. The simulated flow velocities in Zone 1 in the two test cases (involving flood discharges of 1700 and 6100 m^3/s separately) satisfied the critical scour velocity criterion (1.5 m/s) for incipient motion of both fine sediment and coarse sediment. Moreover, the simulated water depths ranged from 1.5 to 4 m in Zone 1 of the replenishment area in the two test cases. The height of the filter structure was 2 m, and that of the replenished material was 1.6 m in the replenishment area, which were determined to be subject to flooding. Consequently, Zone 1 was selected as the location for sediment replenishment. The feasible replenishment area—Zone 1—was determined to be on the right-hand-side floodplain (Figure 6). Moreover, for the experimental river reach for the physical model, we considered the reach between the Shihmen afterbay weir (Section 91 as upstream boundary) and Section 90 (downstream boundary) with a narrow cross section.

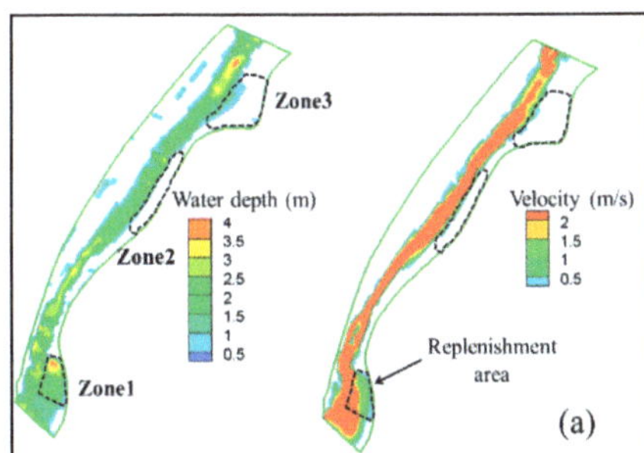

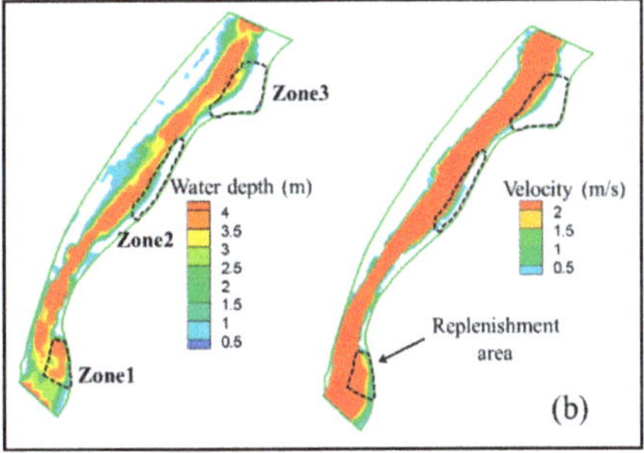

Figure 7. Simulated water depth and flow velocity distributions at flood discharges of (**a**) 1700 and (**b**) 6100 m^3/s.

4. Physical Model Layout and Scaling

4.1. Layout of Physical Model

On the basis of the results of the numerical simulations that were performed for determining candidate replenishment areas, we selected the replenishment area located on the floodplain just downstream of the afterbay weir (Figure 6). For an undistorted model, the geometrical ratio between the prototype and model was set to 64:1 in both the vertical and horizontal directions due to space constraints at the construction site.

The model layout is sketched in Figure 8. The upstream boundary was set at the cross section of the afterbay weir, and the downstream boundary was set at Section 90 with a relatively straight reach and a narrow cross section further downstream of the replenishment area. Water was recirculated using a pump at various desired steady flow rates and was controlled by the valve of the pipeline, which was connected to the head tank. A sediment-settling basin was constructed at the downstream end to settle part of the outflow sediment, and the inflow sediment concentration was monitored to control the background sediment quantity. Two monitoring points were set at the downstream end for measuring outflow sediment concentrations, from which the average was derived. According to the prototype dimension, the volume of replenished sediment was nearly 100×10^3 m^3, which was estimated to be one-fourth of the annual volume of dredged sediment. Accordingly, the thickness of replenished sediment was approximately 1.6 m. To confine the dredged materials inside the selected replenishment area, a permeable barrier with a height of 2 m was constructed. The permeable barrier functioned as a filter to prevent the replenished material containing very fine sediment from polluting the water in the main channel during low-flow periods. Moreover, the barrier was composed of coarse sediment materials with relatively large particle sizes (Figure 2); the materials were obtained from the field dry excavation site near the Lofu gauge station at the upstream end of the Shihmen Reservoir.

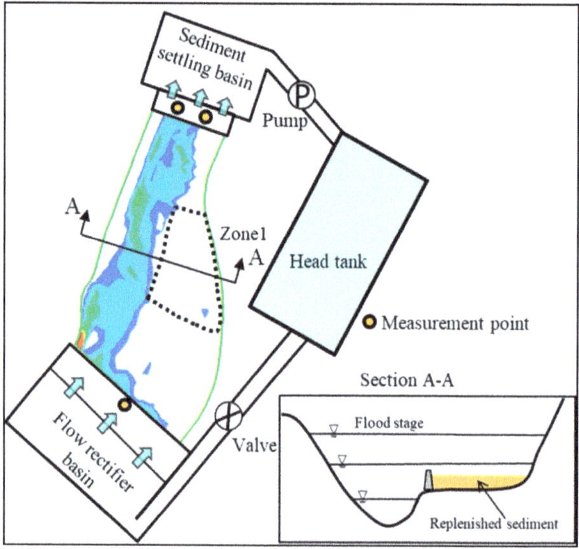

Figure 8. Layout of the physical model.

4.2. Similitude for Model Scaling

A physical model should satisfy the similarity theory of geometrical, kinematic, and dynamic conditions. In open-channel flow, flow patterns under a free-surface condition follow the Froude number (F_r) similarity. Therefore, the following relationship is involved in the scaling of model dimensions for flow patterns: $\lambda_{F_r} = \lambda_u / \sqrt{\lambda_g \lambda_h} = 1$, where λ is the ratio of the model to the prototype; λ_u is the velocity scale ratio; λ_h is the vertical length scale ratio; and g is the acceleration of gravity, with $\lambda_g = 1$. Hence, the flow velocity ratio of the model to the prototype can be derived as follows: $\lambda_u = \lambda_h^{1/2}$. Note that the flow discharge scale is $\lambda_Q = \lambda_u \lambda_A = \lambda_h^{5/2}$, and the time scale is $\lambda_t = \lambda_h^{1/2}$. The subscripts u, Q, A, and t denote the velocity dimension, discharge dimension, cross-sectional area dimension, and time dimension, respectively.

4.2.1. Initiation of Sediment Movement

Similarities in sediment transport mechanisms engender difficulties in combining coarse and fine materials because of the difference in sediment particle sizes. In this study, criteria for the similitude of sediment movements were required to assess coarse sediment for modeling the permeable barrier and to assess cohesive sediment for the dredged material placed in the replenishment area. For coarse sediment, we could adopt the Shields diagram for the condition involving incipient riverbed particle motion [29].

For satisfying the similitude of model scaling, the dimensionless shear parameter in the Shields diagram was employed and can be expressed as follows:

$$\frac{\lambda_{\tau_{cr}}}{\lambda_{(\gamma_s-\gamma)}\lambda_{d_c}} = 1 \tag{8}$$

where τ_{cr} is the critical shear stress ($\lambda_{\tau_{cr}} = \lambda_\rho \lambda_g \lambda_R \lambda_s$), γ_s is the specific density of sediment, γ is the specific density of water, d_c is the grain size of coarse sediment, ρ is the density of water, R is the hydraulic radius, and S is the energy slope. The energy slope ratio λ_s is equal to 1. If we substitute $\lambda_{\tau_{cr}}$ with $\lambda_\rho \lambda_g \lambda_R \lambda_s$ and substitute λ_R with the vertical length scale ratio λ_h, Equation (8) becomes $\lambda_{d_c} = \lambda_\rho \lambda_h \lambda(\rho_s - \rho)^{-1}$, where ρ_s is the sediment density. If we use the same sediment material in the physical model as that in the field, the density scale would be equal to 1. Then, the scale ratio for the particle size of coarse sediment can be expressed as $\lambda_{d_c} = \lambda_h$.

As the scouring process acts on the surface of replenished materials, fine sediment with cohesiveness can be washed away when picked up by running flow, which may behave as a wash load [30]. The scour potential with respect to the flow shear strength can be related to the dry density of cohesive deposits in exponential form, and this relationship has been derived by several researchers [31]. Although this relationship is rather an approximate and is site specific, it is useful for estimating the critical shear stress generated by a flood flow to characterize the structures of replenished sediment as deposits with cohesiveness. Based on flume experiments, the regressed relationship between critical velocity u_c and dry density ρ_d can be expressed as follows:

$$u_c = a{\rho_d}^b \text{ or } \rho_d = (u_c/a)^{1/b} \tag{9}$$

where the coefficient a and exponent b can be determined by conducting a flume experiment [32]. According to the preceding expressions regarding the similarities in the scouring mechanisms of fine sediment and similarities in the Froude number of flows, the scale ratio of dry density can be expressed as follows:

$$\lambda_{\rho_d} = \lambda_{u_c}^{1/b} = \lambda_h^{1/2b} \tag{10}$$

Using sediment sampled from the Agongdian Reservoir located in Southern Taiwan, Lai (1998) [33] conducted experiments by setting the value of a to 1.65 and the value of exponent b to 1.96. As shown in Figure 2, the particle size distribution of the sediment sample (with 22% of clay) from the Agongdian Reservoir is highly similar to the particle size distribution of the sediment sample from the Shihmen Reservoir. Thus, the relationship presented in Equation (9) for fine sediment from the Agongdian Reservoir is analogous to relationship for fine sediment from the Shihmen Reservoir. From the field data regarding dredged sediment properties provided by the Northern Region Water Resources Office (2010) [20], the water content of dredged material ranges from 39.9% to 45.2% at 1 year after deposition and from 81.8% to 98.5% at 16 days after deposition. Therefore, this study adopted water content levels of 40% and 80% to conduct tests in the physical model for sediment replenishment. Based on dredged sediment sample property, the dry densities corresponding to water content levels of 40% and 80% were 1.2 and 0.84 t/m^3, respectively.

4.2.2. Sediment Concentration and Transport Time

Flood flow from the reservoir generates a bed shear force that acts on the surface of replenished sediment. The scouring depth over time for the entire amount of replenished sediment can be denoted as $\rho_d V_S$, where V_S is the volume of sediment eroded. The total amount of eroded sediment in the water column is denoted as SV, where S represents the sediment concentration in water of volume V. The amount of sediment scoured from the riverbed approximately equals that suspended in the water column, that is, $SV = \rho_d V_S$. Because of the requirement of dynamic similarity during scouring, the sediment concentration ratio of the model to the prototype λ_s can be written as follows:

$$\lambda_S = \frac{\lambda_{\rho_d}\lambda_{V_S}}{\lambda_V} = \frac{\lambda_{\rho_d}\lambda_L^3}{\lambda_L^3} = \lambda_{\rho_d} \tag{11}$$

Equation (11) indicates that the scale ratio of the model to the prototype for the sediment concentration is equal to the scale ratio of dry density.

Bed elevation variations during river scouring can be described by the sediment continuity equation:

$$\frac{\partial(qS)}{\partial x} + \rho_d\frac{\partial z_b}{\partial t} = 0 \tag{12}$$

where q is the discharge per unit width, x is the distance along the river, z_b is the bed elevation, and t is the time required for bed variation. On the basis of model similarity, Equation (13) can be derived from Equation (12) and expressed as follows:

$$\frac{\lambda_q\lambda_S}{\lambda_x} = \lambda_{\rho_d}\frac{\lambda_{z_b}}{\lambda_t} \tag{13}$$

where $\lambda_x = \lambda_{z_b} = \lambda_h$ and $\lambda_q = \lambda_u\lambda_h = \lambda_h^{3/2}$ for the undistorted model. Based on the relationship between Equations (10), (12), and (13), the time scale ratio of bed elevation variations due to scouring is $\lambda_t = \lambda_h^{1/2}$. In particular, the ratio is the same as the time scale of flow similarity derived from the Froude number. This indicates that the scouring time scale of flood flow is equal to that of sediment transport.

According to the physical model ratio of the prototype to the model (64:1), the model scale ratios could be derived from similarity theory (Table 1). The installation of replenished sediment and the filter structure to serve as the permeable barrier in the physical model is illustrated in Figure 9. Experimental results and analysis are described on the prototype scale in the following section.

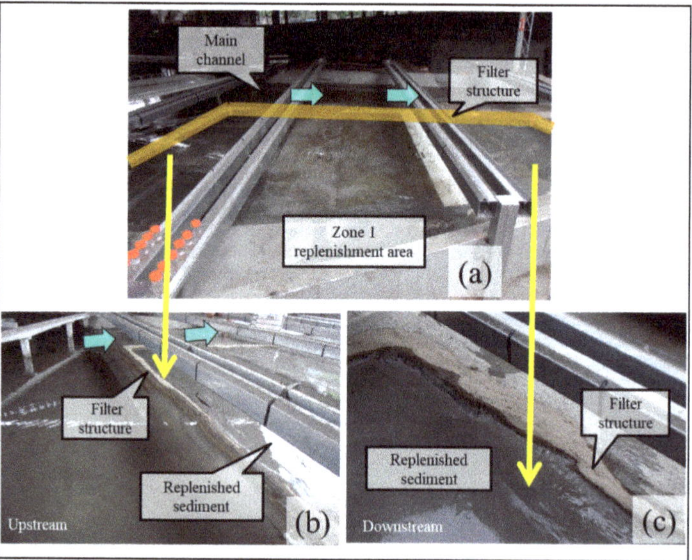

Figure 9. Installation of replenished sediment and filter structure (serving as a permeable barrier): (**a**) Zone 1 in the physical model, (**b**) view from the upstream side, and (**c**) view from the downstream side of the replenishment area.

Table 1. Scale ratio of the flow patterns and sediment properties.

Item	Length (m)	Velocity (m/s)	Discharge (m^3/s)	Time (s)	Coarse Sediment Size (m)	Dry Density (kg/m^3)	Sediment Concentration (mg/L)
Scale	λ_h	${\lambda_h}^{1/2}$	${\lambda_h}^{5/2}$	${\lambda_h}^{1/2}$	λ_h	${\lambda_h}^{1/3.92}$	${\lambda_h}^{1/3.92}$
Ratio	64	8	32768	8	64	2.89	2.89

5. Experimental Results and Analysis

Flood discharges from a reservoir and flood duration are key factors for investigating scouring processes in the sediment replenishment area. On the basis of data regarding flood discharges from the Shihmen Reservoir in the past 10 years, we observed that the flood discharges were higher than 500 and 1000 m^3/s for 38 and 22 h, respectively, in all flood events. Apart from 2-, 10-, and 20-year return period flood discharges, we considered lower discharges of 500 and 1000 m^3/s as the inflow discharges for practical purposes in the physical model tests. These discharges generated overbank flow to flood over the replenishment area on the floodplain.

Figure 10 illustrates the outflow sediment concentration measured against time in test cases involving various inflow discharges (from 500 to 6100 m^3/s) at the downstream end (Section 85) of the physical model. In this figure, the hydrographs in the first 1.5-h period are enlarged to enhance the visualization of changes in the measured sediment concentration. In the test case that involved a Q value of 6100 m^3/s, the peak sediment concentration (12,000 mg/L) was observed at 0.07 h because a higher discharge could generate a higher flow velocity to scour most of the replenished sediment. For comparison, in the test case that involved a Q value of 500 m^3/s, the peak sediment concentration (24,000 mg/L) occurred at 0.7 h. Although a higher discharge could generate a higher flow velocity to scour the replenished sediment, the concentration of suspended sediments may not have increased further due to the larger water volume. The peak sediment concentration for the test case that involved the Q value of 6100 m^3/s was lower than that for the test case that involved the Q value of 500 m^3/s. However, the peak sediment discharge (product of flow discharge and sediment concentration) should be considerably increased due to higher flow discharge. Moreover, the hydrographs revealed that major variations obviously occurred in the first 1.5 h. After 1.5 h, the measured sediment concentrations of all the test cases were lower than 6000 mg/L. This concentration can be managed by the Bansin water treatment plant. Consequently, only the first 1.5 h would be required for complete scouring in the replenishment area.

The volume of scoured sediment could be obtained from the sediment concentration measured at the downstream boundary in the physical model. The scour ratio could be defined as the ratio of the cumulative volume of the scoured sediment to the total amount of replenished sediment. As plotted in Figure 11 for each given discharge, except for 6100 m^3/s, the variations in the scour ratio exhibited a linear trend approximately after 3 h. This indicates that the replenished sediment was scoured at a constant rate with time. According to the observed linear trend, the regression equation could be derived (Figure 11), and the corresponding R-squared value revealed good agreement. By using the regression equation for each given discharge, we estimated the time required for complete scouring (i.e., 100% scour ratio) of replenished sediment (with water content of 40%); the estimation results are listed in Table 2. The results indicated that the scour ratio increased with the discharge; complete scouring was achieved within 8 h. Moreover, when the Q value was 6100 m^3/s, the entire amount of the replenished sediment was scoured (i.e., scour ratio = 100%) within 3 h. These results demonstrate that the scour ratio depends on the flood discharge and scour duration.

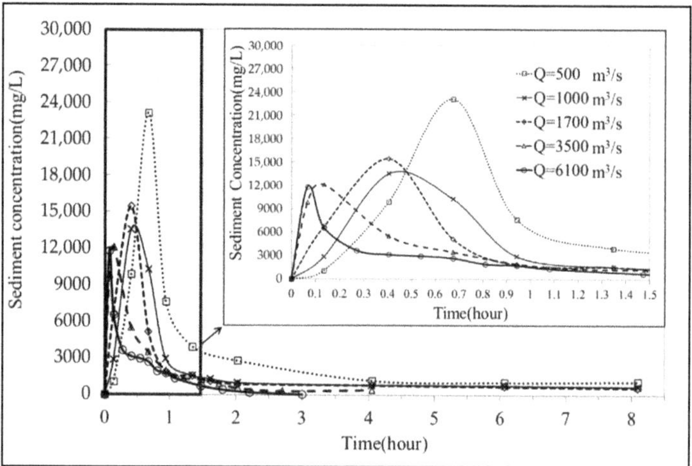

Figure 10. Hydrographs of concentrations of sediment with a water content of 40% at various reservoir flood discharges.

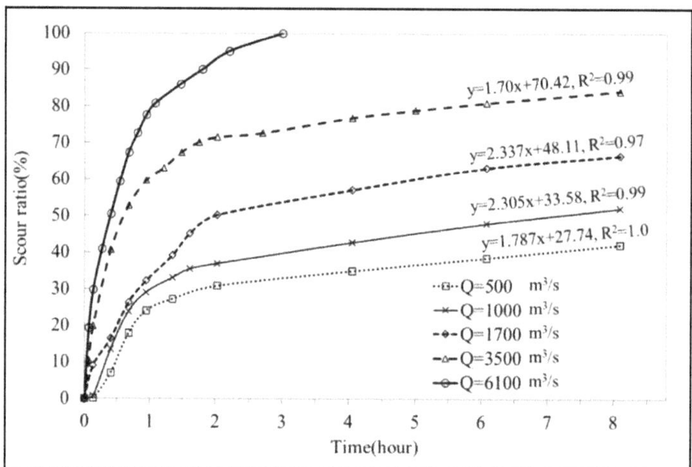

Figure 11. Trend of the scour ratio against time for sediment with the water content of 40% at various reservoir flood discharges.

Table 2. Scour ratio, experimental duration, and estimated duration for achieving a scour ratio of 100% for each given discharge at different water contents.

Discharge (m^3/s)	**500**		**1000**		**1700**		**3500**		**6100**	
Water content (%)	40	80	40	80	40	80	40	80	40	80
Scour ratio (%)/Experimental duration (h)	42/8	57/8	52/8	63/8	67/8	74/8	84/8	98/4	100/3	-
Estimated duration at scour ratio (100%) (h)	40.44	28.55	28.82	26.51	22.20	20.38	17.40	4.25	3.00	-

Figure 12 illustrates hydrographs of the measured sediment concentration against time for the test cases that involved replenished sediment with a water content level of 80%. Because of the lower erodibility of the replenished sediment, the experimental duration was too short to execute the physical model for a Q value of 6100 m^3/s. Therefore, the test cases were conducted under only discharge values ranging from 500 to 3500 m^3/s. In the test case that involved a Q value of 3500 m^3/s, the peak sediment concentration (14,500 mg/L) was observed at 0.07 h. For comparison, in the test case that involved a Q value of 500 m^3/s, the peak sediment concentration (42,000 mg/L) was observed at 0.7 h. Similarly, the peak sediment concentration observed in the test case that involved the Q value of 3500 m^3/s was lower than that observed in the test case that involved the Q value of 500 m^3/s; however, the peak sediment discharge in the test case of 3500 m^3/s was higher. Furthermore, as revealed in the hydrographs, major variations obviously occurred in the first 1.5 h. After 1.5 h, the measured sediment concentrations for all test cases were lower than 3000 mg/L.

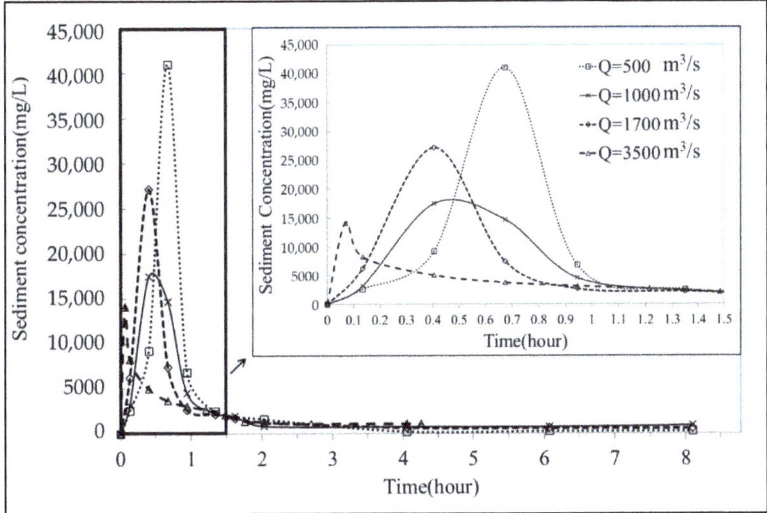

Figure 12. Hydrographs of the concentrations of sediment with a water content of 80% at various reservoir flood discharges.

As plotted in Figure 13, the variations in the scour ratio after approximately 2 h presented a linear trend, indicating the steady rate of the scour ratio at each given discharge with time. Accordingly, the regression equation at each given discharge could be obtained, and the corresponding R-squared values demonstrated good agreement with the measured data. By applying the regression equation to each given discharge value, we estimated the time required for complete scouring of the replenished sediment (with water content of 80%); the estimation results are presented in Table 2. Except for the test case that involved the Q value of 3500 m^3/s, the experiments were conducted within 8 h, and the scour ratio increased with the discharge. When Q was 3500 m^3/s, 4.25 h was required to scour the entire amount of the replenished sediment (i.e., scour ratio = 100%). The experimental results reveal that a higher water content was associated with a higher scour ratio. The estimated time required to achieve a scour ratio of 100% was also longer when the water content was lower. On the basis of these experimental results, we can conclude that the scour ratio depends on the flood discharge, flood duration, and water content of replenished sediment.

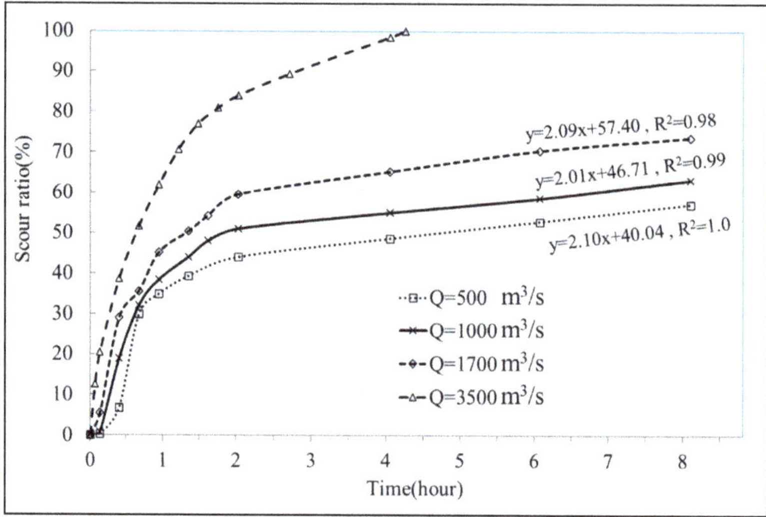

Figure 13. Trends of scour ratio against time for sediment with a water content of 80% at various reservoir flood discharges.

6. Influence of Scouring on Downstream Water Intake

Scouring replenished sediment may influence the water quality at the downstream water intake section. Using the derived experimental data regarding outflow sediment concentrations, we investigated the influence of scouring on downstream water intake by conducting a 2D numerical simulation of suspended sediment transport in the downstream river. The Yuanshan weir is located downstream the Shihmen Reservoir, which is approximately 16.5 km away from the replenishment area. Water is typically withdrawn from the water intake section at the Yuanshan weir to the Bansin water treatment plant, which has a concentration limit of 6000 mg/L. According to the experimental results, the outflow sediment concentrations measured in all the test cases when the water content was 40% were rather high during the first 1.5 h. This may affect the suspended sediment concentration at the water intake and exceed the concentration limit stipulated by the Bansin water treatment plant. Considering the facilities of the treatment plant, the water turbidity can be treated if the sediment concentration decreases to 6000 mg/L or lower [34].

To understand the effect of the scouring of replenished sediment on the downstream water intake, the reduction ratio of sediment concentration should be investigated. By using field measurement data obtained from 2007 to 2016, we derived the relationship between the suspended sediment concentrations at the Yuanshan weir and afterbay weir (nearby replenishment area), and the results are plotted in Figure 14. For modelling suspended sediment transport along the downstream river reach, we applied the 2D numerical model to solve the shallow water equations coupled with the advection–diffusion equation. According to the peak sediment concentration observed for each test case, we simulated the transport of suspended sediment and dissipation of replenished sediment under steady-flow conditions within the river reach, as described in Section 3.2. Figure 14 shows a plot of the simulated sediment concentration in the water intake section of the Yuanshan weir against the concentration in the replenishment area; the figure also shows a plot of the regressed relationship between the field data measured during flood discharge at the Yuanshan weir and the afterbay weir (around the replenishment area). By comparing the regression equations, we observed the simulated results to be close to the measured results. Clearly, the sediment concentration reduction ratio ranged from 91% to 92%. Thus, the reduction ratio of the concentration of suspended sediment scoured from the replenishment area to the concentration of suspended sediment at the Yuanshan weir was approximately 90%. According to the suspended sediment concentration limit (6000 mg/L) imposed at the Bansin water treatment plant, the outflow sediment concentration from the replenishment area was limited to 6667 mg/L.

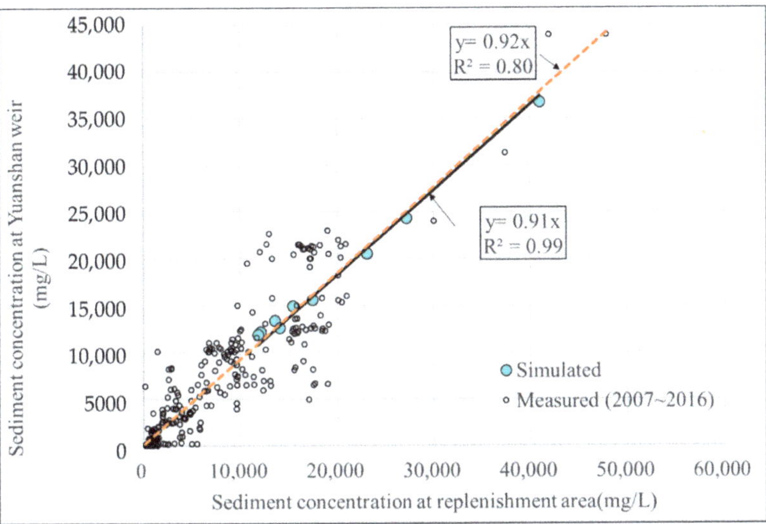

Figure 14. Relationship between sediment concentration in replenishment area and that at the downstream water intake at Yuanshan weir.

7. Summary and Conclusions

Many reservoir management strategies and countermeasures have been implemented for reducing sedimentation effects. Dredging is a commonly used measure to remove mechanically deposited sediment. However, the disposal of dredged sediment can be costly because of engineering, permission, or placement site problems. Nevertheless, dredged sediment can be considered a beneficial resource that can be added to downstream rivers to compensate for dam-induced interruptions of sediment supply. The sediment replenishment method is an experimentally determined strategy in which sediment dredged or excavated from a reservoir can be used for riverbed replenishment to prevent downstream riverbed degradation, reduce coastline erosion, and execute estuary restoration for aquatic habitats. Nevertheless, high concentrations of fine replenished sediment may be scoured during reservoir flood discharges, which may deteriorate the water quality at the water intake in downstream river reaches.

This study selected the Shihmen Reservoir located in Northern Taiwan to investigate the feasibility of reservoir replenishment using fine sediment. Previous studies on sediment replenishment in field or laboratory experiments have mainly focused on coarse materials, with few studies providing examples for fine sediment. In this study, candidate replenishment areas were first analyzed using a validated 2D numerical model to simulate flow velocity fields and water depths. To simulate suspended sediment transport in the model, the 2D shallow water equations along with the advection–diffusion equation was solved numerically in the framework of the finite volume method. On the basis of the analysis of the critical scour velocity, we adopted a flow velocity of 1.5 m/s along with a sufficient water depth to evaluate the incipient motion of both very fine and coarse replenished sediment. Although the concept of critical scour velocity is used herein for illustrative purposes, it indeed provided the preliminary criterion to assess the most feasible replenishment area. By comparing the simulated velocity fields and water depths in the candidate replenishment areas through hydraulic assessment, we identified the most feasible replenishment area, namely Zone 1, as the experimental domain (Figure 6) for the physical model.

For satisfying the similitude of model scaling, the physical model was developed to mimic the scouring process of replenished sediment in the downstream river reach of Shihmen dam. As mentioned, similarities in sediment transport mechanisms engender difficulties in combining coarse and fine materials because of the difference in sediment particle sizes. The dimensionless shear parameter in the Shields diagram was employed for scaling the incipient particle motion by using Equation (8) for coarse sediment in the filter structure. For determining the scouring similarity of the fine replenished sediment, the relationship between flow-critical velocity and sediment dry density was regressed through flume experiments by using Equation (10). Using different water content levels (40% and 80% corresponding to dry densities of 1.2 and 0.84 t/m^3, respectively) of the dredged fine sediment, we performed experiments in the physical model to investigate the incipient motion of replenished sediment with cohesiveness. On the basis of dynamic similarity during scouring, the sediment concentration scale ratio was equal to the dry density scale ratio presented in Equation (11). The time scale ratio of the scouring process was the same as that of the flow similarity derived from the Froude number. This indicates that the scouring time scale of flood flow is equal to that of sediment transport.

According to flood discharge data and frequency analysis data available for the Shihmen Reservoir, five inflow discharge values ranging from 500 to 6100 m^3/s were used in the physical model tests as the upstream boundary conditions, which generated overbank flow during a flood in the replenishment area on the floodplain. The hydrographs obtained from the test cases that involved the water content of 40% (Figure 10) revealed that the variations in the sediment concentration apparently occurred in the first 1.5 h. After 1.5 h, the measured sediment concentrations for all test cases were lower than 6000 mg/L. This concentration value was determined to be suitable for the downstream water intake stipulated by the Bansin water treatment plant. For the test cases that involved the water content of 80% (Figure 12), the measured sediment concentrations after 1.5 h were lower than 3000 mg/L for all test cases. Thus, for all test cases, the entire scouring operation took less than 1.5 h in the replenishment area to affect the downstream water treatment plant.

In the scouring process, the scour ratio was defined as the cumulative amount of scoured volume to the total amount of replenished sediment. Through experiments, the scour ratio was analyzed using relevant factors, including flood discharge and duration and water content of the replenished sediment. The scour ratio presented a linear trend, which provided the basis for obtaining the regression equation for each given discharge value (Figures 11 and 13). Using the regression equation, we estimated the time required to scour the entire amount of the replenished sediment (i.e., scour ratio = 100%), as presented in Table 2. As the scour ratio increased, the inflow discharge increased, and the time required to scour the entire amount of the replenished sediment decreased.

According to the experimental results obtained for the study site, high concentrations of replenished sediment in the outflow could affect the suspended sediment concentrations at the water intake and exceed the concentration limit stipulated by the Bansin water treatment plant. The reduction ratio of sediment concentration was investigated by establishing the relationship between the suspended sediment concentration data obtained at the Yuanshan weir and those obtained at the afterbay weir near the replenishment area (Figure 14). Through the use of the 2D numerical model, the peak sediment concentration in each test case was provided as the input to obtain the simulated sediment concentration at the water intake of the Yuanshan weir. By comparing the regression equations, it was found that the sediment concentration reduction ratio ranged from 91% to 92%. The reduction ratio of the concentration of the suspended sediment scoured from the replenishment area to the concentration of the sediment at the Yuanshan weir was approximately 90%. Thus, the concentration of the outflow sediment produced at the replenishment area was limited to 6667 mg/L, thus matching the concentration limit of 6000 mg/L stipulated by the Bansin water treatment plant.

Author Contributions: F.-Z.L. developed the study methodology, conducted the experiments, and wrote the manuscript. J.-S.L. provided research directions and comments and revised the manuscript. W.-D.G. assisted in the numerical simulations and result analysis. T.S. was responsible for data collection regarding sediment replenishment and for providing practice suggestions.

Funding: This study was funded by the Ministry of Science and Technology, Taiwan, under grant numbers MOST 107-2625-M-002-002, MOST 105-2221-E-002-063-MY3, and MOST 105-2625-M-002-025-MY2.

Acknowledgments: The authors acknowledge the Northern Region Water Resources Office, Water Resources Agency, and Ministry of Economic Affairs, Taiwan, for the valuable data provided for this study. The authors also appreciate the Hydrotech Research Institute of National Taiwan University for providing facilities and technical support.

Conflicts of Interest: The authors declare no conflict of interest.

References

1. Annandale, G.W.; Morris, G.L.; Karki, P. *Extending the Life of Reservoirs: Sustainable Sediment manAgement for Dams and Run-of-River Hydropower*; World Bank Group: Washington, DC, USA, 2016.
2. Kondolf, G.M. Hungry water: Effects of dams and gravel mining on river channels. *Environ. Manag.* **1997**, *21*, 533–551. [CrossRef]
3. Kondolf, G.M.; Gao, Y.X.; Annandale, G.W.; Morris, G.L.; Jiang, E.H.; Zhang, J.H.; Cao, Y.T.; Carling, P.; Fu, K.D.; Guo, Q.C.; et al. Sustainable sediment management in reservoirs and regulated rivers: Experiences from five continents. *Earth's Future* **2014**, *2*, 256–280. [CrossRef]
4. Shields, F.D.; Simon, A.; Steffen, L.J. Reservoir effects on downstream river channel migration. *Environ. Conserv.* **2000**, *27*, 54–66. [CrossRef]
5. Wilcock, P.R.; Kondolf, G.M.; Matthews, W.V.; Barta, A.F. Specification of sediment maintenance flows for a large gravel-bed river. *Water Resour. Res.* **1996**, *32*, 2911–2921. [CrossRef]
6. Sumi, T.; Kantoush, S.A. Comprehensive sediment management strategies in Japan: Sediment bypass tunnels. In Proceedings of the 34th IAHR World Congress, Brisbane, Australia, 26 June–1 July 2011; pp. 1803–1810.
7. Sumi, T.; Kantoush, S.A. Sediment replenishing measures for revitalization of Japanese rivers below dams. In Proceedings of the 34th IAHR World Congress, Brisbane, Australia, 26 June–1 July 2011; pp. 2838–2846.
8. Wang, H.W.; Kondolf, G.M.; Desiree, T.; Kuo, W.C. Sediment management in Taiwan's reservoirs and barriers to implementation. *Water* **2018**, *10*, 1034. [CrossRef]
9. Stähly, S.; Franca, M.J.; Robinson, C.T.; Schleiss, A.J. Sediment replenishment combined with an artificial flood improves river habitats downstream of a dam. *Nat. Sci. Rep.* **2019**, *9*, 5176. [CrossRef] [PubMed]
10. Sumi, T.; Lee, F.Z.; Lai, J.S.; Tan, Y.C.; Huang, C.C. Flushing replenished sediment on floodplain downstream of a reservoir. In Proceedings of the International Symposium on Dams for a Changing World, Kyoto, Japan, 5 June 2012.
11. Battisacco, E.; Franca, M.J.; Schleiss, A.J. Sediment replenishment: Influence of the geometrical configuration on the morphological evolution of channel-bed. *Water Resour. Res.* **2016**, *52*, 8879–8894. [CrossRef]
12. Central Region Water Resources Office. *Report of Implementation and Efficiency of Replenishment in Shigang Dam Downstream River*; Water Resources Agency, Ministry of Economic Affairs: Taichung, Taiwan, 2009. (In Chinese)
13. Kantoush, S.A.; Sumi, T.; Kubota, A. Geomorphic Response of Rivers Below Dams by Sediment Replenishment Technique. In Proceedings of the River Flow 2010, Braunschweig, Germany, 8–10 September 2010; Dittrich, A., Koll, K., Aberle, J., Geisenhainer, P., Eds.; Japan Water Agency: Tokyo, Japan, 2010.
14. Okano, M.; Kikui, M.; Ishida, H.; Sumi, T. Reservoir sedimentation management by coarse sediment replenishment below dams. In Proceedings of the Ninth International Symposium on River Sedimentation, Yichang, China, 18–21 October 2004; pp. 1070–1078.
15. Seto, K.; Sakamoto, T.; Suetsugi, T. Sediment control measures and improvement effects of physical condition and environment by sediment flushing—A case study in the Yahagi dam. In Proceedings of the International Congress on Large Dams, Brasilia, Brazil, 22–29 May 2009; Q.89-R.3.
16. Sumi, T.; Kobayshi, K.; Yamaguchi, K.; Takata, Y. Study on the applicability of the asset management for reservoir sediment management. In Proceedings of the International Congress on Large Dams, Brasilia, Brazil, 22–29 May 2009; Q.89-R.4.

17. Ock, G.; Sumi, T.; Takemon, Y. Sediment replenishment to downstream reaches below dams: Implementation perspectives. *Hydrol. Res. Lett.* **2013**, *7*, 54–59. [CrossRef]
18. Guo, W.D.; Lai, J.S.; Lin, G.F.; Lee, F.Z.; Tan, Y.C. Finite-volume multi-stage scheme for advection-diffusion modeling in shallow water flow. *J. Mech.* **2011**, *27*, 415–430. [CrossRef]
19. Environmental Protection Administration (EPA). Water Pollution Control Act, Taiwan. 2018. Available online: https://law.moj.gov.tw/ENG/LawClass/LawAll.aspx?pcode=O0040001 (accessed on 13 June 2018).
20. Northern Region Water Resources Office. *Research of Arrangement and Planning on Deposition Sediment in Shihmen Reservoir*; Water Resources Agency, Ministry of Economic Affairs: Taoyuan, Taiwan, 2010. (In Chinese)
21. Ziegler, C.K.; Nisbet, B.S. Long-term simulation of fine-grained sediment transport in large reservoir. *J. Hydraul. Eng.* **1995**, *121*, 773–781. [CrossRef]
22. Teisson, C. Cohesive suspended sediment transport: Feasibility and limitations of numerical modeling. *J. Hydraul. Res.* **1991**, *29*, 755–769. [CrossRef]
23. Liu, W.C.; Hsu, M.H.; Kuo, A.Y. Modelling of hydrodynamics and cohesive sediment transport in Tanshui river estuarine system, Taiwan. *Mar. Pollut. Bull.* **2002**, *44*, 1076–1088. [CrossRef]
24. Guo, W.D.; Lai, J.S.; Lin, G.F. Finite-volume multi-stage schemes for shallow-water flow simulations. *Int. J. Numer. Methods Fluids* **2008**, *57*, 177–204. [CrossRef]
25. Hjulstrom, F. Decay of Pressure Fluctuation in the Hyporheic Zone around a Cylinder. *World J. Mech.* **1935**, *6*, 159–168.
26. Fortier, S. Permissible canal velocities. *Proc. Am. Soc. Civil Eng.* **1923**, *51*, 1398–1414.
27. Garey, W.P.; Simon, A. *Physical Basis and Potential Estimation Basis and Potential Estimation Techniques for Soil Erosion Parameters in the Precipitation-Runoff Modeling System (PRMS)*; Water-Resources Investigations Report 84-4218; U.S. Geological Survey: Reston, VA, USA, 1984.
28. Graf, W.H. *Hydraulics of Sediment Transport*; Water Resources Publications, LLC.: Highlands Ranch, CO, USA, 1984.
29. Ettema, R.; Arndt, R.; Roberts, P.; Wahl, T. *Hydraulic Modeling: Concepts and Practice. ASCE Manuals and Reports on Engineering Practice*; American Society of Civil Engineers: Reston, VA, USA, 2000.
30. Wang, Z.; Zhang, S. Model study on emptying flushing in a reservoir with cohesive deposits. *J. Sediment Res.* **1989**, *2*, 62–68. (In Chinese)
31. Mehta, A.J.; Hayter, E.J.; Parker, W.R.; Krone, R.B.; Teeter, A.M. Cohesive sediment transport. I: Process description. *J. Hydraul. Eng.* **1989**, *115*, 1076–1093. [CrossRef]
32. Lai, J.S.; Chang, F.J. Physical modeling of hydraulic desiltation in Tapu Reservoir. *Int. J. Sediment Res.* **2001**, *16*, 363–379.
33. Lai, J.S. Research of threshold velocity on fine sediment in Agongdian Reservoir. *J. Taiwan Water Conserv.* **1998**, *46*, 76–83.
34. Lee, F.Z.; Lai, J.S.; Guo, W.D.; Huang, C.C.; Chang, F.C.; Tan, Y.C. The Impacts of water withdraw at Yuanshanyan during sediment venting in Shihmen Reservoir. *J. Taiwan Agric. Eng.* **2012**, *58*, 80–94.

Article

Case Study: Model Test on the Effects of Grade Control Datum Drop on the Upstream Bed Morphology in Shiting River

Xudong Ma, Lu Wang *, Ruihua Nie, Kejun Yang and Xingnian Liu

State Key Laboratory of Hydraulics & Mountain River Engineering, Sichuan University, Chengdu 610065, China; maxd@scu.edu.cn (X.M.); nierh@scu.edu.cn (R.N.); yangkejun@scu.edu.cn (K.Y.); liuxingnian@126.com (X.L.)
* Correspondence: wanglu@scu.edu.cn

Received: 28 June 2019; Accepted: 6 September 2019; Published: 11 September 2019

Abstract: This paper conducted an undistorted scaled model test (geometric scale λ_L = 1:80; the others are derived scales based on Froude similitude) of a 1.3 km-long river reach in Shiting River, China, investigating the impacts of the grade control datum (GCD, defined as the crest elevation of the grade control structure) drop on the upstream bed morphology. Three GCDs and six flood events (occurrence probability 1–50%, discharge = 600–4039 m^3/s) were tested on the model. Experimental results indicate that, for a constant GCD, the increase in discharge deepens and widens the upstream river bed. For a lower GCD, the increase in channel depth and width caused by the increasing discharge is greater. For each discharge, the decrease in GCD induces a lower and steeper upstream river bed, widening the upstream main channel. For lower discharge, the GCD drop induces a head cut erosion area upstream of the grade control structure and the head cut erosion area is filled by the upstream sediment when the flow discharge is high. Experimental data also indicate that the maximum general scour depth at the 105th Provincial Highway Bridge is approximately independent of discharge for a constant GCD. For a lower GCD, the general scour depth at the 105th Provincial Highway Bridge increases slightly with discharge.

Keywords: bed morphology; bridge scour; general scour; grade control

---

1. Introduction

Scour is a main design concern for hydraulic structures such as bridge or wind turbine foundations [1–4], buried pipelines [5–7], weirs and sills [8–12], etc. The scour at instream structures can be classified into three types: general scour, constriction scour and local scour [13–15]. The general scour is normally caused by the evolution of the river bed and is independent of the instream structure. The constriction scour occurs where an instream structure narrows the flow section. The local scour is caused directly by the change of flow pattern induced by the existence of the instream structure.

Bed degradation usually occurs when the bed erosion rate is greater than the upstream sediment replenishment rate. It can induce general scour at the foundations of instream structures and destabilize riverbanks, threatening the security of both the public and private properties [16]. Grade control structures (GCSs) such as submerged weirs, bed sills and check dams are common countermeasures for bed degradation [9–11,17–19]. They can raise the upstream water level and reduce the flow capacity for sediment transport, preventing the upstream river bed from being excessively degraded. Properly designed GCSs play a role as a grade control datum (GCD) for the upstream river reach [16,20]. The GCD is usually the crest elevation of the GCS which is lower than the upstream bed level.

However, the flow over GCSs can cause local scour [8–12], and the downstream general scour may continue if there are no further downstream GCSs, leading to structural damages or failures (Figure 1). The failure of a GCS induces a drop in GCD, accelerating the adverse impacts they are initially built to prevent [16]. Thus, it is important to understand the scour process at GCDs for safe design. Also, understanding the impacts of GCD drop on the upstream bed morphology can help to assess the stability of the upstream riverbanks and instream structures if the GCS fails.

Figure 1. Scour at the grade control structure (GCS) downstream of the 105th Provincial Highway Bridge after the flood event on 9 July 2013 (discharge $Q = 2710$ m^3/s).

The local scours at GCSs have been extensively studied and many empirical equations have been proposed [8,10–12,21–38]. Some studies also investigated the impacts of GCSs on the river bed profile [17,18]. Although very important, no studies have investigated the effects of GCS failure (i.e., GCD drop) on the upstream river reach. This paper conducted a scaled model test based on a 1.3 km-long river reach upstream of a GCS in Shiting River, China, investigating the effects of GCD drop on the upstream riverbed morphology.

2. Background

Shiting River is the first tributary of Tuo River. It originates from the Longmen mountain located in Sichuan Province, China, having a length of 131.7 km and a basin area of 1501 km^2. The mountainous reach (upstream reach) of the Shiting River is 61.1 km long with a gradient between 92.3–10.4‰. The gradient of the river reach downstream of the mountain decreases gradually and ranges between 12 and 2.5‰ before it joins the Tuo River. Since the Wenchuan Earthquake (Ms 8.0) in 2008, serious bed degradation occurred in the piedmont reach of Shiting River, significantly exposing and endangering the foundations of instream infrastructures [39,40]. Thus, grade control structures (GCSs) were extensively used in this river reach as countermeasures for bed degradation.

The studied reach is about 13 km downstream of the mountainous area with a gradient of around 4‰, in which the 105th Provincial Highway Bridge is located (Figure 2). The studied reach is about 1.3 km upstream of the 105th Provincial Highway bridge, and has a channel width between 300–400 m. This reach is a quasi-straight channel and has no instream structures upstream of the 105th Provincial Highway Bridge. Thus, the impacts of stream curve and other instream structures on the bed morphology are negligible. As mentioned previously, significant bed degradation occurred in the piedmont reach of Shiting River after 2008 [39,40]. Thus, an 18 m high grade control structure was built upstream of the studied reach (2 km upstream of the 105th Provincial Highway Bridge) for stabilizing the upstream riverbed. The flood events during the period of 2009–2012 degraded the river bed at the 105th provincial Hyghway Bridge to an elevation of 539 m (544 m before the Wenchuan Earthquake). In 2012, for protecting the upstream river reach and the bridge, a grade control structure (GCS) with a downstream stilling basin (for minimizing the downstream local scour) was built downstream of the 105th Provincial Highway Bridge with a crest at z = 539 m (i.e., grade control datum, GCD). However, as the bed degradation downstream of the GCS continued after the flood event in July 2013 (Q = 2710 m^3/s), the bed elevation downstream of the GCS became 527 m, exposing the GCS foundation and causing structural failure (Figure 1). In order to assess the impacts of GCD drop (or GCS failure) on the upstream bed morphology, as well as the general scour depth at the 105th Provincial Highway Bridge, a scaled model test is conducted in this study.

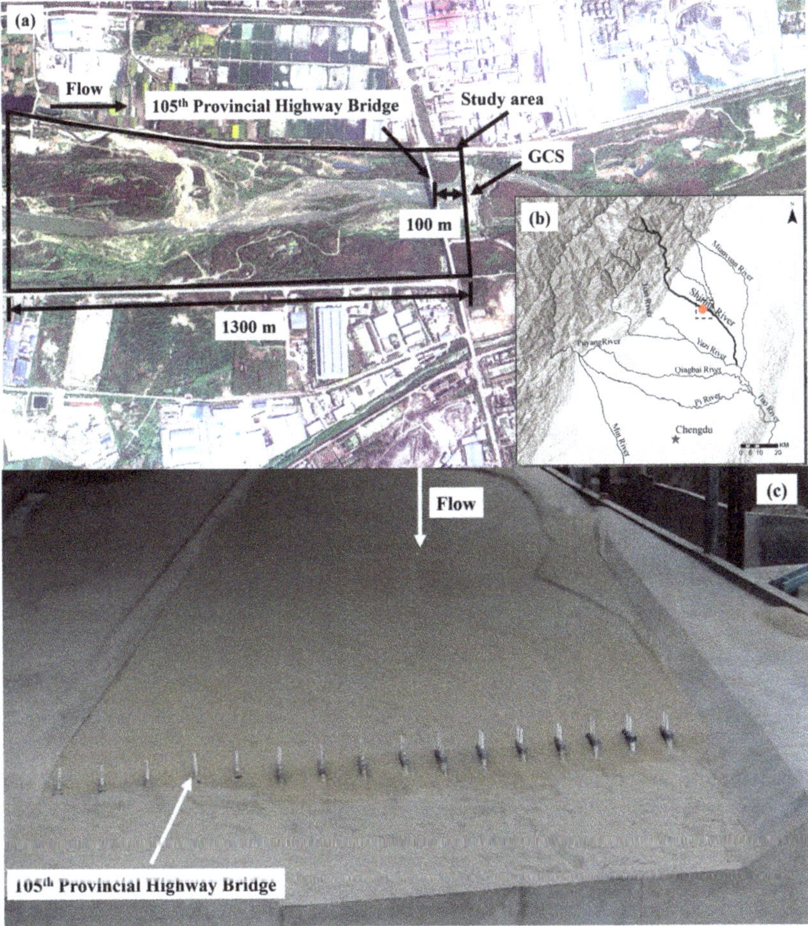

Figure 2. Plan view of the studied reach of Shiting River (**a**); location of Shiting River (**b**); a photo of the model (**c**).

3. Experimental Setup

An undistorted model based on the prototype shown in Figure 2 was built with a geometry scale λ_L = 1:80 in the State Key Laboratory of Hydraulics and Mountain River Engineering, Sichuan University, China. The scaled model was implemented into a water-recirculating system. The bank revetments were built in concrete and had a 0.5 m-deep non-cohesive sediment bed. The flow passed through a triangular weir for flowrate measurement before entering the model, and exited the model through the concrete grade control datum as a free flow which is the same as that of the prototype.

Froude similitude [41] was adopted for the flow motion. Thus, the velocity scale is $\lambda_U = \lambda_L{}^{0.5} = 8.94$, the discharge scale is $\lambda_Q = \lambda_L{}^{2.5} = 1{:}57243$ and the time scale is $\lambda_t = \lambda_L{}^{0.5} = 8.94$. In order to achieve the similitude of sediment motion, U/U_c in the prototype and model should be the same (U_c is the critical velocity of the sediment entrainment). The Shamov formula (Equation (1)) is commonly used for calculating U_c in a scale model test, as it is simple to use and can provide reliable estimations [42].

$$\frac{U_c}{\sqrt{gd}} = 1.47\left(\frac{h}{d}\right)^{1/6} \tag{1}$$

where g is gravity acceleration, h is approach flow depth, d is sediment size. Based on Equation (1), the sediment size scale is $\lambda_d = \lambda^2{}_{Uc} = \lambda^2{}_U = 1{:}80$. The model sediment was scaled down by $\lambda_d = 1{:}$ 80 from the prototype sediment size distribution based on a field survey in the studied river reach (Figure 3). As there is an 18 m-high grade control structure with a stilling basin (for minimizing the downstream local scour) located about 2 km upstream of the 105th Provincial Highway Bridge, the upstream sediment is blocked by the grade control structure from entering the studied river reach. Therefore, the upstream sediment replenish rate was considered as zero and no sediment was fed during the test.

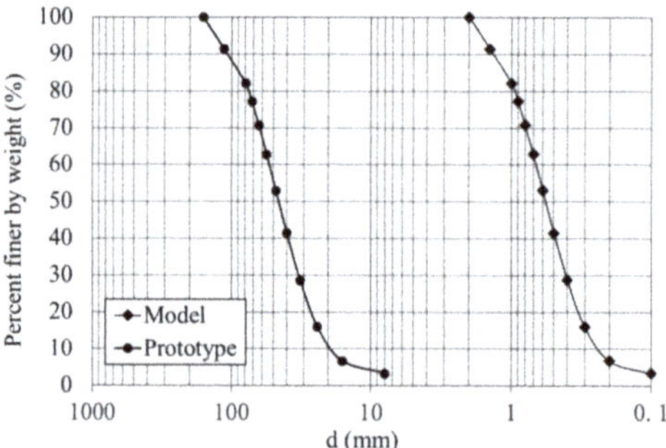

Figure 3. Grain size distribution of the prototype and model sediment.

Six flood events (discharge Q = 600–4039 m^3/s, occurrence probability P = 1%–50%) were tested for three GCDs (z = 527 m, 533 m, 539 m). Among which, z = 539 m is the crest of the current GCS, z = 533 m is the crest elevation of the new GCS design plan, z = 527 m is the bed level downstream of the GCS after the flood event on 9 July 2013.

The tests of each z commenced with an initially flattened sediment bed and the smallest discharge Q. The gradient of the initial flat bed was set at 4‰, which is the average bed gradient of the studied reach after the flood event in July, 2013. The test stopped and the water dried gradually after the scaled flood duration t. Then, the bed profile was measured using a Total Station (Nikon, Japan, DTM-352C). The next test, with a higher discharge, commenced without flattening of the sediment bed.

Before the formal tests, a preliminary test based on the flood event on 9 July 2013 (Q = 2710 m^3/s) was conducted for model calibration. As this model test is aimed to assess the impacts of GCD drop on the upstream bed morphology, only the bed profile upstream of the GCS was measured for calibration. Based on a field survey after the flood event on July 9 2013, the average gradient of the studied river reach was 4‰. The measured model talweg profile of the calibration test has an average gradient of 4.3‰ (multiple correlation coefficient R^2 = 0.91), which is close to that of the prototype. The discrepancy of bed profile between the model and prototype is within ± 0.8 m (prototype vaule), which is acceptable for the large scale prototype of this study. As the scour design for instream structures normally adds a safe value greater than 1 m to the estimated scour depth, the model data is reliable for engineering use.

4. Results and Discussions

The experimental results and conditions are summarized in Table 1. In Table 1, P is the probability of a flood event; Q is the prototype flood event discharge; t is the prototype flood event duration; z is the prototype grade control datum elevation; i is the measured average bed slope (model) upstream of the GCS; d_p is the maximum general scour depth at the site of 105th Provincial Highway Bridge in terms of z = 544 m (i.e., the bed elevation before the Wenchuan Earthquake). In this study, as the detailed hydrograph is not available, the peak flood rate was applied during each flood event. As the flood in this area is caused by intense rain, the flood rising and recession periods are very short and can be neglected. In order to avoid the impacts of local scour and constriction scour on the bed profile, the general scour depth at bridge site was measured upstream of the scour area of the bridge foundation.

Table 1. Summary of the experimental conditions and results.

Group Number	P (%)	Q (m^3/s)	z (m)	i (‰)	d_p (m)	t (h)
1	50	600	539	4.5	4.9	54
2	20	1225	539	4.4	5.2	36
3	10	2095	539	4.3	5.3	36
4	5	2664	539	4.3	5.3	36
5	3.3	2935	539	3.8	5.5	24
6	1	4039	539	3.8	5.7	24
7	50	600	533	8.5	9.4	54
8	20	1225	533	7.7	9.9	36
9	10	2095	533	6.3	10.2	36
10	5	2664	533	5.5	10.6	36
11	3.3	2935	533	5.3	10.7	24
12	1	4039	533	4.9	10.9	24
13	50	600	527	12.6	14.8	54
14	20	1225	527	10.2	15.5	36
15	10	2095	527	8.2	16	36
16	5	2664	527	7.2	16.5	36
17	3.3	2935	527	6.2	17.2	24
18	1	4039	527	4.9	17.6	24

4.1. Talweg Profile Upstream of GCS

Figure 4 shows the talweg profile upstream of the GCS of each GCD z for different flood events. Figure 4 indicates that, for each test, the bed elevation immediately upstream of the GCS is approximately equal to the GCS crest (i.e., GCD). For each z, the upstream talweg elevation is lower with a lower P (higher discharge, Q). This is because a higher flowrate has a greater capacity for sediment transport, causing more erosion on the bed. For z = 539 m, the bed in -150 m $< x < 0$ is flatter with a higher Q, as some of the sediment driven by the approach flow is blocked by the GCS, resulting in an aggraded bed near the GCS. For z = 533 m and z = 527 m, the aggradation in -150 m $< x < 0$ disappears. This is because, for a lower GCD, the sediment above the GCD is flushed downstream over the GCS without any blockage. Figure 4 also indicates that, for a lower z, the difference in the talweg profile caused by increasing Q is larger. The bed incision due to GCD drop can be affected by two factors: (i) the approach flow capacity for sediment transport; (ii) GCS blockage. The bed is stabilized when these two factors reach a balance. As the GCS only blocks the sediment when the GCD is above the bed, for a lower GCD, factors (i) and (ii) reach the balance at a lower bed level.

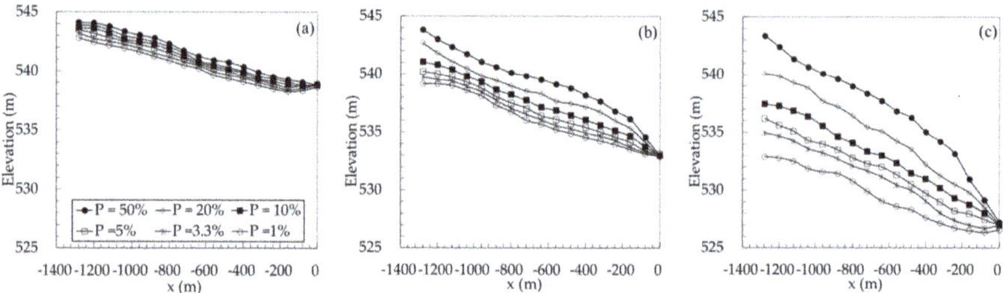

Figure 4. Upstream talweg profile after different flood events for $z = 539$ m (**a**), $z = 533$ m (**b**) and $z = 527$ m (**c**).

Figure 5 indicates that for each Q, the upstream talweg elevation decreases with decreasing z. For $P = 50\%$, there is an abrupt steepening in the talweg profile for $z = 533$ m and 527 m at $x \approx 200$ m and $x \approx 300$ m, respectively. For $P \leq 10\%$, the abrupt steepening in the talweg profile for $z = 533$ m and 527 m disappears. This is because the bed degradation induced by the drop in GCD begins as a head cut erosion process from $x = 0$. For small discharges ($P \geq 20\%$), the sediment transport rate is slow and is unable to fill the head cut erosion area during the flood event. For high discharges ($P \leq 10\%$), the upstream sediment transport rate is high enough to fill the head cut erosion area within the duration of the flood. Figure 5 also indicates that for $P \leq 3.3\%$, the talweg profile of -200 m $< x <0$ is flatter than that of $x < -200$ m for $z = 527$ m. This is because the grade control datum protrudes high enough above the upstream river bed to block the sediment from the upstream, inducing an aggraded bed near the GCS.

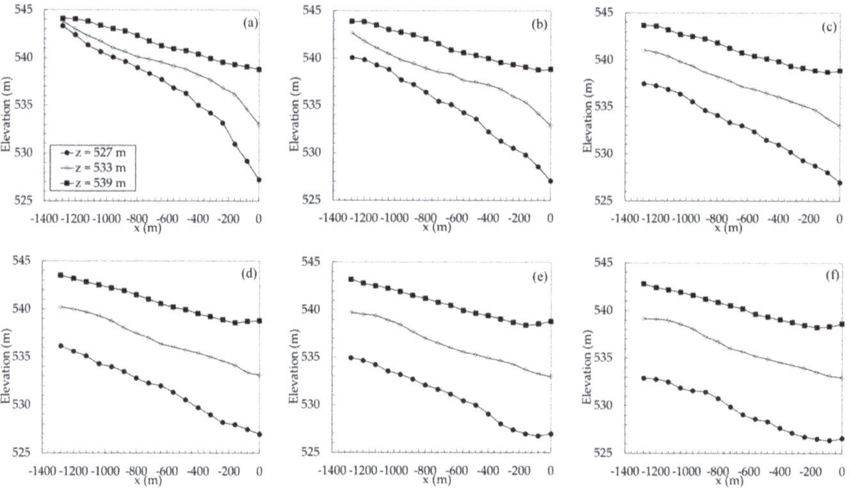

Figure 5. Upstream talweg profile of each GCDs z for $P = 50\%$ (**a**), $P = 20\%$ (**b**), $P = 10\%$ (**c**), $P = 5\%$ (**d**), $P = 3.3\%$ (**e**) and $P = 1\%$ (**f**).

Figure 6 highlights the dependence of the average upstream bed gradient i on the flood discharge Q. Figure 6 shows that i is greater with a lower z. As the GCD drop enlarges the elevation difference between $x = 1300$ m and $x = 0$, the bed tends to develop to be steeper. Figure 6 also indicates that i is approximately independent of Q for $z = 539$ m, but decreases with Q for $z = 533$ m and $z = 527$ m. As shown in Figures 4 and 5, the final bed elevation at $x = 0$ is fixed at the GCD. Thus, the increasing Q erodes more sediment from the upstream bed, resulting in a lower upstream bed level and a smaller average gradient.

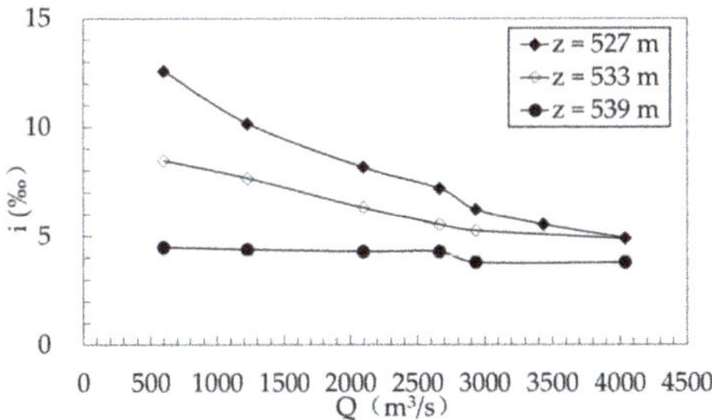

Figure 6. Dependence of average talweg gradient i on discharge Q for different GCDs z.

4.2. Transverse Profile Upstream of GCS

Figure 7 indicates that, for each z, the channel cross-section is deeper and wider with a higher Q. This is because the increase in Q increases the sediment transport capacity of the flow, resulting in greater erosion of the riverbed and riverbanks. Figure 7 also shows that, for a lower z, the difference in the cross-section depth and width caused by increasing Q is greater. More specifically, as shown in Figure 8, the channel cross-section of each discharge is wider and deeper for a lower z. The GCD drop can cause significant bed incision as shown in Figures 4 and 5, inducing mass failures in the riverbank as the incised bed can not support the bank material. Our experimental observations also confirm that, for a lower GCD, more mass failure occurs in the bank.

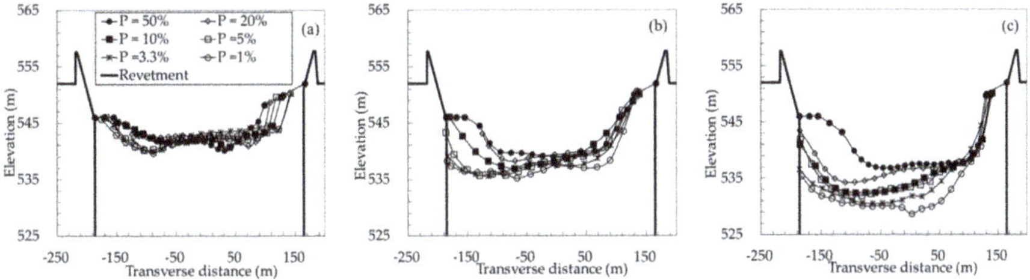

Figure 7. Transverse profile of different flood events (P = 1%–50%) at x = −800 m for z = 539 m (**a**), z = 533 m (**b**) and z = 527 m (**c**).

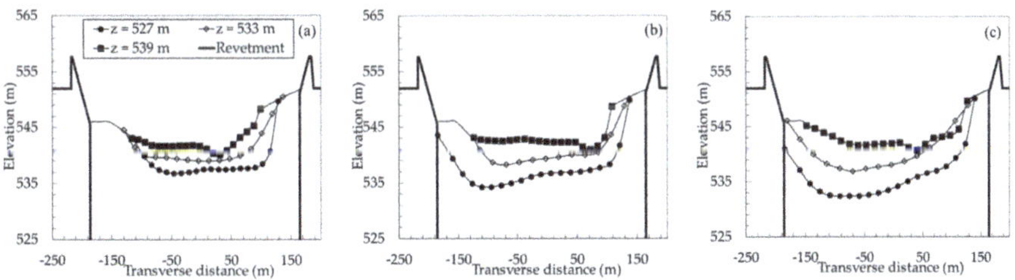

Figure 8. *Cont.*

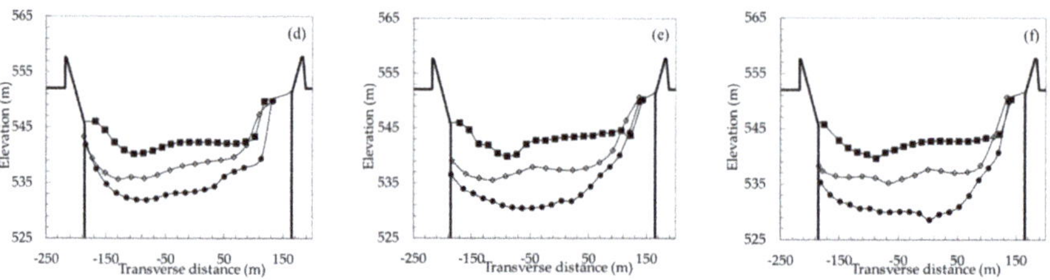

Figure 8. Transverse profile of different GCDs z for $P = 50\%$ (**a**), $P = 20\%$ (**b**), $P = 10\%$ (**c**), $P = 5\%$ (**d**), $P = 3.3\%$ (**e**) and $P = 1\%$ (**f**).

4.3. General Scour Depth at the 105th Provincial Highway Bridge

Figure 9 highlights the dependence of the maximum general scour depth d_p at the 105th Provincial Highway Bridge on the discharge Q for different GCDs. In Figure 9, for $z = 539$ m, with the help of the GCS, d_p is approximately independent of Q. For a lower z, even very small Q ($P = 50\%$) can cause serious general scour at the bridge site ($d_p \approx 9.4$ m and 14.8 m). For $z = 527$ m, i.e., the GCS is completely removed, the d_p can reach up to 17.6 m. For the new GCS design plan $z = 533$ m, the minimum d_p is 9.4 m. Thus, this study suggests building a new GCS with $z \geq 539$ m or a new bridge with much deeper foundations.

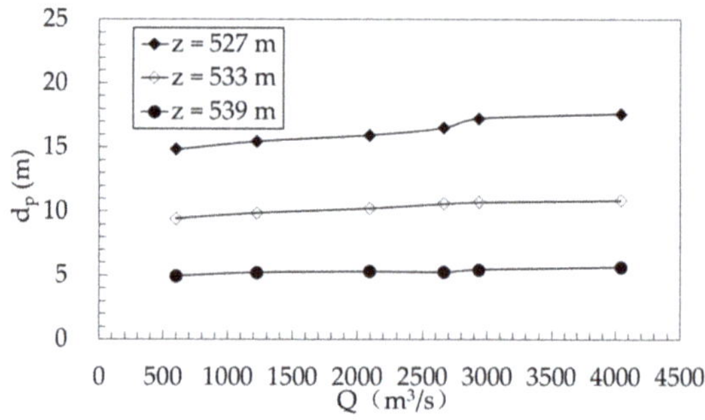

Figure 9. Dependence of the maximum general scour depth d_p at the 105th Provincial Highway Bridge on discharge Q for different GCDs z.

5. Conclusions

Bed degradation is a common river process and can cause general scour depth at instream infrastructures, leading to structural damages or failures. Grade control structures (GCSs) are conventional countermeasures for the general scour at instream structures. However, improper scour design for the GCS can destabilize both the GCS and the upstream river reach. This study conducts a 1:80 scaled model test based on a 1.3 km-long reach of Shiting River, China, investigating the effects of GCS failure (drop in GCD) on the upstream river bed morphology.

The experimental results indicate that, for each GCD z, the upstream river bed elevation decreases with increasing discharge Q. For a lower z, the difference caused by increasing Q is greater. For $z = 533$ m and $z = 527$ m, a head cut erosion area exists upstream of the GCS for $P \geq 20\%$. For $P \leq 10\%$, the head cut erosion area of $z = 533$ m and $z = 527$ m vanishes. The average upstream bed gradient i is approximately independent of Q for $z = 539$ m; i decreases with increasing Q for $z = 533$ m and $z = 527$ m. The experimental results also indicate that, for each z, the increase in Q can widen and deepen the channel. The difference in the channel width and depth caused by increasing Q is greater for a lower z. For each Q, the decrease in z induces a wider and deeper channel.

The maximum general scour depths d_p at the 105th Provincial Highway Bridge are analyzed. For $z = 539$ m, d_p is approximately independent of Q. For a lower z, d_p increases slightly with increasing

Q. For the new GCS design plan (z = 533 m), the general scour depth at the bridge site is still significant (d_p = 9.4–10.9 m). Thus, this study suggests building a GCS with $z \geq 539$ m or a new bridge with much deeper foundations.

Author Contributions: X.M. did all the experiments, data process and part of result analysis. L.W. wrote the paper and did most of the results analysis. R.N., K.Y., X.L. proofread the paper, and provided a lot of valuable advice on experimental operation, result presentation and writing.

Funding: This research is supported by the National Key Research and Development Program of China (2016YFC0402302), the National Natural Scientific Foundation of China (51809187) and the Fundamental Research Funds for the Central Universities (YJ201935).

Acknowledgments: The authors would like to thank the editors of the Special Issue of "Experimental, Numerical and Field Approaches to Scour Research" for their invitations.

Conflicts of Interest: The authors declare no conflict of interest.

References

1. Yang, Y.; Melville, B.W.; Sheppard, D.M.; Shamseldin, A.Y. Live-Bed Scour at Wide and Long-Skewed Bridge Piers in Comparatively Shallow Water. *J. Hydraul. Eng.* **2019**, *145*, 06019005. [CrossRef]
2. Yang, Y.; Melville, B.W.; Sheppard, D.M.; Shamseldin, A.Y. Clear-Water Local Scour at Skewed Complex Bridge Piers. *J. Hydraul. Eng.* **2018**, *144*, 04018019. [CrossRef]
3. Guan, D.; Chiew, Y.-M.; Melville, B.W.; Zheng, J. Current-induced scour at monopile foundations subjected to lateral vibrations. *Coast. Eng.* **2019**, *144*, 15–21. [CrossRef]
4. Guan, D.; Chiew, Y.-M.; Wei, M.; Hsieh, S.-C. Characterization of horseshoe vortex in a developing scour hole at a cylindrical bridge pier. *Int. J. Sediment Res.* **2019**, *34*, 118–124. [CrossRef]
5. Guan, D.; Hsieh, S.-C.; Chiew, Y.-M.; Low, Y.M. Experimental study of scour around a forced vibrating pipeline in quiescent water. *Coast. Eng.* **2019**, *143*, 1–11. [CrossRef]
6. Chiew, Y.M. Mechanics of Local Scour Around Submarine Pipelines. *J. Hydraul. Eng.* **1990**, *116*, 515–529. [CrossRef]
7. Wu, Y.; Chiew, Y.-M. Mechanics of Pipeline Scour Propagation in the Spanwise Direction. *J. Waterw. Port Coast. Ocean Eng.* **2015**, *141*, 04014045. [CrossRef]
8. Guan, D.; Melville, B.W.; Friedrich, H. Local scour at submerged weirs in sand-bed channels. *J. Hydraul. Res.* **2016**, *54*, 172–184. [CrossRef]
9. Wang, L.; Melville, B.W.; Whittaker, C.N.; Guan, D. Scour estimation downstream of submerged weirs. *J. Hydraul. Eng.* **2019**. [CrossRef]
10. Wang, L.; Melville, B.W.; Guan, D. Effects of Upstream Weir Slope on Local Scour at Submerged Weirs. *J. Hydraul. Eng.* **2018**, *144*, 04018002. [CrossRef]
11. Wang, L.; Melville, B.W.; Guan, D.; Whittaker, C.N. Local Scour at Downstream Sloped Submerged Weirs. *J. Hydraul. Eng.* **2018**, *144*, 04018044. [CrossRef]
12. Guan, D.; Melville, B.W.; Friedrich, H. Live-Bed Scour at Submerged Weirs. *J. Hydraul. Eng.* **2015**, *141*, 04014071. [CrossRef]
13. Breusers, H.; Raudkivi, A.J. *Scouring*; A.A. Balkema: Rotterdam, The Netherlands, 1991.
14. Hoffmans, G.J.; Verheij, H.J. *Scour Manual*; CRC Press: Boca Raton, FL, USA, 1997; Volume 96.
15. Melville, B.W.; Coleman, S.E. *Bridge Scour*; Water Resources Publication: Highlands Ranch, CO, USA, 2000.
16. Radspinner, R.; Diplas, P.; Lightbody, A.; Sotiropoulos, F. River Training and Ecological Enhancement Potential Using In-Stream Structures. *J. Hydraul. Eng.* **2010**, *136*, 967–980. [CrossRef]
17. Martín-Vide, J.P.; Andreatta, A. Disturbance Caused by Bed Sills on the Slopes of Steep Streams. *J. Hydraul. Eng.* **2006**, *132*, 1186–1194. [CrossRef]
18. Martín-Vide, J.P.; Andreatta, A. Channel degradation and slope adjustment in steep streams controlled through bed sills. *Earth Surf. Process. Landf.* **2009**, *34*, 38–47. [CrossRef]
19. Wang, L.; Melville, B.W.; Whittaker, C.N.; Guan, D. Effects of a downstream submerged weir on local scour at bridge piers. *J. Hydro Environ. Res.* **2018**, *20*, 101–109. [CrossRef]
20. Jeon, J.; Lee, J.Y.; Kang, S. Experimental Investigation of Three-Dimensional Flow Structure and Turbulent Flow Mechanisms Around a Nonsubmerged Spur Dike with a Low Length-to-Depth Ratio. *Water Resour. Res.* **2018**, *54*, 3530–3556. [CrossRef]

21. Ben Meftah, M.; Mossa, M. Scour holes downstream of bed sills in low-gradient channels. *J. Hydraul. Res.* **2006**, *44*, 497–509. [CrossRef]
22. Pagliara, S.; Kurdistani, S.M. Scour downstream of cross-vane structures. *J. Hydro Environ. Res.* **2013**, *7*, 236–242. [CrossRef]
23. Pagliara, S.; Mahmoudi Kurdistani, S. Clear water scour at J-Hook Vanes in channel bends for stream restorations. *Ecol. Eng.* **2015**, *83*, 386–393. [CrossRef]
24. Tregnaghi, M.; Marion, A.; Bottacin-Busolin, A.; Tait, S.J. Modelling time varying scouring at bed sills. *Earth Surf. Process. Landf.* **2011**, *36*, 1761–1769. [CrossRef]
25. Tregnaghi, M.; Marion, A.; Coleman, S. Scouring at Bed Sills as a Response to Flash Floods. *J. Hydraul. Eng.* **2009**, *135*, 466–475. [CrossRef]
26. Tregnaghi, M.; Marion, A.; Coleman, S.; Tait, S. Effect of Flood Recession on Scouring at Bed Sills. *J. Hydraul. Eng.* **2010**, *136*, 204–213. [CrossRef]
27. Tregnaghi, M.; Marion, A.; Gaudio, R. Affinity and similarity of local scour holes at bed sills. *Water Resour. Res.* **2007**, *43*. [CrossRef]
28. Bormann, N.E.; Julien, P.Y. Scour Downstream of Grade-Control Structures. *J. Hydraul. Eng.* **1991**, *117*, 579–594. [CrossRef]
29. D'Agostino, V.; Ferro, V. Scour on Alluvial Bed Downstream of Grade-Control Structures. *J. Hydraul. Eng.* **2004**, *130*, 24–37. [CrossRef]
30. Lenzi, M.A.; Marion, A.; Comiti, F. Local scouring at grade-control structures in alluvial mountain rivers. *Water Resour. Res.* **2003**, *39*, 1176. [CrossRef]
31. Scurlock, S.M.; Thornton, C.I.; Abt, S.R. Equilibrium Scour Downstream of Three-Dimensional Grade-Control Structures. *J. Hydraul. Eng.* **2011**, *138*, 167–176. [CrossRef]
32. Lenzi, M.A.; Marion, A.; Comiti, F.; Gaudio, R. Local scouring in low and high gradient streams at bed sills. *J. Hydraul. Res.* **2002**, *40*, 731–739. [CrossRef]
33. Gaudio, R.; Marion, A.; Bovolin, V. Morphological effects of bed sills in degrading rivers. *J. Hydraul. Res.* **2000**, *38*, 89–96. [CrossRef]
34. Marion, A.; Tregnaghi, M.; Tait, S. Sediment supply and local scouring at bed sills in high-gradient streams. *Water Resour. Res.* **2006**, *42*, W06416. [CrossRef]
35. Marion, A.; Lenzi, M.A.; Comiti, F. Effect of sill spacing and sediment size grading on scouring at grade-control structures. *Earth Surf. Process. Landf.* **2004**, *29*, 983–993. [CrossRef]
36. Lenzi, M.A.; Marion, A.; Comiti, F. Interference processes on scouring at bed sills. *Earth Surf. Process. Landf.* **2003**, *28*, 99–110. [CrossRef]
37. Lu, J.-Y.; Hong, J.-H.; Chang, K.-P.; Lu, T.-F. Evolution of scouring process downstream of grade-control structures under steady and unsteady flows. *Hydrol. Process.* **2012**, *27*, 2699–2709. [CrossRef]
38. Guan, D.; Melville, B.W.; Friedrich, H. Flow Patterns and Turbulence Structures in a Scour Hole Downstream of a Submerged Weir. *J. Hydraul. Eng.* **2014**, *140*, 68–76. [CrossRef]
39. Fan, N.; Nie, R.; Wang, Q.; Liu, X. Dramatic undercutting of piedmont rivers after the 2008 Wenchuan Ms 8.0 Earthquake. *Sci. Rep.* **2016**, *6*, 37108. [CrossRef]
40. Nie, R.; Wang, X.; Liu, F.; Wang, Q.; Fan, N.; Liu, X. Study on Fluvial Processes of Piedmont Rivers Damaged by Strong Earthquakes. *Adv. Eng. Sci.* **2018**, *50*, 105–111. [CrossRef]
41. ASCE. *Hydraulic Modeling: Concepts and Practices*; ASCE: Reston, VA, USA, 2000. [CrossRef]
42. Chien, N.; Wan, Z. *Mechanics of Sediment Transport*; ASCE Press: Reston, VA, USA, 1999. [CrossRef]

Article

Vortex Evolution within Propeller Induced Scour Hole around a Vertical Quay Wall

Maoxing Wei [1], Nian-Sheng Cheng [1,*], Yee-Meng Chiew [2] and Fengguang Yang [3]

[1] Ocean College, Zhejiang University, Zhoushan 316021, China
[2] School of Civil and Environmental Engineering, Nanyang Technological University, Singapore 639798, Singapore
[3] State Key Laboratory of Hydraulics and Mountain River Engineering, Sichuan University, Chengdu 610065, China
* Correspondence: nscheng@zju.edu.cn

Received: 24 May 2019; Accepted: 8 July 2019; Published: 25 July 2019

Abstract: This paper presents an experimental study on the characteristics of the propeller-induced flow field and its associated scour hole around a closed type quay (with a vertical quay wall). An "oblique particle image velocimetry" (OPIV) technique, which allows a concurrent measurement of the velocity field and scour profile, was employed in measuring the streamwise flow field (jet central plane) and the longitudinal centerline scour profile. The asymptotic scour profiles obtained in this study were compared with that induced by an unconfined propeller jet in the absence of any berthing structure, which demonstrates the critical role of the presence of the quay wall as an obstacle in shaping the scour profile under the condition of different wall clearances (i.e., longitudinal distance between propeller and wall). Moreover, by comparing the vortical structure within the asymptotic scour hole around the vertical quay wall with its counterpart in the case of an open quay (with a slope quay wall), the paper examines the effect of quay types on the formation of the vortex system and how it determines the geometrical characteristic of the final scour profile. Furthermore, the temporal development of the mean vorticity field and the vortex system are discussed in terms of their implications on the evolution of the scour hole. In particular, comparison of the circulation development of the observed vortices allows a better understanding of the vortex scouring mechanism. Energy spectra analysis reveals that at the vortex centers, their energy spectra distributions consistently follow the −5/3 law throughout the entire scouring process. As the scour process evolves, the turbulent energy associated with the near-bed vortex, which is responsible for scouring, is gradually reduced, especially for the small-scale eddies, indicating a contribution of the dissipated turbulent energy in excavating the scour hole. Finally, a comparison of the near-bed flow characteristics of the average kinetic energy (AKE), turbulent kinetic energy (TKE), and Reynolds shear stress (RSS) are also discussed in terms of their implications for the scour hole development.

Keywords: propeller jet; confined propeller scour; energy spectra; turbulence; vortex system

---

1. Introduction

During the berthing and deberthing processes, the quay structure is often present in close proximity to the ship propeller, and the resulting local scour hole that forms the base of the quay wall is a growing concern, as it may cause structural instability or even failure. The Permanent International Association of Navigation Congresses [1] reported that ship propeller-induced jet flow has been recognized as the main cause of scour around the quay structures. With reference to the impact of propeller jets, Sumer and Fredsøe [2] stated that the quay structures can be categorized into two principal types, closed and open quays. The former is characterized by a vertical solid wall that is constructed in the berth front to resist the horizontal load from the landfill and any other live loads. The latter, on the other hand, consists of a slope foundation, above which the quay slab is supported by a group of piles, columns or lamellar walls.

The scour problems caused by propeller jets in both types of quay have received extensive attention over the past decades. For the closed quay, Stewart [3] and Hamill et al. [4] systematically investigated the scouring action related to a stern propeller whose axis is perpendicular to the quay wall. Their results suggested that the maximum scour depth was significantly augmented when compared to that in the absence of any quay structures. Furthermore, they reported that the maximum scour depth exhibited a monotonically decreasing trend with the increasing wall clearance (=the longitudinal distance between the propeller and quay wall). Following a similar experimental configuration to that of Stewart [3] and Hamill et al. [4], Ryan [5] further extended their work by introducing the effect of a rudder in his study. As for the open quay, Sleigh [6] and Cihan et al. [7] experimentally investigated propeller-induced scour around an open quay with the focus on erosion on the quay slope itself. More recently, Wei and Chiew [8] and Wei et al. [9] examined the characteristics of the local scour hole around the toe of the quay slope, showing that the maximum scour depth first increases and then decreases with increasing toe clearance (=longitudinal distance between propeller and slope toe) until the quay effect becomes insignificant. To date, various empirical equations have been proposed to determine the maximum scour depth for both the closed and open quays. Although of important practical use, these studies have shed limited light on the underlying scouring mechanisms due to a lack of detailed flow field data within the scour hole, which is crucial to understanding the effect of the flow structure and the type of quay on the development of the scour.

In the case of the open quay, Wei and Chiew [10] made the first attempt at investigating the mean flow and turbulence characteristics within the developing scour hole, in which the scouring mechanism was found to be subjected to the jet diffusion and quay obstruction effects. Their relative dominance, which is dependent on the magnitude of the toe clearance, dictates the characteristics of the final scour hole. For the closed quay, Wei and Chiew [11] detailed the flow properties associated with the impingement behavior between the propeller jet and a vertical quay wall but in the absence of an erodible sand bed. Their results evidence the complex flow nature of the impinging jet. Thus far, few, if any, studies have been carried out to quantify how an impinging propeller jet flow induces the local scour hole around a vertical quay wall, and therefore a comprehensive understanding of the underlying physics remains elusive. For this reason, the main objective of the current study is to furnish a complementary investigation on this subject, which is of great importance for both the fundamental understanding and potential practical application to scour protection.

In what follows, the experimental setup and methodology are first introduced. Then, the asymptotic scour profiles and the associated flow patterns are discussed with four different wall clearances. Moreover, the temporal development of the vortex system within the developing scour hole is presented, together with a comparison of the circulation associated with each individual vortex. Furthermore, an energy spectra analysis of the time series of velocity fluctuations is conducted to examine the turbulent energy dissipation and its implication for the scouring process. Finally, near-bed flow characteristics are also discussed in terms of their correlation with the scouring development.

2. Experimental Setup and Methodology

The experiments were conducted in a straight flume that is 11 m long, 0.6 m wide, and 0.7 m deep, in the Hydraulic Modeling Laboratory, Nanyang Technological University. The glass-sided flume walls enabled optical observations through the use of PIV techniques. The test section was located at a distance of 7 m downstream from the flume entrance. A five-bladed propeller with an overall diameter of D_p = 7.5 cm and a hub diameter of D_h = 1.0 cm was used in this study. The propeller rig was mounted on an appropriately designed movable carriage that spanned transversely across the flume and could be moved along the longitudinal rail installed on the sidewalls of the flume. In this way, the propeller was able to operate at different clearances from the model quay wall, namely the wall clearance X_w, which is defined as the longitudinal distance between the propeller face and vertical wall.

The quay model located downstream of the propeller was built using an acrylic plate with a dimension of 60 cm (height = water level) × 60 cm (width = flume width) × 2 cm (thickness). A 10 cm thick sand bed with uniform sediment of d_{50} (=0.45 mm) was placed on the bottom of the flume. The still water depth above the sand bed was 0.5 m, which is reasonably deep, such that the effect of the free surface is negligible. Before the commencement of each test, the sand bed was carefully prepared and leveled to minimize the compaction difference among different test runs. The experimental setup is shown in Figure 1, in which a right-handed coordinate system (*o-xyz*) is adopted, with the origin located at the undisturbed bed level and directly beneath the propeller face plane. The x-axis is streamwise-oriented along the bottom centerline, the y-axis spanwise-oriented toward the starboard, and the z-axis along the upward vertical. Accordingly, the mean velocity components in the direction of (x, y, z) are represented by (u, v, w). The PIV system comprised a 5-W air-cooling laser with a wavelength of 532 nm as the light source and a high-speed camera. The laser was placed on top of the flume and the beam emitted from the laser source passed through the optics, resulting in a laser light fan of 1.5 mm thickness being cast down into the water. The laser sheet was set to align with the propeller axis of rotation, which is in the vertical plane of symmetry of the flume (see Figure 1a). In this way, the streamwise flow data were collected from the sectional view of the jet central plane (see Figure 1b). Meanwhile, using GetData Graph Digitizer Software, the centerline scour profile was determined from the illuminated line where the laser sheet intersects the sand bed. The high-speed camera used (Phantom Miro 320 with Nikkor 50 mm f/1.4 lens) had a maximum resolution of 1920 × 1200 pixels, 12-bit depth, and more than 1200 frames per second (fps) sampling rate. A sampling rate of 300 fps was used in this study to ensure that the particle displacement was within 50% overlap between adjacent interrogation windows for cross-correlation analyses. Aluminium particles with d_{50} of 10 μm and specific density of 2.7 were used as seeding particles. The settling velocity of the aluminium particles was estimated to be 92.6 μm/s using Stoke's law, which is negligible compared with the propeller jet velocity. The same particle has been extensively used in previous studies (e.g., Lin et al. [12], Hsieh et al. [13] and Wei et al. [14]) and validated as a satisfactory seeding particle in PIV applications.

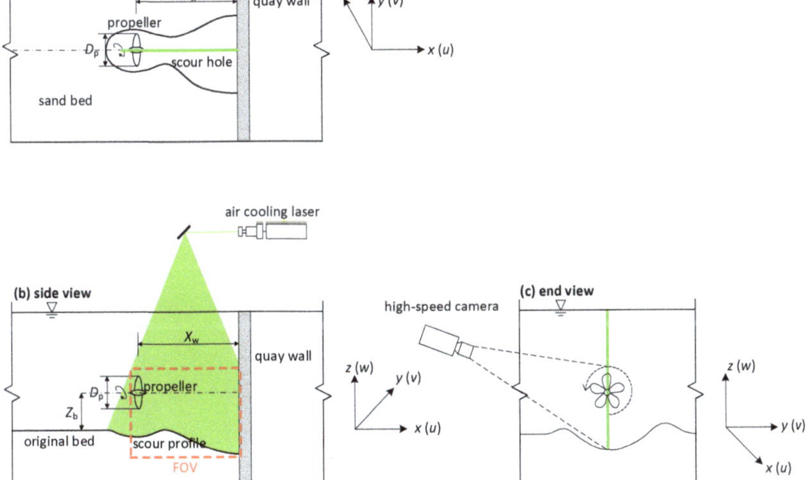

Figure 1. Experimental setup of PIV system, (**a**) top view; (**b**) side view; (**c**) end view, where FOV = field of view.

For a three-dimensional scour hole induced by a propeller jet, one may find it difficult to capture the flow field within the scour hole, since the lateral sediment deposition could block the optical access in a normal PIV operation where the camera is 90° to the laser sheet. For this reason, Wei and Chiew [10] adopted an oblique particle image velocimetry (OPIV) method, in which the camera was tilted with a depression angle, and thus could capture a complete view of the flow field inside the scour hole. By performing an image correction procedure, their error analysis confirmed a reasonable accuracy of the OPIV method in measuring the flow field within a developing three-dimensional scour hole. Guan et al. [15] have also successfully applied the same protocol for measuring the horseshoe vortex evolution of a pier-scour hole. By following a similar approach as in those studies, the camera in this study was set at a depression angle of 20° to cover a rectangular field of view (FOV) in front of the quay wall as shown in Figure 1b,c. To ensure that the entire FOV was in focus, the lens aperture was set to $f = 5.6$ to achieve a sufficient depth of field for the oblique viewing. Before capturing the particle seeded flow field, a calibration plate (with a regularly spaced grid of markers) was placed at the position of the laser sheet in still water, and a calibration image was obtained as shown in Figure 2a, which shows that the coordinates of the markers were distorted from their actual positions due to the oblique viewing. To correct this distortion, a third-order polynomial transformation function was employed, in which the calibration coefficients were obtained by fitting the position of the distorted dots to the regular grid as shown in Figure 2b. Thereafter, the same calibration parameters were applied to correct all the raw PIV images during the postprocessing stage. A comparison of sample images before and after correction is exemplified in Figure 2c,d. In addition, a Butterworth high pass filter [16] was applied to filter out undesirable light reflections (low-frequency) and highlight the seeding particles (high-frequency), as illustrated in Figure 2c,d. The velocity vector fields were then calculated by using the Davis 8.4.0 software, in which a multi-pass iteration process was adopted with the interrogation windows starting from 64×64 pixels to 32×32 pixels.

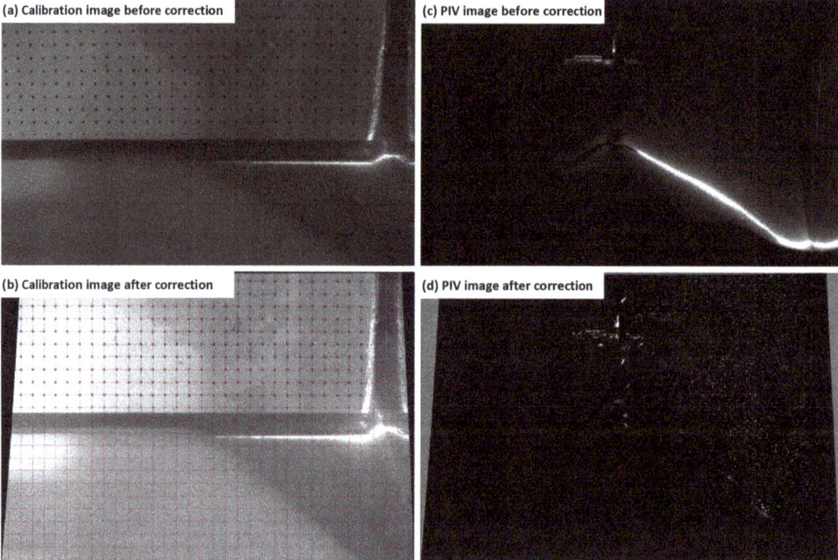

Figure 2. Comparison of calibration and PIV images before and after correction.

To examine the influence of wall clearance on the final scour profile, the tests were conducted at four wall clearances, i.e., $X_w = 1D_p$, $2D_p$, $3D_p$, and $4D_p$, of which the asymptotic scour profiles and their associated flow fields were measured using OPIV. To further examine the temporal development of the flow and scour subject to the wall confinement, a small clearance of $X_w = 2D_p$ was selected as a typical case, for which PIV measurements were carried out during the entire scouring process from the initial instant to the asymptotic state at predetermined time intervals of t = 0, 0.5, 2, 12, 24 h. The other variables were kept as constant, including the clearance height $Z_b = D_p$, and propeller rotational speed n = 300 rpm (revolution per minute), where Z_b is the vertical distance between the propeller axis and the undisturbed sand bed; and n is the propeller rotational speed. The specific test conditions are tabulated in Table 1, in which U_o is the efflux velocity obtained as the maximum mean velocity along the initial efflux plane (i.e., the propeller disk) [5]; and F_o is the densimetric Froude number calculated as $U_o/\sqrt{(\frac{\rho_s-\rho}{\rho})gd_{50}}$, with ρ_s denoting the density of sediment grains, ρ the density of water, and g the gravitational acceleration. For each case, 3683 images were captured and then used for data analysis. According to a convergence analysis, the maximum residual within the FOV of the mean velocity fields was calculated as 0.00028 m/s. Furthermore, to assess the PIV measurement error, the uncertainty calculation was performed in DaVis software, which quantifies the differences between the two interrogation windows mapped onto each other by the computed displacement vector. In the case of the current study ($X_w = 2D_p$), the velocity vector uncertainty inside the scour hole ranges from 0.002 m/s to 0.023 m/s, which is far less than the target flow velocity. Moreover, a detailed error analysis associated with the OPIV method can be found in Wei and Chiew [10].

Table 1. Test conditions.

Test No.	D_p (cm)	d_{50} (mm)	n (rpm)	Z_b/D_p	X_w/D_p	U_o (cm/s)	F_o	t (hr)	Spatial Resolution (cm)	FOV (cm^2)
1	7.5	0.45	300	1	1	36.4	4.31	24	0.27	9 × 22
2	7.5	0.45	300	1	2	36.4	4.31	0–24	0.27	16.5 × 22
3	7.5	0.45	300	1	3	36.4	4.31	24	0.27	24 × 22
4	7.5	0.45	300	1	4	36.4	4.31	24	0.27	31.5 × 22

3. Effects of Wall Clearance and Type of Quay

Figure 3 depicts the comparison of the asymptotic scour profiles between the unconfined ($X_w = \infty$) and confined cases at different wall clearances ($X_w = 1D_p, 2D_p, 3D_p$, and $4D_p$). Also superimposed in the figure is the unconfined profile that was obtained under the same test conditions, but without any quay wall [10]. Hong et al. [17] observed that an unconfined propeller scour hole comprises a small scour hole directly beneath the propeller (due to the ground vortex), a primary scour hole (due to jet diffusion) and a deposition mound. In contrast, Figure 3 shows that the confined scour profiles around a vertical quay wall were significantly altered, as the primary scour hole was truncated in length but enlarged in depth. In general, with the increasing wall clearance, the development of the scour profile exhibits a trend approaching that of its unconfined counterpart. Specifically, Figure 3a–c show that the scour profiles are featured by a single primary scour hole when $X_w \leq 3D_p$. As the wall clearance further increases to $X_w = 4D_p$, the wall effect is less pronounced, and the scour profile evolves into a combination of a primary scour hole and a small scour hole immediately in front of the wall (see Figure 3d). A similar behavior was also observed by Wei and Chiew [8], who experimentally investigated toe clearance effects on the propeller jet induced scour hole around a sloping quay and found that the asymptotic scour profiles could be classified into three types in terms of the toe clearance, namely, near field (featured by a single toe scour hole), intermediate field (featured by a primary and a toe scour holes), and far field (resembling the unconfined case). However, in terms of the development of the maximum scour depth, Figure 3 shows that its magnitude appears to decrease monotonically as the wall clearance increases, which agrees well with what was observed in Hamill et al. [4]. This further confirms the marked difference of the in scouring mechanisms associated with the closed and open quay, as was already pointed out in the introduction section.

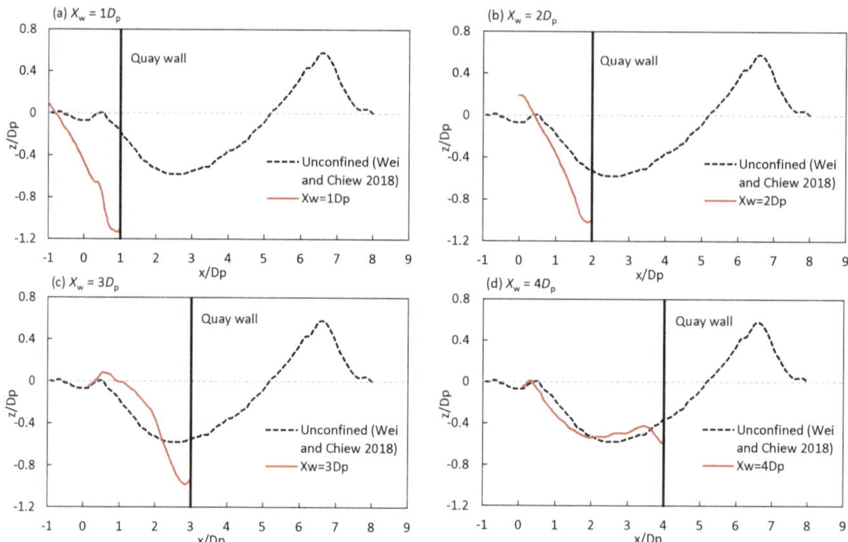

Figure 3. Comparison of asymptotic scour profiles between unconfined and confined cases: (**a**) $X_W = 1D_p$; (**b**) $X_W = 2D_p$; (**c**) $X_W = 3D_p$; (**d**) $X_W = 4D_p$.

To examine the underlying mechanism associated with the observed scour characteristics at different wall clearances, it is useful at this stage to qualitatively discuss the features of the flow pattern within an asymptotic scour hole, which is most explicitly described in the streamline plots as shown in Figure 4. Corresponding to Figure 3, Figure 4a–d depict a comparison of the results obtained in the present study for the four wall clearances, $X_W = 1D_p$, $2D_p$, $3D_p$, and $4D_p$. Additionally, Figure 4e,f, which present the flow fields around two typical open quay cases with toe clearances of $X_t = 2D_p$ and $4D_p$, are also included in the figure for illustrating the effect of the type of quay. Please note that the open quay cases were conducted by Wei and Chiew [10] under the same test conditions (e.g., propeller configurations, bed material, etc.) as those in the current study but with an inclining quay.

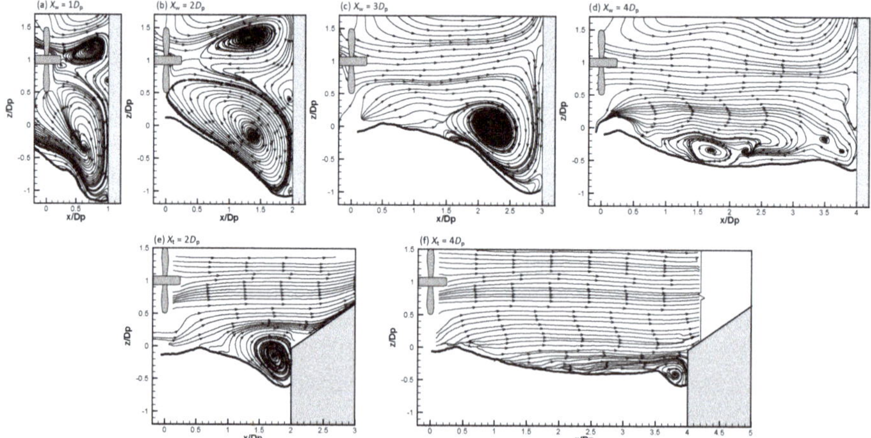

Figure 4. Comparison of streamline plots of the asymptotic state at different wall clearances (vertical quay wall) and toe clearances (slope quay wall): (**a**) $X_W = 1D_p$; (**b**) $X_W = 2D_p$; (**c**) $X_W = 3D_p$; (**d**) $X_W = 4D_p$; (**e**) $X_t = 2D_p$; (**f**) $X_t = 4D_p$. Note: (**e**) and (**f**) are reproduced from Wei and Chiew [10].

In general, Figure 4a,b reveal that the dimension of the asymptotic scour hole is closely related to the size of vortex that resides in it. Specifically, Figure 4a,b show that with small wall clearances, in other words, intense wall effects, there exists a complex vortex system comprising three vortices in front of the wall, in which the primary vortex completely fills the entire scour hole, signifying its dominating role in shaping the scour hole. As the wall effect decreases with the increase in wall clearance, only one vortex that is responsible for scouring persists at $X_w = 3D_p$ (see Figure 4c). When $X_w = 4D_p$, Figure 4d reveals that only a feeble vortex remains, with the formation of a small scour hole at the base of the quay wall because the majority of jet energy has already been dissipated before the jet impinges onto the wall. Based on this observation, one can reasonably infer that in the case of the closed quay, the confinement effect of the wall plays a crucial role in generating and stabilizing the vortex, which in turn facilitates the scouring development. An implication of this phenomenon is that the formation of a well-established vortex, i.e., the driving force of the scouring action, is positively related to the wall effect, which provides an explanation for the observed inverse correlation between the maximum scour depth and wall clearance. On the other hand, in the case of the open quay, only one vortex can be observed around the slope toe, even with the small toe clearance of $X_t = 2D_p$ (see Figure 4e). This is because the sharp edge of the slope toe fixes the separation point at the toe, thus limiting the size of the vortex. Moreover, the open type quay, as its name implies, provides an open space above the slope for jet energy dissipation, which allows much of the jet energy to be deflected upwards along the slope upon the flow separation takes place at the toe. This is especially true for the cases with the small clearance, i.e., the near field scenario defined in Wei and Chiew [10], in which the maximum scour depth is positively correlated with the toe clearance, thus highlighting a significant difference between the scouring mechanism in open and closed quay.

For larger wall/toe clearances, Figure 4d,f show that for the upstream portion, where the jet essentially is unconfined, the primary scour hole is subject to the direct impact associated with the radial expansion of the jet, in which the underlying scouring mechanism resembles that of an unconfined propeller scour. Given the similarities of the scour profile and the associated flow structures shown in Figure 4d,f, one may conclude that as the wall/slope effect decreases with the increasing wall/toe clearance, the type of the quay, either closed (with vertical wall) or open (with slope wall) types, does not matter as much as in the small clearance cases.

4. Temporal Development of Mean Flow Structure

4.1. Evolution of Vortex System

To further examine how the vortex system evolves during the scouring process, the temporal development of the flow field at a near wall clearance case of $X_t = 2D_p$ is exemplified in this section. Figure 5 compares the flow patterns at different scouring phases (t = 0, 0.5, 2, 12, and 24 h) in terms of the mean vorticity contour (left column) and the streamline plot (right column). The out-of-plane vorticity is calculated by using $\omega = \frac{\partial w}{\partial x} - \frac{\partial u}{\partial z}$ and normalized with the propeller diameter D_p and efflux velocity U_o. It should be noted that the mean velocity vectors are also superimposed in the vorticity contour, in which the magnitude and direction of the velocity vectors are calculated as $(u^2 + w^2)^{1/2}$ and $\tan^{-1}(w/u)$, respectively. For the convenience of discussion, the three vortices identified in the streamline plots are hereinafter referred to as V1, V2, and V3, whose center locations are denoted as VC1, VC2, and VC3, respectively.

Before a close examination of Figure 5, it would be helpful to revisit the flow structures associated with the free expanding jet and the confined jet in the presence of a quay wall alone. Several previous studies have shown that a free expanding propeller jet features an iconic double-stream flow structure due to the presence of the hub at the center of the propeller disk [14,18,19]. Wei and Chiew [11] reported that when the propeller jet is placed in the vicinity of a vertical wall, its two streams, namely, upper and lower streams, would be forced to spread out and deflected upwards and downwards along the wall, resulting in a pair of symmetrical vortices in between.

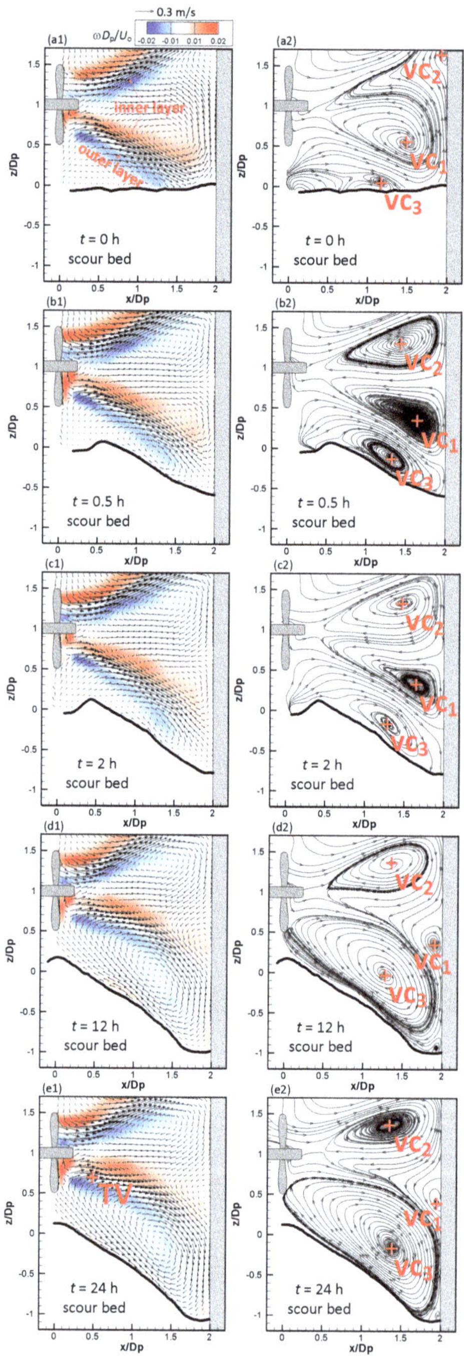

Figure 5. Temporal development of flow patterns within developing scour hole at $X_w = 2D_p$: (**a1**,**a2**) t = 0 h; (**b1**,**b2**) t = 0.5 h; (**c1**,**c2**) t = 2 h; (**d1**,**d2**) t = 12 h; (**e1**,**e2**) t = 24 h.

In the context of the current study with a scour hole, Figure 5 depicts the spread-out features of the two jet streams, similar to that observed in Wei and Chiew [11]. More explicitly, both the upper and lower streams are characterized by a pair of outer and inner shear layers with opposite signs, as denoted in Figure 5a1. A side-by-side comparison between Figure 5a1,a2 reveal that V1 and V2 essentially reside within the triangular region between the spread-out jet streams. The zero-vorticity layer (in white) between the positive (in red) and negative (in blue) shear layers, in fact, reflects a layer of zero-shear-stress which determines the separation line that envelops V1 and V2. Accordingly, the counterclockwise V1 and clockwise V2 are associated with the lower shear layer (with positive vorticity) and upper shear layer (with negative vorticity), respectively, which is consistent with those observed in Wei and Chiew [11]. This comparison is important because it indicates that the formation mechanism of V1 and V2 is exclusively related to the deflection effect that is associated with the jet impingement on the wall and has nothing to do with the presence of the bed, although they are no longer symmetrical due to the confinement effect of the latter. On the contrary, a near-bed vortex, V3, which is absent in Wei and Chiew [11], is directly emanated from the flow separation that occurred at the bed.

As the scour hole evolves, it is interesting to note that the shear layer structure associated with the spread-out jet streams seems to be constant and steady during the entire scouring process, although the developing scour hole, to some extent, allows the lower stream to be expanded farther downwards (Figure 5b1,e1). It therefore can be concluded that the presence and development of the scour hole have little impact on the spread-out flow structure, which primarily is dependent on the wall effect. In contrast, the streamline plots on the right column of Figure 5 depicts a considerable variation in the overall structure of the vortex system from the initial to asymptotic state. Specifically, at the initial instant of $t = 0$, Figure 5a2 shows that the presence of the flatbed prevents the formation of the downward flow, which would otherwise occur when the lower stream impinges on the vertical wall in the absence of the bed. Instead, a strong upward flow can be found along the vertical wall, which enhances the strength of V1 and squeezes V2 to the upper right corner of the FOV. On the other side, V3 is still in its embryonic phase. Thus, one could surmise that at the initial instant, it is V1 that is the driving force for the onset of scouring, during which the bed sediment particles are entrained in a counter-clockwise manner and transported to the lateral sides by the vortex tube formed in front of the wall. As soon as an initial local scour hole is excavated around the base of the quay wall, the original "confinement effect" associated with the flatbed diminishes, allowing the jet flow (lower stream) to be deflected downwards along the wall. As a result, a clockwise vortex is expected to be formed inside the scour hole. This is exactly what is shown at the subsequent time of $t = 0.5$ h in Figure 5b2, in which a well-established V3 is present, resembling the horseshoe vortex in a pier scour hole. This vortex, in turn, facilitates the subsequent scouring process, during which the bed sediment would be driven to the upstream by the clockwise flow, finally depositing at the upstream of the propeller or being carried away with the oncoming jet flow. Meanwhile, Figure 5b2 also shows that without the effect of the previously observed upward flow at $t = 0$ h, V1 and V2 appear to be quasi-symmetrical about the propeller axis, similar to those observed by Wei and Chiew [11]. From then on, Figure 5b2–e2 simply show that V3 is exclusively responsible for the scouring action, during which the sediments are swept in the clockwise fashion from the bed bottom and carried upstream largely in the form of bedload (visual observation). More interestingly, the enlargement of the scour hole does not support a further development of V1 and V2 as one could have envisioned. On the contrary, the enhancement of the primary vortex (V3) within the scour hole appears to overwhelm the growth of V1 and V2. In particular, V1 exhibits an evident shrinking trend from $t = 0.5$ h to $t = 12$ h, while V2 seems to be able to maintain its size. Thereafter, the vortex system appears to be stabilized by the presence of the large scour hole. At $t = 24$ h, both the size and position of all the three vortices are found to reach an asymptotic state when the scouring process ceases.

4.2. Comparison of Vortex Strength

To further explore the strength evolution of each individual vortex within the evolving vortex system, their normalized circulation is plotted against time in Figure 6a, in which the magnitude of circulation is computed from the vorticity distributions (see left column of Figure 5), as follows,

$$\Gamma = \iint_A \omega dA \tag{1}$$

where A is the enclosed area of the vortex. In this study, the vortex center and boundary were determined by using the vortex identification algorithms developed by Graftieaux et al. [20]. The so-obtained coordinates of the vortex centers confirm a reasonable agreement with those inferred from the visualized vortices shown in the streamline plots (see VC1, VC2, and VC3 in the right column of Figure 5). It is also noted that both V2 and V3 exhibit a clockwise rotation (negative circulation). To compare their relative strength with V1, Figure 6a shows the absolute value of their magnitudes. For easy reference, the temporal development of the maximum scour depth ($d_{s,t}$) is also plotted against time in Figure 6b.

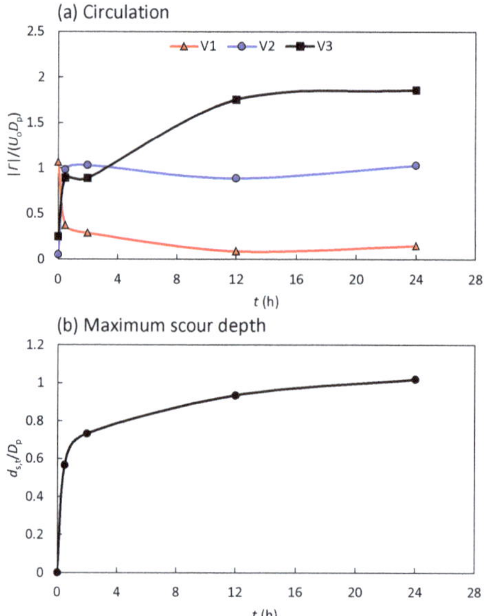

Figure 6. Temporal development of (**a**) vortex circulation of V1, V2, and V3; (**b**) maximum scour depth at $X_w = 2D_p$.

Figure 6a clearly shows that during the initial scouring phase (t = 0–0.5 h), V1 undergoes a rapid decrease; meanwhile, V2 and V3 reveal a synchronous increase with a comparable increasing rate. During the developing phase (t = 0.5–12 h), V1 and V2 exhibit a similar decreasing trend, whereas V3 still retains a relatively high increasing rate, which is consistent with that of the scour depth development. During the stabilizing phase (t = 12–24 h), all three vortices approach an asymptotic state in both size and location. When studying the horseshoe vortex evolution at pier scour, Baker [21] suggested that there exists a constancy of the vortex strength during the scouring process. On the other hand, Muzzammil and Gangadhariah [22] reported that during the scouring process the strength of the horseshoe vortex at a cylinder pier experiences an initial increase, followed with a decreasing trend as the scour hole continues to enlarge. However, these observations are different from that of a propeller-induced scour around a vertical wall. In the present case, Figure 5 reveals that the vortex circulation undergoes a monotonical increase as the scour hole evolves until the asymptotic state is reached around $t = 24$ h. A possible reason may be the fact that pier scour is often caused by multiple horseshoe vortices that formed around the scour hole [23], while in the present study, it is the single primary vortex (V3) that directly shapes and completely fills the scour hole (see Figure 5). As a result, the vortex size exhibits a concurrent growth with the developing scour hole.

5. Energy Spectra Analysis

To provide a further insight into the interaction between the turbulent characteristics of the vortex system and developing scour hole, this section presents an energy spectra analysis of the time series of velocity fluctuations at four representative points. Graftieaux et al. [20] suggested that the velocity fluctuations are their maximum near the mean vortex center within the region of a well-established vortex. For this reason, the vortex centers (marked as VC1, VC2, and VC3 in the right column of Figure 5) were selected to examine the turbulent energy cascade associated with the vortex system. Additionally, a characteristic point located in the shear layer in the vicinity of the propeller (marked as TV in Figure 5e1) was also selected for the sake of comparison. The TV position is located within the shedding path of the tip vortex originating from the propeller blades, as observed by Felli et al. [24] and Wei and Chiew [11], thus serving as a reference for the flow nature that is exclusively associated with the free jet diffusion.

Figures 7 and 8, respectively, illustrate the energy spectra distributions of the velocity fluctuations at point TV and the three vortex centers. The left column (S_u) and right column (S_w) in the figures denote the spectral energy density of the horizontal and vertical velocity components, respectively. Since the shear layer structure close to the propeller possesses a steady attribute throughout the entire scouring process (see Figure 5), Figure 7 only presents the power spectra of point TV at the asymptotic state ($t = 24$ h), whereas Figure 8 includes the power spectra at the vortex centers associated with the evolving vortex system at different time intervals of $t = 0, 0.5, 2, 12, 24$ h. Figure 7 reveals that in the near wake region of the propeller, both S_u and S_w exhibit two obvious spectra peaks superimposed on a broadband spectrum. The former is conjectured to be associated with the dominant frequencies of the tip vortex shedding originating from the rotating propeller blade. The latter reveals a less steep slope than the well-known Kolmogorov energy cascade slope of −5/3 as the turbulent energy is enhanced at small scale eddies (larger frequencies). One may, therefore, expect a local unbalance between turbulence production and dissipation close to the propeller, which is not surprising since the turbulence there is neither isotropic nor fully developed. Moreover, a side-by-side comparison between Figure 7a,b show that the turbulent energy associated with the horizontal velocity fluctuation appears to be greater than its vertical counterpart for the entire frequency band, confirming the dominant role of the axial (horizontal) velocity within a propeller jet.

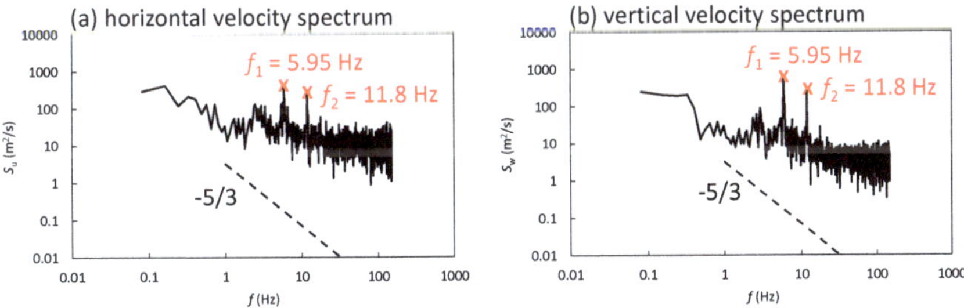

Figure 7. Power spectra of horizontal and vertical velocity fluctuations at a typical point of tip vortex shedding ($x = 0.5D_p$, $z = 0.7D_p$) at $X_w = 2D_p$.

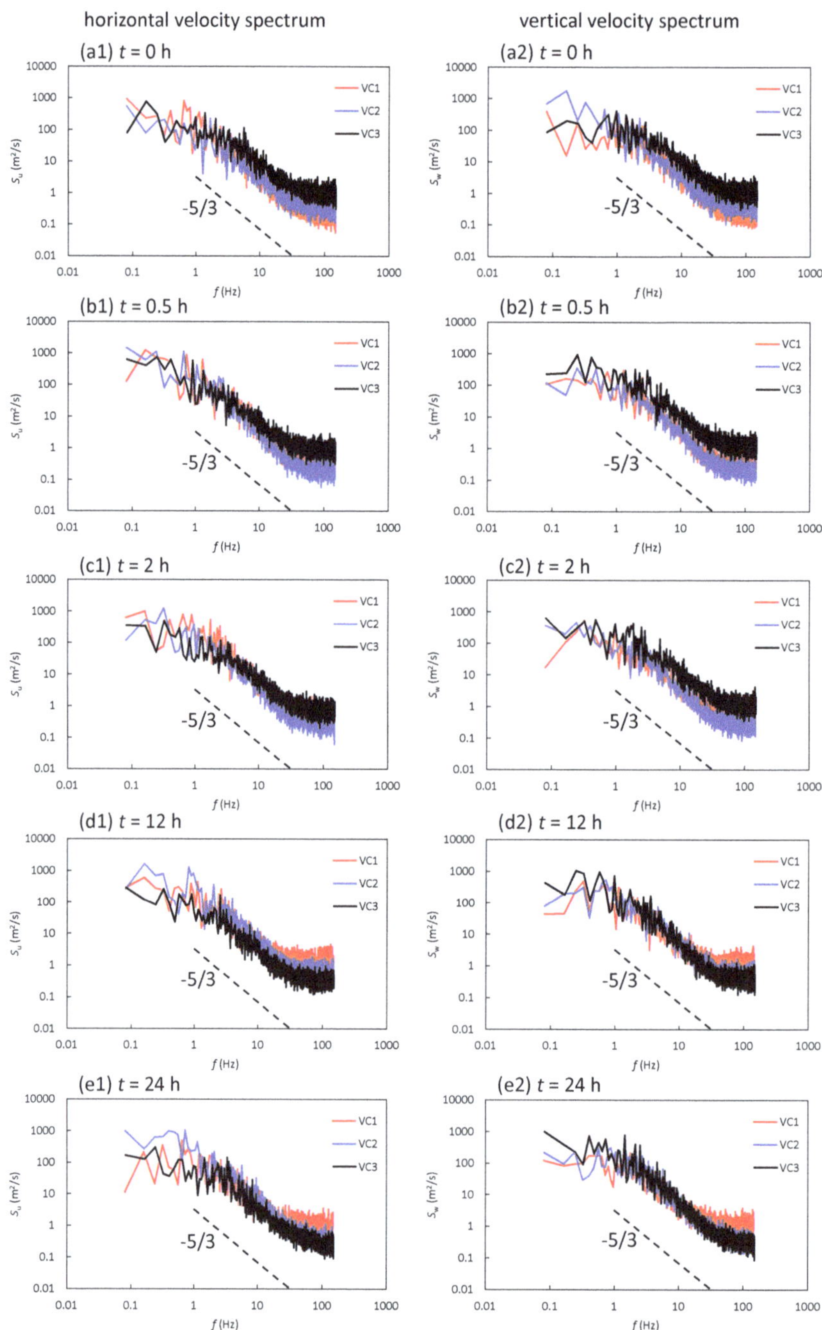

Figure 8. Power spectra of horizontal and vertical velocity fluctuations at vortex core centers of VC1, VC2, and VC3 at $X_w = 2D_p$: (**a1**,**a2**) $t = 0$ h; (**b1**,**b2**) $t = 0.5$ h; (**c1**,**c2**) $t = 2$ h; (**d1**,**d2**) $t = 12$ h; (**e1**,**e2**) $t = 24$ h.

On the other hand, away from the near wake region of the propeller, Figure 8 shows that no perceptible peak frequency can be observed from the spectra distributions associated with the vortex centers, indicating that the periodicity emanated from the propeller rotation has been attenuated. In the meantime, for all three vortices, Figure 8 reveals that both S_u and S_w exhibit a comparable power spectrum distribution which obeys the −5/3 law within the inertial range defined by Kolmogorov [25], implying a fully developed turbulence and an asymptotic between production and dissipation. This means that unlike the near wake region (e.g., point TV), where the small-scale eddies hold a significant amount of turbulent energy, it is the large-scale turbulent eddies that matter in governing the formation of the coherent structures, i.e., the well-established vortices (V1, V2, and V3). Furthermore, it can be observed that, compared to V1 (in red line) and V2 (in blue line), the energy spectra of V3 (in black line) gradually decreases with the development of the scour hole. Given that V3 is directly responsible for enlarging the scour hole as already pointed out, this observation implies a direct coupling between the vortex evolution and the developing scour hole, that is, an energy transfer from the turbulent energy dissipation to the scouring action. On the other hand, V1, which initiated the onset of scour process, reveals an opposite scenario as its energy spectra in small-scales appears to be enhanced and exceed its two counterparts during t = 12–24 h. One may surmise that under the confinement effect exerted by the two larger vortices (V2 and V3) on both sides, V1 is subject to a more intense velocity fluctuation and thus has higher turbulent kinetic energy, albeit in a smaller size.

6. Comparison of Near-Bed Flow Characteristics

The near-bed flow characteristics are always of great importance in understanding the interaction between the flow and scouring bed. According to the adopted resolution of the PIV measurement (as shown in Table 1), this study was able to extract the near-bed data at a distance of 2.7 mm (i.e., the location of the nearest data point close to the bed boundary) above the scoured bed boundary. The so-obtained results are plotted in Figure 9, in which AKE is the averaged kinetic energy (= $\frac{1}{2}(u^2+w^2)/U_o^2$); TKE is the turbulent kinetic energy (= $\frac{1}{2}(\overline{u'^2}+\overline{w'^2})/U_o^2$); and RSS is the Reynolds shear stress (= $-\overline{u'w'}/U_o^2$), in which u', w' are the fluctuating velocity components in the horizontal and vertical directions, respectively. It should be stated that these three parameters only consider the in-plane components of the three-dimensional jet flow due to the nature of the planer PIV measurement. For the sake of clarity, the scour bed profile is also superimposed in the figure whose scale follows the secondary ordinate axis on the right.

At the onset of the scouring action, Figure 9a reveals a dominant TKE (blue filled circle) prevailing along the initial flatbed with a local maximum value between x/D_p = 1–1.5. This location is reasonably correlated with the core of the near-bed vortex (V3) as shown in Figure 5a2. As the scour hole develops, the overall TKE profile undergoes a significant reduction (see Figure 9b–d), which reflects a weakening of the scouring capacity associated with the scour driving vortex. This inference is in conformity with the inference that one may deduce from the turbulent energy spectra plots. The local increase in TKE (around x/D_p = 2 at the quay wall), on the other hand, is likely attributed to the presence of the corner vortex that formed at the junction between the scour bed and the quay wall. At the asymptotic state (t = 24 h), when the scour process almost ceases altogether, the near-bed TKE distribution drops to approximately zero (see Figure 9e).

Conversely, the development of the near-bed AKE distribution (black filled square) reveals an opposite trend to that of its turbulent counterpart. Initially, the flatbed acts as a blockage and restrict jet expansion; this constraint facilitates an energy transfer from the mean flow to the turbulent flow field, thus resulting in a relatively small AKE value along the initial bed (see Figure 9a). Thereafter, during t = 0.5–2 h, Figure 9b,c show that the AKE reveals a local increase a round x/D_p = 1–1.5, which is probably caused by the growing near-bed vortex (V3). Entering the stabilizing phase (t = 12–24 h), the larger scour hole seems to be able to stabilize the well-established near-bed vortex, which, in turn, conserves the mean flow energy as the AKE is found to exceed its TKE counterpart (see Figure 9d,e). In other words, being different from the initial state of the scouring process, less energy is extracted from the mean flow to the turbulence during the later scouring phases, which eventually retards the scouring process.

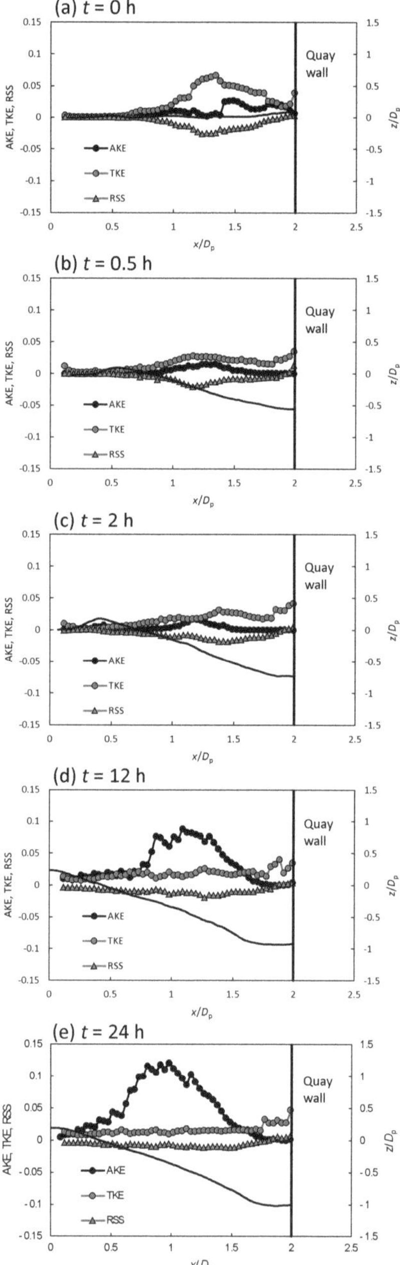

Figure 9. Comparison of AKE, TKE, RSS distributions along the developing scour bed at $X_w = 2D_p$: (**a**) $t = 0$ h; (**b**) $t = 0.5$ h; (**c**) $t = 2$ h; (**d**) $t = 12$ h; (**e**) $t = 24$ h.

As for the RSS profile, Figure 9a,b show that during the initial scouring stage the near-bed RSS exhibits a negative value due to the reverse flow associated with the clockwise vortex (V3). As the scour evolves, Figure 9c–e reveal that the RSS distribution gradually decreases to a near-zero value. On the whole, in contrast to the distributions of AKE and TKE, the RSS exhibits a relatively insignificant change during the scouring process, which may indicate its less important role in driving the scour development.

7. Conclusions

This study presents an experimental study of the temporal evolution of the propeller jet flow within a developing scour hole around a vertical quay wall. The asymptotic scour profiles under the condition of four wall clearances ($X_w = 1D_p, 2D_p, 3D_p$, and $4D_p$) were compared with their counterpart without any confinement; the results highlight the crucial role of the presence of the quay wall in shaping the final scour hole. The flow patterns within the asymptotic scour hole for the same four wall clearance cases were also compared to examine the vortex scouring mechanisms. In the small wall clearance cases ($X_w = 1D_p, 2D_p$), where the wall effect is more pronounced, a vortex system comprising three vortices was observed with the primary vortex being responsible for the scouring action. With increasing wall clearances and thus decreasing wall effects, the former three-vortices system transforms into a single vortex residing in the scour hole at $X_w = 3D_p$; when $X_w = 4D_p$, the single vortex further diminishes. This transformation provides an explanation for the observed inverse relationship between the scour depth and wall clearance in the case of the closed type quay. Moreover, similarities and differences were discussed with respect to the vortical structure in the case of the closed quay and that of the open quay. Such comparison elucidates the distinct influence of different quay types on the formation of flow patterns and the resulting scour profiles, which furnishes a laboratory reference for the maintenance work of quay structure in terms of propeller induced scour damage.

To investigate the evolution of the vortex system and its interaction with the developing scour hole, the temporal development of the flow patterns at a typical case of $X_w = 2D_p$ was discussed. The results highlight the significant difference in the vortex scouring mechanism at the initial instant with a flatbed and that in the presence of a developing scour hole. During the scouring process, although the shear layer structure associated with the propeller jet streams seems to remain steady over time, the vortex system undergoes a considerable change, for which the relative dominance of each individual vortex was analyzed in terms of their circulation developments. As the scour process develops, the near-bed vortex (V3) gradually overwhelms its two counterparts (V1 and V2) and exhibits a positive correlation with the increasing scour depth. In addition, the energy spectra analysis was conducted to examine the turbulent energy cascade and its implication for the associated scouring process. The −5/3 law was confirmed for the energy spectra at the centers of all the three vortices throughout the entire scouring process, indicating an asymptotic balance between turbulence production and dissipation within large-scale eddies associated with the vortex system. Furthermore, a comparison of the overall spectra over the entire frequency band of the three vortices reveals that the turbulent kinetic energy associated with the scour-driving vortex (V3) gradually decreases as the scour hole evolves. This phenomenon, to some extent, links the energy transfer from the turbulence dissipation to the scour excavating action.

Finally, the near-bed flow characteristics (i.e., AKE, TKE, RSS), which are indicative of erosive flow mechanisms, are discussed. The results show that the near-bed TKE gradually decreases with the development of the scour hole, confirming the inference from the energy spectra analysis. Meanwhile, the near-bed AKE appears to be enhanced when the vortex is stabilized by the presence of a large scour hole. The increase of AKE and decrease of TKE with time shows that less energy is transferred from the mean to turbulent flow fields, which is the reason the scouring process progressively diminishes. The RSS distribution, on the other hand, undergoes an insignificant change during the scouring process, indicating its comparatively less important role in affecting scour.

Author Contributions: Conceptualization, M.W., N.-S.C. and Y.-M.C.; Data curation, M.W.; Formal analysis, M.W.; Supervision, N.-S.C. and Y.-M.C.; Writing–original draft, M.W.; Writing–review & editing, M.W., N.-S.C. and Y.-M.C. and F.Y.

Funding: The National Key Research and Development Program of China (2016YFC0402302; 2017YFC1502504).

Conflicts of Interest: The authors declare no conflict of interest.

References

1. PIANC. *Guidelines for Protecting Berthing Structures from Scour Caused by Ships*; PIANC: Brussels, Belgium, 2015.
2. Sumer, B.M.; Fredsøe, J. *The Mechanics of Scour in the Marine Environment*; World Scientific: River Edge, NJ, USA, 2002; p. 536.
3. Stewart, D.P.J. Characteristics of a Ship's Screw Wash and the Influence of Quay Wall Proximity. Ph.D. Thesis, Queen's University of Belfast, Belfast, Northern Ireland, 1992.
4. Hamill, G.; Johnston, H.; Stewart, D. Propeller wash scour near quay walls. *J. Waterw. Port Coast. Ocean Eng.* **1999**, *125*, 170–175. [CrossRef]
5. Ryan, D. Methods for Determining Propeller Wash Induced Scour in Harbours. Ph.D. Thesis, Queen's University of Belfast, Belfast, Northern Ireland, 2002.
6. Sleigh, A. A Laboratory Study into the Problem of Slope Erosion by Propeller Wash. Master's Thesis, University of Strathclyde, Glasgow, UK, 1981.
7. Cihan, K.; Ozan, A.Y.; Yüksel, Y. The effect of slope angle on propeller jet erosion near quays. *Proc. ICE-Marit. Eng.* **2012**, *165*, 81–92. [CrossRef]
8. Wei, M.; Chiew, Y.M. Influence of toe clearance on propeller scour around an open-type quay. *J. Hydraul. Eng.* **2017**, *143*, 04017012. [CrossRef]
9. Wei, M.; Chiew, Y.M.; Guan, D.W. Temporal development of propeller scour around a sloping bank. *J. Waterw. Port Coast. Ocean Eng.* **2018**, *144*, 06018005. [CrossRef]
10. Wei, M.X.; Chiew, Y.M. Characteristics of propeller jet flow within developing scour holes around an open quay. *J. Hydraul. Eng.* **2018**, *144*, 04018040. [CrossRef]
11. Wei, M.; Chiew, Y.M. Impingement of propeller jet on a vertical quay wall. *Ocean Eng.* **2019**, *183*, 73–86. [CrossRef]
12. Lin, C.; Hsieh, S.C.; Lin, I.J.; Chang, K.A.; Raikar, R.V. Flow property and self-similarity in steady hydraulic jumps. *Exp. Fluids* **2012**, *53*, 1591–1616. [CrossRef]
13. Hsieh, S.C.; Low, Y.M.; Chiew, Y.M. Flow characteristics around a circular cylinder subjected to vortex-induced vibration near a plane boundary. *J. Fluids Struct.* **2016**, *65*, 257–277. [CrossRef]
14. Wei, M.; Chiew, Y.M.; Hsieh, S.C. Plane boundary effects on characteristics of propeller jets. *Exp. Fluids* **2017**, *58*. [CrossRef]
15. Guan, D.; Chiew, Y.M.; Wei, M.; Hsieh, S.-C. Characterization of horseshoe vortex in a developing scour hole at a cylindrical bridge pier. *Int. J. Sediment Res.* **2019**, *34*, 118–124. [CrossRef]
16. Sciacchitano, A.; Scarano, F. Elimination of PIV light reflections via a temporal high pass filter. *Meas. Sci. Technol.* **2014**, *25*, 084009. [CrossRef]
17. Hong, J.H.; Chiew, Y.M.; Cheng, N.S. Scour Caused by a Propeller Jet. *J. Hydraul. Eng.* **2013**, *139*, 1003–1012. [CrossRef]
18. Hamill, G.A. Characteristics of the Screw Wash of a Manoeuvring Ship and the Resulting Bed Scour. Ph.D. Thesis, Queen's University of Belfast, Belfast, Northern Ireland, 1987.
19. Hsieh, S.C.; Chiew, Y.M.; Hong, J.H.; Cheng, N.S. 3-D Flow Measurements of a Swirling Jet Induced by a Propeller by Using PIV. In Proceedings of the 35th IAHR World Congress, Chengdu, China, 8–13 September 2013; pp. 5141–5150.
20. Graftieaux, L.; Michard, M.; Grosjean, N. Combining PIV, POD and vortex identification algorithms for the study of unsteady turbulent swirling flows. *Meas. Sci. Technol.* **2001**, *12*, 1422. [CrossRef]
21. Baker, C.J. Vortex Flow Around the Bases of Obstacles. Ph.D. Thesis, University of Cambridge, Cambridge, UK, 1978.
22. Muzzammil, M.; Gangadhariah, T. The mean characteristics of horseshoe vortex at a cylindrical pier. *J. Hydraul. Res.* **2003**, *41*, 285–297. [CrossRef]

23. Baker, C. The position of points of maximum and minimum shear stress upstream of cylinders mounted normal to flat plates. *J. Wind Eng. Ind. Aerodyn.* **1985**, *18*, 263–274. [CrossRef]
24. Felli, M.; Di Felice, F.; Guj, G.; Camussi, R. Analysis of the propeller wake evolution by pressure and velocity phase measurements. *Exp. Fluids* **2006**, *41*, 441–451. [CrossRef]
25. Kolmogorov, A.N. The local structure of turbulence in incompressible viscous fluid for very large Reynolds numbers. *Cr. Acad. Sci. URSS* **1941**, *30*, 301–305. [CrossRef]

Article

Scour Induced by Single and Twin Propeller Jets

Yonggang Cui [1], Wei Haur Lam [1,2,*], Tianming Zhang [1], Chong Sun [1] and Gerard Hamill [3]

1 State Key Laboratory of Hydraulic Engineering Simulation and Safety, Tianjin University, Tianjin 300072, China; cui_yonggang@tju.edu.cn (Y.C.); zhangtianming@tju.edu.cn (T.Z.); chong@tju.edu.cn (C.S.)
2 First R&D Services, A-08-16 M Suites, 283 Jalan Ampang, Kuala Lumpur 50450, Malaysia
3 School of Natural and Built Environment, Architecture, Civil & Structural Engineering and Planning, Queen's University Belfast, David Keir Building, Stranmillis Road, Belfast BT9 5AG, UK; g.a.hamill@qub.ac.uk
* Correspondence: joshuawhlam@hotmail.com or wlam@tju.edu.cn

Received: 27 April 2019; Accepted: 23 May 2019; Published: 25 May 2019

Abstract: Single and twin ship propeller jets produce scour holes with deposition dune. The scour hole has a maximum depth at a particular length downstream within the propeller jet. Existing equations are available to predict maximum scour depth and the corresponding scour length downstream. Experiments conducted with various physical propeller models, rotational speeds, propeller-to-propeller distances and bed clearances are presented. The measurements allowed a better understanding of the mechanism of temporal scour and deposition formation for scour caused by single-propeller and twin-propeller. Results show that the propeller jet scour profiles can be divided into three zones, which are the small scour hole, primary scour hole and deposition dune. An empirical 2D scour model is proposed to predict the scour profile for both a single-propeller and twin-propeller using a Gaussian normal distribution.

Keywords: ship twin-propeller; jet; scour; 3D printing

---

1. Introduction

Ships are getting larger and faster to meet the needs of world economic development. Engine power has been increased for bigger ships leading to the use of the twin-propeller system. Kim et al. [1] and PIANC [2] stated that the twin-propeller system improved the ship power system and its handling ability. A twin-propeller ship produces a complicated jet affecting the navigation safety in waterways and damaging the seabed condition. Seabed scour is widely discussed due to the direct impingement of a high-velocity jet to cause excessive sediment transport in harbour and waterways. Hamill [3] and Gaythwaite [4] stated that the impingement of the ship's propeller jet can damage the seabed.

Albertson et al. [5] firstly proposed the efflux velocity within the plain water jet from orifice being predicted using axial momentum velocity. Blaauw and van de Kaa [6] and Verhey et al. [7] suggested the use of a theoretical foundation of a plain water jet in Albertson et al. to investigate the ship propeller jet induced seabed scour in the harbour without consideration of the berth structure. Hamill et al. [8], Lam et al. [9] and Hong et al. [10] continued the investigations in laboratories by using a physical single propeller model to observe the scouring condition. Hamill et al. [11] include the effects of the quay wall in the seabed scour prediction. The aforementioned researchers emphasised on the estimation of maximum scour depth induced by ship propeller wash. Wang et al. [12], Sun et al. [13] and Ma et al. [14] implemented the foundation of propeller jet induced scour to predict the wake of tidal turbine and the tidal turbine induced scour and Jiang et al. [15] proposed the theoretical structure of ship twin propeller jet.

The impingement of this rotating jet forms the scour pit with the scoured sediments being transported downstream to form dune deposited surrounding the scour pit on the seabed when the watercraft or vessels entering a port. The accumulated sediment forms a dune reducing the water depth and increasing the risk of ship collision in shallow waterways. A ship propeller can be damaged when hitting the seabed. The understanding of both the scour pit and deposition dune are important to ensure navigation safety. Twin-propeller jet produces the scour profiles with deposited dune downstream, which is different compared to the single-propeller. Yew et al. [16] and Hong et al. [10] stated that the deposition is an M-shaped distribution with the largest accumulated height on the longitudinal axis of both propellers. The height of deposited dunes decreased along the side of two propellers surrounding scour hole.

In this works, a purpose-built power train system was developed to rotate the physical propeller model at the desired rotational speeds in the laboratory in order to investigate the ship twin-propeller jet induced scour. Previous researchers used the maximum scour depth without available equations to predict the entire cross-section of the scour profile. The current experimental data were used to establish the 2D scour model by proposing the scour depth equation and Gaussian normal distribution to estimate the entire scour profile for both single-propeller and twin-propeller systems.

2. Propeller Jet and Scour

2.1. Propeller Jet

Hamill et al. [3] started the research on propeller jets according to the axial momentum theory and plain jet theory. Hamill et al. [17] found that the flow structure of the propeller jet is symmetrical along the rotational axis. The jet can be divided into two development stages, which are the zone of flow establishment and zone of established flow and the whole jet is conical in 3D view. The zone of flow establishment is the initial development stage of flow velocity with more complicated nature compared to the second zone downstream. Low-velocity region was found near to the propeller jet outflow plane due to the hub effects with no contribution of velocity from blades. The maximum lateral velocity distribution occurred in the middle of the blade, where the maximum blade width occurred. The velocity distribution is symmetrical along the propeller axis. For the zone of established flow, the maximum velocity is lesser compared to the initial zone due to water entrainment to reduce the velocity gradient within the jet. The high-velocity flow is mixed with the low-velocity flow to balance the velocity distribution.

Jiang et al. [15] proposed the flow structure of the twin-propeller jet has two zones as a single propeller, as shown in Figure 1. The zone of flow establishment can be divided to be non-interference zone (ZFE-TP-NI) and inference zone (ZFE-TP-I). Two propeller jets in twin system have no contact in the non-interference zone and expand separately to meet at one point where the interference zone starts. The zone of flow establishment consists of the four-peak region (ZFE-TP-4P) and two-peak region (ZFE-TP-2P) according to the lateral velocity section. A mixing point happens downstream within the twin-propeller jet. The twin-propeller jet is the same as two separate jets in the upstream region of the mixing point. Four maximum speeds were found on the cross-section, which corresponds to the two velocity peaks for every single propeller. The four maximum velocities combined to be two maximum velocities located at the rotational axis of each propeller when two single jets mixed together. In the zone of established flow, only one maximum velocity was found on the symmetrical plane.

2.2. Propeller Induced Scour

Chiew and Lim [18] studied the scouring problem by using a submerged circular jet to simulate the rotating propeller instead of the complicated propeller jet. The scour hole dimensions produced by the circular wall jet were highly dependent on the densimetric Froude number (F_0). The maximum equilibrium scour depth increased with an increase in F_0. Chiew and Lim introduced the offset height (G) of the circular wall jet as an additional parameter governing the diffusion characteristics of the jet. The maximum scour depth reduced with an increase in the offset height.

Hamill et al. [11] investigated the scour induced by a propeller jet both in the absence of the port structure and near the quay walls. A predictive equation for the time-dependent maximum scour depth as a function of the Froude number (F_0), propeller size (D_p), sediment bed material size (d_{50}), and distance between propeller tip and seabed was proposed. Seabed scour around a marine current turbine was also studied by Chen and Lam [19].

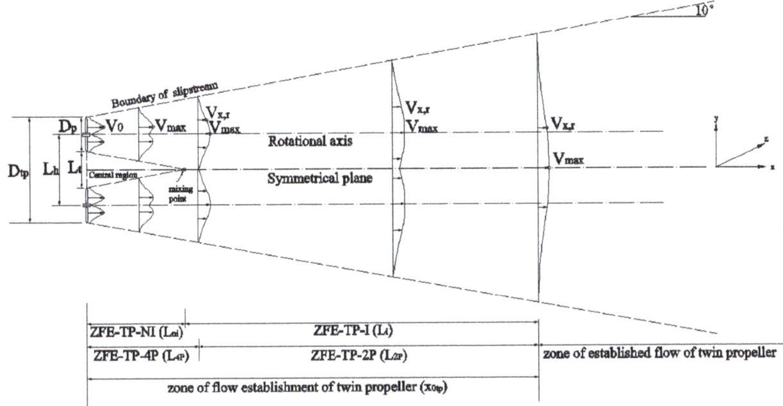

Figure 1. Flow structure of a twin-propeller jet [15].

Hong et al. [10] conducted an experimental study of a single propeller jet. Limited works from Mujal-Colilles et al. [20] and Yew [16] were found to discuss the twin-propeller induced scour. All these researchers stressed the importance of estimating the maximum scouring depth caused by ship propeller. The equations to predict the scour depth proposed by the previous researches are shown in Table 1.

Table 1. Summarised scour depth induced by the propeller.

Source	Scour Type	Equations
Chiew [18]	Circular jet	$\varepsilon_{max} = 0.21 d_0 F_0$
Hamill [8]	Single propeller	$\varepsilon_{max} = \Omega[ln(t)]^G$ $\Omega = 6.9 \times 10^{-4} \times \left(\frac{C}{d_{50}}\right)^{-4.63}\left(\frac{D_p}{d_{50}}\right)^{3.58} F_0{}^{4.535}$ $\Gamma = 4.113 \times \left(\frac{C}{d_{50}}\right)^{0.742}\left(\frac{D_p}{d_{50}}\right)^{-0.522} F_0{}^{-0.628}$
Hamill [11]	Single propeller	$\varepsilon_{max} = 38.97\Omega[ln(t)]^G$ $\Gamma = \left(\frac{C}{d_{50}}\right)^{0.94}\left(\frac{D_p}{d_{50}}\right)^{-0.48} F_0{}^{-0.53}$ $\Omega = \Gamma^{-6.38}$
Hong [10]	Single propeller	$\frac{\varepsilon_{max}}{D_p} = k_1\left[log_{10}\left(\frac{U_0 t}{D_p}\right) - k_2\right]^{k_3}$ $k_1 = 0.042 * F_0{}^{1.12}\left(\frac{C}{D_p}\right)^{-0.4}\left(\frac{C}{d_{50}}\right)^{-0.17}$ $k_2 = 1.882 * F_0{}^{-0.009}\left(\frac{C}{D_p}\right)^{2.302}\left(\frac{C}{d_{50}}\right)^{-0.441}$ $k_3 = 2.477 * F_0{}^{-0.073}\left(\frac{C}{D_p}\right)^{0.53}\left(\frac{C}{d_{50}}\right)^{-0.045}$
Yew [16]	Twin propeller	$\varepsilon_{twin} = k(\text{logt})^{0.0231}$ $k = \left(\frac{C}{D_p}\right)^{-0.488}\left(\frac{U_0 t}{C}\right)^{0.241}$
Cui	Twin propeller	$\varepsilon_{\text{twin}} = \Omega_t[ln(t)]^{\Gamma_t}$ $\Omega_t = 0.2526 \times \left(\frac{d_p}{d_{50}}\right)^{-0.859}\left(\frac{C}{d_{50}}\right)^{-4.63}\left(\frac{D_p}{d_{50}}\right)^{3.58} F_0{}^{4.535}$ $\Gamma_t = 1.389 \times \left(\frac{d_p}{d_{50}}\right)^{0.1571}\left(\frac{C}{d_{50}}\right)^{0.742}\left(\frac{D_p}{d_{50}}\right)^{-0.522} F_0{}^{-0.628}$

3. Experimental Works

Experiments were conducted in the Marine Renewable Energy Laboratory (MREL) at Tianjin University. The scaled twin-propeller system consisted of two single propellers. Two propellers were fixed at the shaft to rotate at the desired rotational speeds in a tank. The experimental setup, measurement method and 3D printed propeller model are discussed in the following sections. The rotational directions of twin-propellers were controlled to rotate internally and externally as internal rotation and external rotation. The propeller at the port side (left from the aft) rotated counter clockwise, while the propeller at starboard side (right from the aft) rotated clockwise to form the external rotation system. Most of the ships used the external rotation system to prevent flow compression in the clearance between two propellers with strong internal mixing jet. The observation of the development of a scour profile with time was made and followed by the measurement of the scour depth.

3.1. Experiment Setup

The experiment was carried out in a water tank of 1.2 m in length, 0.8 m in width, and 0.45 m in height. The sand was evenly laid on the ground with a height of 0.1 m. Scour depth was much smaller than the depth of available laid sand. The sand used was the construction sand screened through 0.1 mm to 0.3 mm diameter filters. The diameter of sand in this experiment was 0.2 mm. The distances between propeller tip and sandbed were set to 5.0 mm, 27.5 mm and 55.0 mm according to Hong et al. [10]. The schematic diagram of the experiment is shown in Figure 2a,b.

3.2. Measurement Methods

The measurement system consisted of a laser rangefinder to measure the scour depth and the traverse system to move the laser rangefinder to the designated position for point measurement within the 1 m × 1 m area, as shown in Figure 2. A laser rangefinder was used to measure the scour depth and deposition height. The laser rangefinder used is made by Sndway Co. Ltd. with a measurement range up to 40 m. The laser rangefinder had an aspherical optical focusing mirror, which could enhance reflected light acceptance. The measurement accuracy was within ± 0.2 mm. The lasers had different refractive indices in water and in air. The experimental data needs to be multiplied by 1.33 to eliminate the differences due to the transmission of the light beam through the air and water as consideration of refractive index.

The propellers were not rotating before the measurement and the designed rotational speeds can be adjusted by using the speed control system. Depth measurement was made on a point by point basis through comparison of the undisturbed reference point and the scoured point to obtain the relative scour depth. The laser rangefinder was initially moved to the undisturbed sandbed point and read the measurement point L_1, as shown in Figure 3. Then, the laser rangefinder was moved to the sandpit measurement point to read the data L_2. The scour depth can be obtained by calculating the $\varepsilon_m = (L_2 - L_1)\ n$, where n is the refractive index of water.

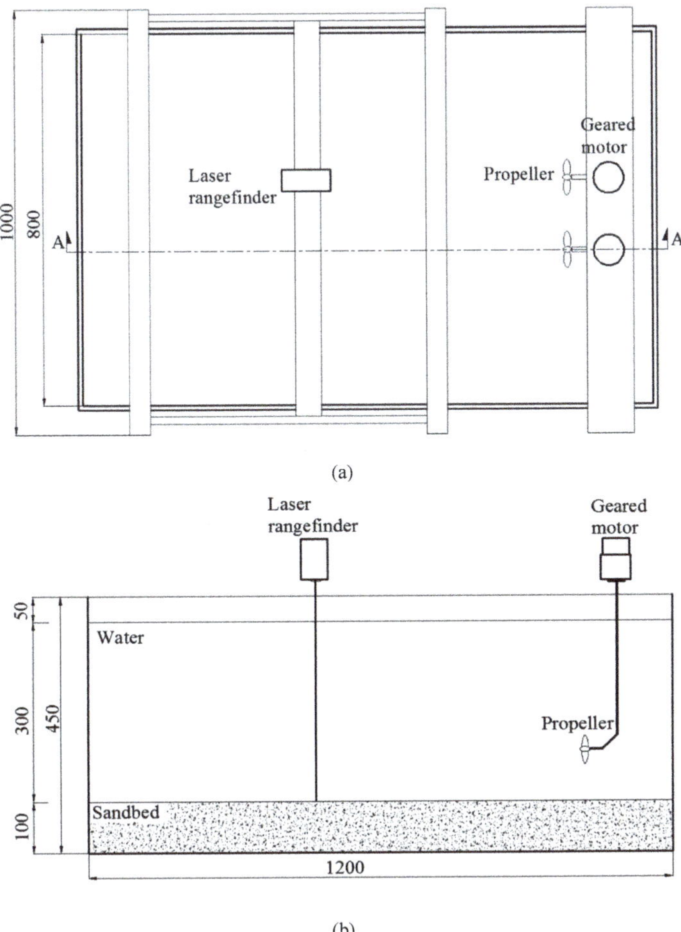

Figure 2. Schematic diagram of the experiment (1.2 m × 0.8 m × 0.45 m). (**a**) Plan view; (**b**) longitudinal cross-section (A-A).

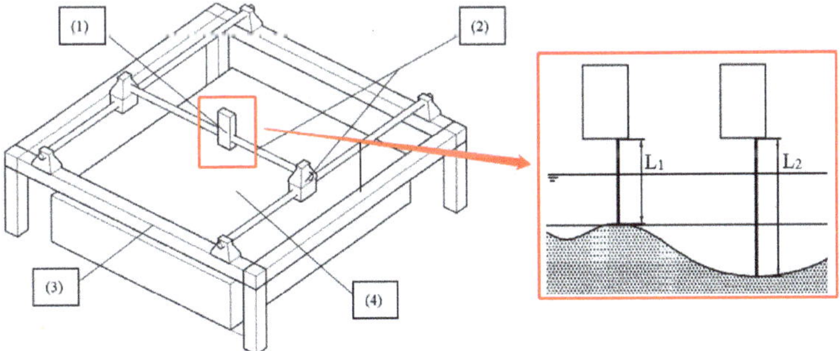

Figure 3. Measurement system: (1) Laser rangefinder; (2) optical axis and sliding block; (3) support; (4) experimental tank.

The support structure for the measurement system was built with four columns and four beams surrounding the experimental tank to accommodate the traverse system. The stands were 0.4 m high, and its length and width were the same as 1 m. Two rods were fixed in the direction of the propeller jet (x-direction) on supporting beams with sliders allowing y-direction movement. An additional rod was installed on top of the two rods with sliders on each side allowing the x-direction motion. A fixed clip was used to fix the laser rangefinder the rod. A laser rangefinder had the optical axis perpendicular to the seabed for depth measurement. The single rod allowed the laser rangefinder to move at any axial distance from the flow field. The double rods allowed the laser rangefinder to measure any position in the y-direction. Therefore, all desired points within the x-y measurement grid can be acquired to provide data for scour depth and deposition height.

3.3. 3D Printed Propeller Model

Hamill et al. [8] studied the characteristics of the propeller wash of a manoeuvring ship and the resulting bed scour based on two propellers, propeller-76 and propeller-131. The propeller characteristics used in experiments are shown in Table 2. The current propeller model used is being termed as propeller-A and propeller-B.

Table 2. Propeller characteristics.

Propeller	N	D_p (mm)	C_t	P′	β	D_h (mm)
A	3	55	0.40	1.00	0.47	11.5
B	6	55	0.56	1.14	0.922	11.5

For the current experiments, the solid propeller geometry was created by using the Solidworks and printed on the 3D printer made by JG Aurora in China using biodegradable polylactic acid filament (PLA) material. The actual propeller model was created through the additive process by printing the propeller layer by layer up to 0.1 mm. Two hours were taken to complete the entire printing process. Figure 4a shown the propeller printed in progress and Figure 4b shows the completed propeller with the temporary supporting structure, which needs to be eliminated in the post-processing.

3.4. Scaling of Experiment

According to Verhey et al. [7], the scaling effects caused by viscosity are negligible if the Reynolds number for flow (R_{flow}) and the propeller (R_{prop}) are greater than 3×10^3 and 7×10^4. The Reynolds numbers can be calculated using Equations (1) and (2).

$$R_{flow} = \frac{V_0 D_P}{\nu} \tag{1}$$

$$R_{prop} = \frac{n L_m D_p}{\nu} \tag{2}$$

where L_m is the length that depends on β as defined by Blaauw and van de Kaa [6] as Equation (3).

$$L_m = \beta D_p \pi \left(2N \left(1 - \frac{D_h}{D_p} \right) \right)^{-1} \tag{3}$$

where V_0 is the efflux velocity (m/s); D_p is the diameter of the propeller (m); D_h is the diameter of the hub (m); ν is the kinematic viscosity of the fluid (8.54×10^{-7} m^2/s); n is the rotational speed (rps); β is the blade area ratio; and N is the number of blades.

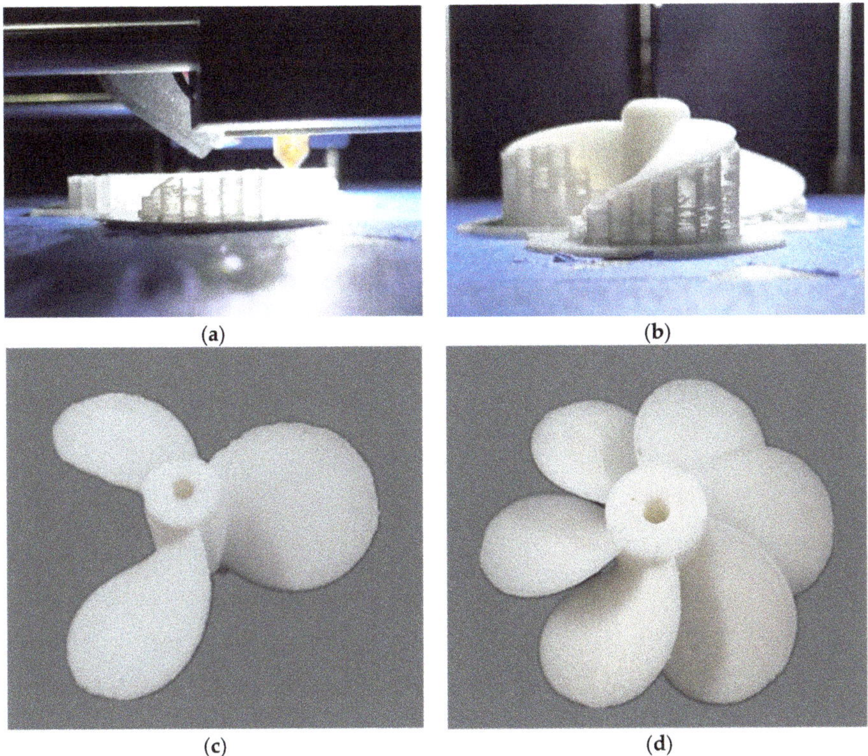

Figure 4. 3D printing for propeller model. (**a**) Printing status; (**b**) print completed; (**c**) propeller-A; (**d**) propeller-B.

Typical propeller sizes and rotational speeds that might cause seabed scouring in British ports and harbours ranged between 1.5 m and 3 m for propeller diameters running at 200–400 rpm according to Qurrain [21]. A typical ship propeller with a diameter of 1.65 m that operated at 200 rpm with a coefficient of thrust (C_t) of 0.35 was used as the prototype for the twin-propeller. The propeller rotation speeds used for this investigation were set to 500 and 700 rpm for observing the resulting scour profiles. Reynolds numbers for the proposed speed ranges were 2.9×10^4 R_{flow} and 1.3×10^4 for R_{prop}. In the current study, the motor speed was limited. R_{prop} was slightly smaller than the specified value. However, Blaauw and van de Kaa [6] and Verhey [7] proposed that these scale effects were insignificant. The Reynolds numbers for the jets were all greater than 3×10^3 satisfying the criteria for Froudian scaling.

The dimensions of the scour profile induced by twin-propeller are shown in Figure 5. Hong et al. [10] stated that scour induced by a propeller can be divided into small scour hole (Zone A), primary scour hole (Zone B) and deposition dune (Zone C). The current research focused on the scour caused by the twin-propellers. Ten different twin-propeller experiments and one single-propeller experiment were set up as summarised in Table 3.

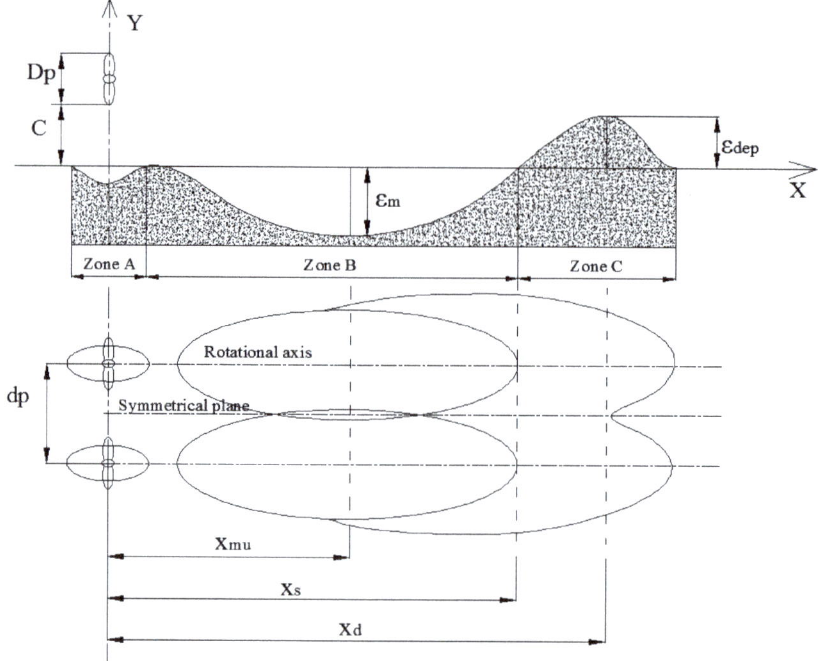

Figure 5. Dimensions of the scour profile induced by twin-propeller.

Table 3. Summary of experiments.

Experiment	Propeller	Distance between Propeller Hubs, d_p	C (mm)	n (rpm)
T-1	Propeller-A	1.5 D_p	27.5	500
T-2	Propeller-A	1.5 D_p	55.0	500
T-3	Propeller-A	2.0 D_p	27.5	700
T-4	Propeller-A	2.0 D_p	55.0	700
T-5	Propeller-B	1.5 D_p	27.5	500
T-6	Propeller-B	1.5 D_p	55.0	700
T-7	Propeller-B	2.0 D_p	27.5	500
T-8	Propeller-B	2.0 D_p	55.0	700
T-9	Propeller-A	3.0 D_p	27.5	500
T-10	Propeller-A	3.0 D_p	55.0	500
T-11	Single-propeller (Propeller-A)	/	27.5	500

4. Twin-Propeller Jet Induced Sandbed Scour

4.1. Temporal Depth of Scour

The experiments found the maximum scour position of a twin-propeller occurred symmetrically at the rotational axis of both two propellers. The scour holes were mirrored for two propellers in the twin system at the port and starboard sides with time, as shown in Figure 6. The depth of the scoured hole increased with time from the observation of various experiments. The motor speed reached 500 rpm in a very short time and the propeller reached a stable rotation speed in 3 s. The scour data were recorded at 60, 120, 300, 600, 1200 and 1800 s in the experiment. The increase of the scour depth was insignificant after 1800 s (30 min), which reached the equilibrium state. Hong et al. [10] stated the scour process includes initial stage, developing stage, stabilisation stage and asymptotic stage.

The scour process for T-1 case in current experiments is shown in Figure 6. The evolution of a typical scouring profile along the longitudinal direction was measured.

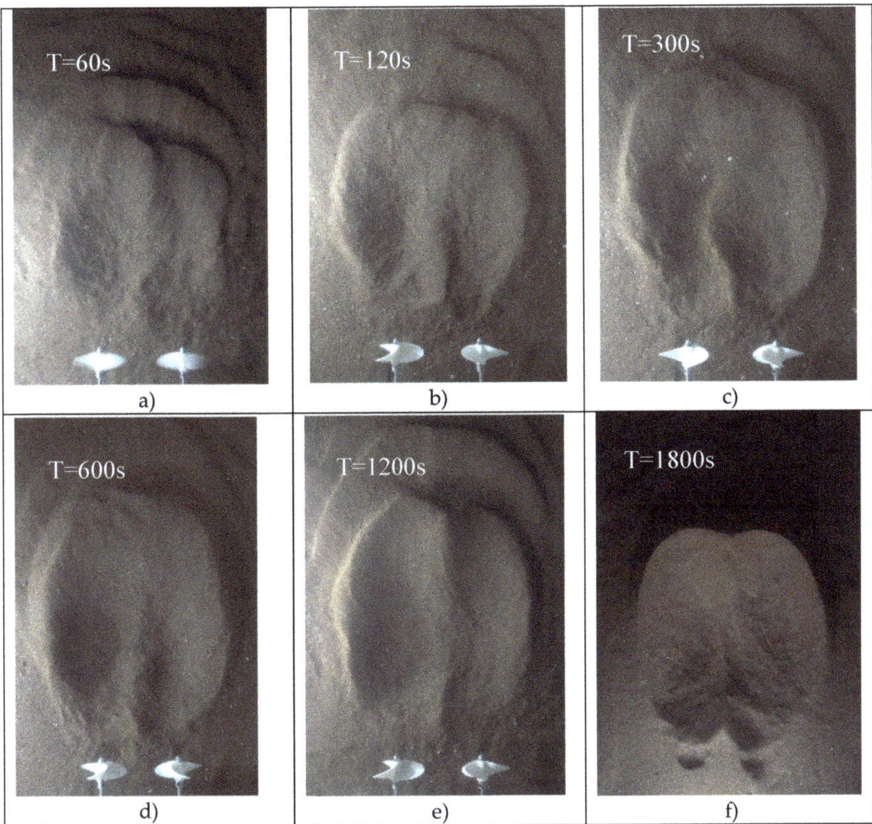

Figure 6. Temporal scour processes for twin-propeller, (**a**) 60 s; (**b**) 120 s; (**c**) 300 s; (**d**) 600 s; (**e**) 1200 s; (**f**) 1800 s.

Figure 7 shows the experimental results obtained from six scour times up to 1800 s (30 min). The external rotating twin-propeller produced the two largest scour holes at the propeller axis and the scour profile was approximately symmetrical. The maximum scour depth in the axial direction of the right single propeller (starboard) was recorded. The x-axis shows the axis distance, while the y-axis indicates scour depth. The main scour hole and deposition appeared downstream of the jet. A shallow hole formed immediately below the propeller. The sediments located on the surface of the sand bed were washed downstream with time due to the jet impingement. The maximum scour depth increased gradually with time. The location of the maximum scour depth did not change significantly with time. The deposition height increased gradually with time, but the position of dune peak moves downstream with time. For the non-dimensionalised scour profile, the x-axis shows the axis distance (X/X_m), where X_m is the length of the maximum scour hole from the propeller. The y-axis indicates scour depth ($\varepsilon_m/\varepsilon_{max}$). The non-dimensionalised scour profiles show the scour profile at different times had high similarity.

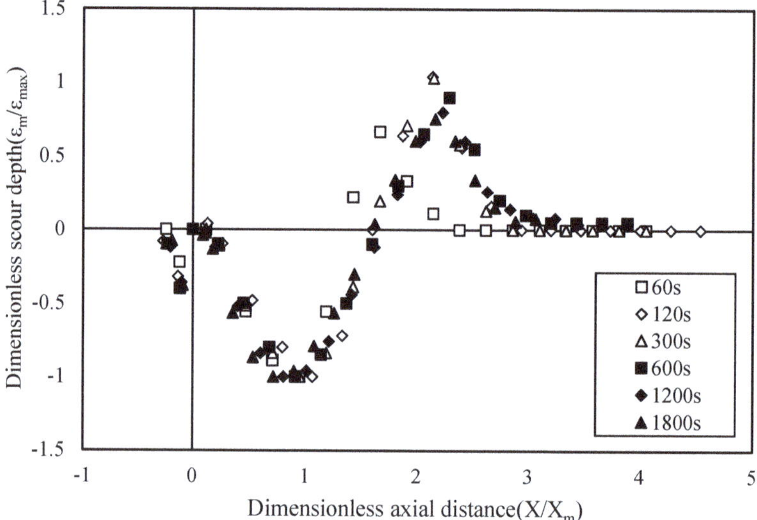

Figure 7. Dimensionless scour depth for a twin-propeller.

4.2. 2D Scour Section for Single-propeller

Previous researchers proposed an empirical formula to predict the maximum scour depth based on the experimental results. The scour structure of the entire scour section has not been studied. The current research studied the 2D scour profile of single-propeller and twin-propeller induced scour. The maximum scour occurred on the rotational axis of single-propeller. Hamill et al. [8] stated that the maximum scour depth location (X_m) can be calculated by Equation (4). The location of the maximum scour depth was proportional to the distance from the tip of the propeller blade to the sandbed.

$$X_m = F_0^{0.94} C \tag{4}$$

For current propeller diameter of 55 mm, speed of 500 rpm, and C of 0.0275 m, F_0 was calculated as 8.135. The maximum scour depth and position was measured, as presented in Figure 8.

The dimensionless scour profiles show that the current research has the same scour profile compared with Hamill et al. [8] and Hong et al. [10]. The scour profile is divided into three zones: small scour hole, primary scour hole and deposition dune in order to describe the whole scour section for single-propeller. The length of each zone was necessary to describe the scour profile. Table 4 summarises the suggested dimensionless length of each zone. The current work suggests the application of a Gaussian probability distribution to represent each scour section, which the scour depth is distributed in the holes. A Gaussian distribution curve was used to fit the three zones, as shown in Figure 9.

Table 4. Dimensionless length of each zone for single-propeller induced scour.

Source	Small Scour Hole (Zone A)	Primary Scour Hole (Zone B)	Deposition Dune (Zone C)
Hamill [8]	No defined	$0 < X/X_m < 1.7$	$1.7 < X/X_m < 2.2$
Hong [10]	$-0.5 < X/X_m < 0$	$0 < X/X_m < 1.83$	$1.83 < X/X_m < 2.9$
Current research	$-0.45 < X/X_m < 0$	$0 < X/X_m < 1.8$	$1.8 < X/X_m < 3$
Suggested length	$-0.5 < X/X_m < 0$	$0 < X/X_m < 1.8$	$1.8 < X/X_m < 3$

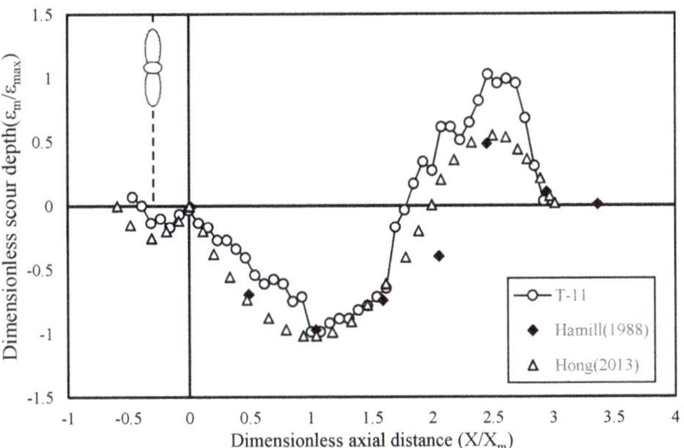

Figure 8. Dimensionless scour profiles for a single-propeller.

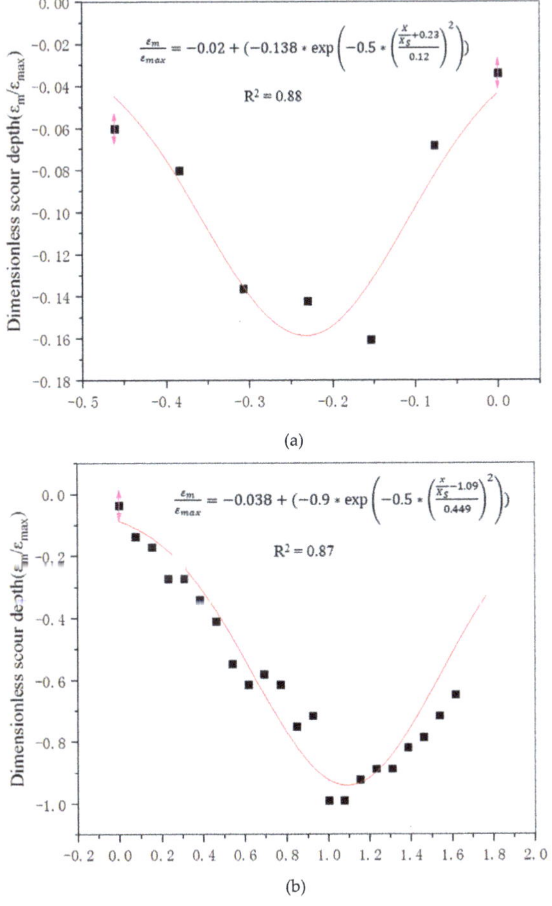

Figure 9. *Cont.*

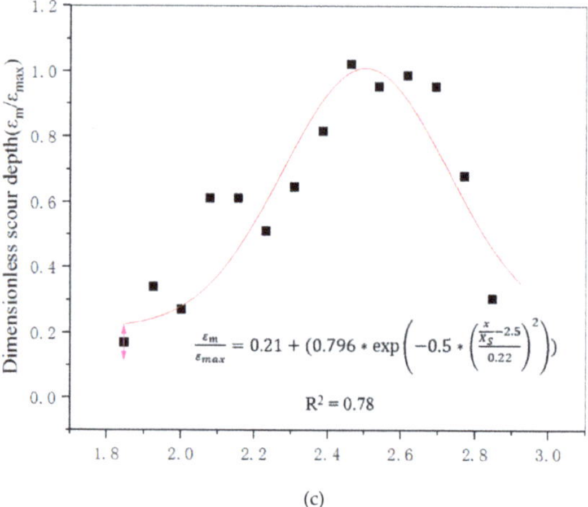

(c)

Figure 9. Dimensionless scour section of different zones. (**a**) Small scour hole ($-0.5 < X/X_m < 0$); (**b**) primary scour hole ($0 < X/X_m < 1.8$); (**c**) deposition dune ($1.8 < X/X_m < 3$).

The entire scour profile can be predicted by using Equations (5)–(7).

$$-0.5 < X/X_m < 0: \frac{\varepsilon_m}{\varepsilon_{max}} = -0.02 + \left(-0.138 * \exp\left(-0.5 * \left(\frac{\frac{x}{X_S} + 0.23}{0.12}\right)^2\right)\right) \quad (5)$$

$$0 < X/X_m < 1.8: \frac{\varepsilon_m}{\varepsilon_{max}} = -0.038 + \left(-0.9 * \exp\left(-0.5 * \left(\frac{\frac{x}{X_S} - 1.09}{0.449}\right)^2\right)\right) \quad (6)$$

$$1.8 < X/X_m < 3: \frac{\varepsilon_m}{\varepsilon_{max}} = 0.21 + \left(0.796 * \exp\left(-0.5 * \left(\frac{\frac{x}{X_S} - 2.5}{0.22}\right)^2\right)\right). \quad (7)$$

The proposed calculation was compared to the experimental results by Hong et al. [10] in order to validate the proposed scour equation to form 2D scour profile. The maximum scour depth was 0.107 m. The location of the maximum scour depth was 0.75 m. The comparison between the experimental results and the predicted equation is shown in Figure 10.

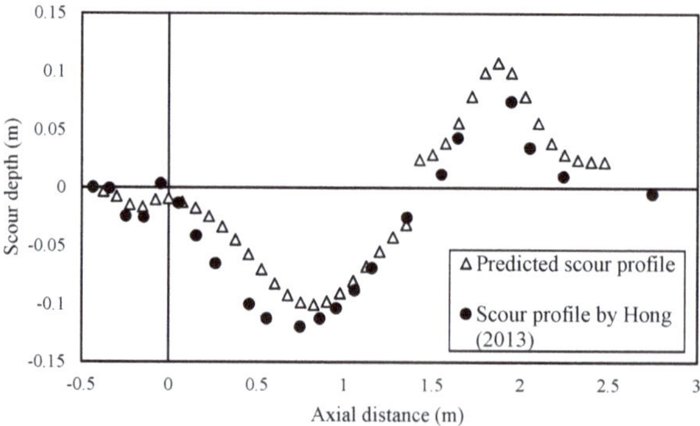

Figure 10. Comparison between the experimental values and the predicted equation.

The results of the predicted scour profile are consistent with the experimental results. The predicted scour depth has an error of less than 20%. The current research suggests that the scour profile of a single propeller can be predicted by using Equations (5)–(7) after the maximum scour depth and position was calculated using the existing equations.

4.3. 2D Scour Section for Twin-Propeller

Scour induced by twin-propeller produced a sand deposition downstream after the primary scour hole. The deposition is in the "M" distribution with two deposition peaks at the rotational axis of both propellers. The deposition decreased towards the two propellers laterally from the M-shaped lateral scour distribution. Previous researchers did not propose a corresponding formula to predict the deposition profile.

For propeller-B, the scour results were recorded at a speed of 700 rpm in T-6 case. The measured twin-propeller scour results in the rotational axis and the symmetrical plane are shown in Figure 11. The maximum scour depth on the rotational axis was significantly larger than the maximum scour depth on the symmetrical plane. The maximum deposition height on the symmetrical plane was higher than the deposition height on the rotational axis. The position of deposition peak on the symmetrical axis was 310 mm from the propeller, but the position of deposition peak on the rotational axis was 350 mm from the propeller. The experimental results show the ridge-like deposition profile with three peaks. The first deposition peak is bigger compared to the two other peaks downstream. The dimensionless scour profile on the rotational axis of twin-propeller (T-1 case) and single-propeller (T-11 case) induced scour are shown in Figure 12.

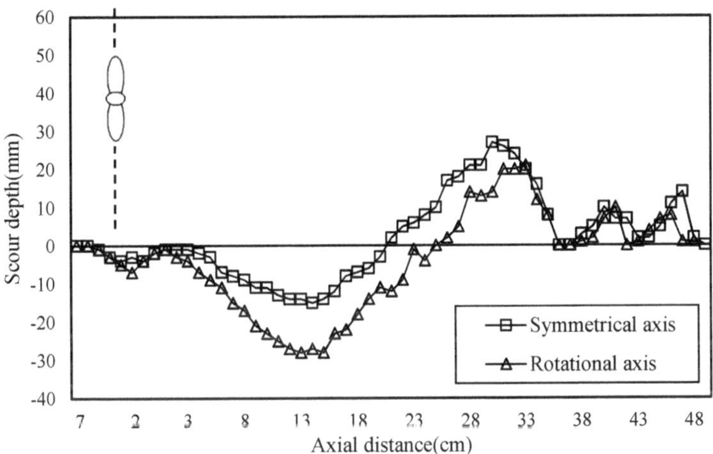

Figure 11. Scour on rotational axis and symmetrical planes for twin-propeller.

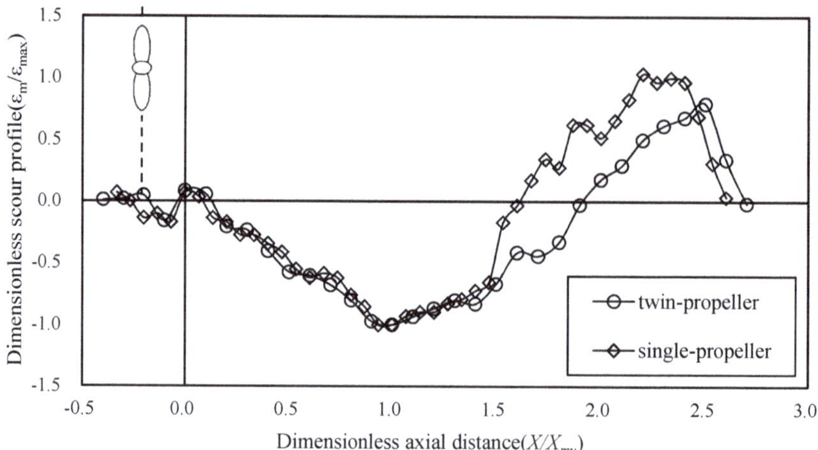

Figure 12. Scour induced by a twin-propeller and single-propeller.

The maximum scour position for twin-propeller can be calculated by Equation (8).

$$X_{m,twin} = 2.1D_p \quad (8)$$

The scour profile of twin-propeller was also divided into three zones including small scour hole, primary scour hole and deposition dune. Table 5 gives the position of the maximum scour depth ($X_{m,twin}$) and the length of primary scour hole ($X_{s,twin}$). The dimensionless length of each zone was also suggested. The three zones were predicted by using the Gaussian distribution curve, as shown in Figure 13.

Table 5. Dimensionless length of each zone for twin-propeller induced scour.

Case	$X_{m,twin}$ (m)	$X_{s,twin}$ (m)	Small Scour Hole (Zone A)	Primary Scour Hole (Zone B)	Deposition Dune (Zone C)
T-1	0.15	0.24	$-0.5 < X/X_m < 0$	$0 < X/X_m < 1.58$	$1.58 < X/X_m < 2.4$
T-2	0.17	0.28	$-0.5 < X/X_m < 0$	$0 < X/X_m < 1.63$	$1.63 < X/X_m < 2.6$
T-3	0.18	0.29	$-0.45 < X/X_m < 0$	$0 < X/X_m < 1.62$	$1.62 < X/X_m < 2.8$
T-4	0.17	0.28	$-0.5 < X/X_m < 0$	$0 < X/X_m < 1.65$	$1.65 < X/X_m < 2.7$
T-5	0.15	0.23	$-0.5 < X/X_m < 0$	$0 < X_m < 1.53$	$1.53 < X/X_m < 2.4$
T-6	0.17	0.27	$-0.5 < X/X_m < 0$	$0 < X/X_m < 1.6$	$1.6 < X/X_m < 2.6$
T-7	0.19	0.33	$-0.5 < X/X_m < 0$	$0 < X/X_m < 1.72$	$1.72 < X/X_m < 3$
T-8	0.18	0.31	$-0.5 < X/X_m < 0$	$0 < X/X_m < 1.71$	$1.71 < X/X_m < 2.9$
T-9	0.14	0.24	$-0.5 < X/X_m < 0$	$0 < X/X_m < 1.68$	$1.68 < X/X_m < 2.8$
T-10	0.15	0.25	$-0.5 < X/X_m < 0$	$0 < X/X_m < 1.64$	$1.64 < X/X_m < 2.8$
Suggested length	/	/	$-0.5 < X/X_m < 0$	$0 < X/X_m < 1.6$	$1.6 < X/X_m < 2.6$

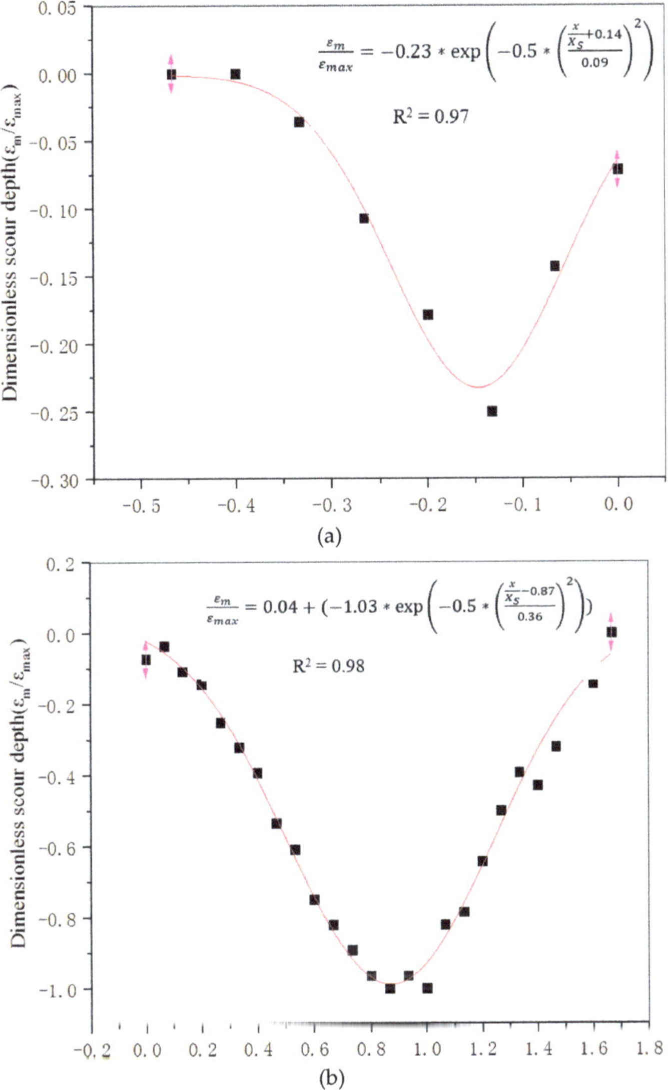

Figure 13. *Cont.*

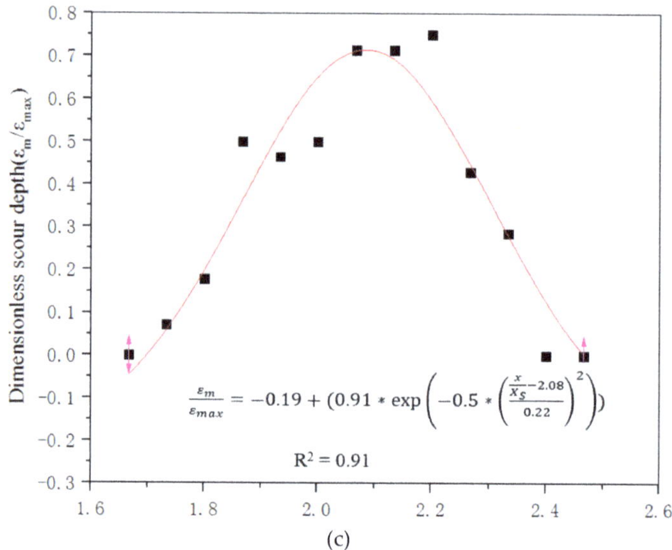

Figure 13. Dimensionless scour section of different zones for twin-propeller. (**a**) Small scour hole ($-0.5 < X/X_{m,twin} < 0$); (**b**) primary scour hole ($0 < X/X_{m,twin} < 1.6$); (**c**) deposit mound ($1.6 < X/X_{m,twin} < 2.6$).

The entire scour profile can be predicted using Equations (9)–(11).

$$-0.5 < X/X_{m,twin} < 0: \ \frac{\varepsilon_m}{\varepsilon_{max}} = -0.23 * \exp\left(-0.5 * \left(\frac{\frac{x}{X_S} + 0.14}{0.09}\right)^2\right) \tag{9}$$

$$0 < X/X_{m,twin} < 1.6: \ \frac{\varepsilon_m}{\varepsilon_{max}} = 0.04 + \left(-1.03 * \exp\left(-0.5 * \left(\frac{\frac{x}{X_S} - 0.87}{0.36}\right)^2\right)\right) \tag{10}$$

$$1.6 < X/X_{m,twin} < 2.6: \ \frac{\varepsilon_m}{\varepsilon_{max}} = -0.19 + \left(0.91 * \exp\left(-0.5 * \left(\frac{\frac{x}{X_S} - 2.08}{0.22}\right)^2\right)\right). \tag{11}$$

The prediction of the scour profile by twin-propeller was compared with the experimental results. The measured maximum scour depth was 28 mm. The location of the maximum scour depth was 140 mm from propeller. The comparison between the predicted equation and previous research is shown in Figure 14.

The predicted scour profile for q twin-propeller proposed by the current research agreed with the scour profile of a single-propeller by Hamill et al. [8] and Hong et al. [10]. The predicted primary scour profile was different from the research provided by Yew [16]. The current research suggests that the scour profile of a twin-propeller can be predicted by Equations (9), (10) and (11). The maximum scour depth and position should be calculated first.

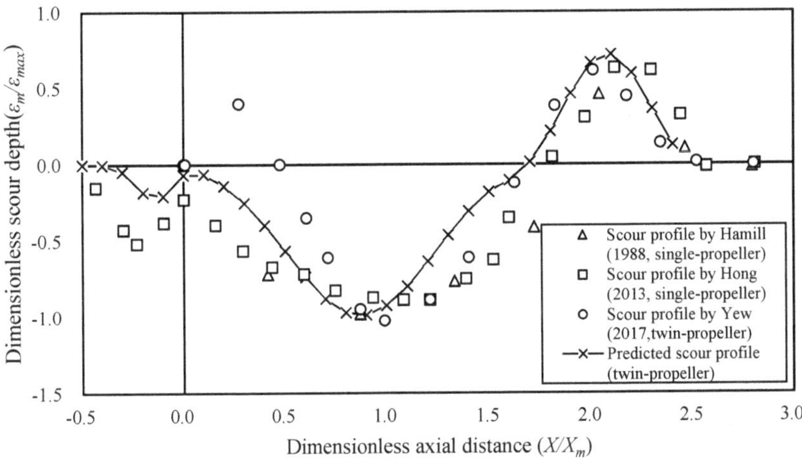

Figure 14. Comparison between the predicted equation and previous research.

5. Conclusions

Scour induced by single and twin propeller jets were investigated to present the scour structure and followed by the proposal of empirical scour equations based on the maximum scour depth and its position. Empirical scour distribution equation was proposed by using Gaussian normal distribution. The scour structure can be divided into three zones, which are the small scour hole, primary scour hole and deposition dune. 2D scour model are proposed enabling the prediction of the scour profiles for both single-propeller induced scour and twin-propeller induced scour.

For the single propeller, Equations (5)–(7) are proposed to predict the entire scour profile to establish a 2D single propeller scour model.

$$-0.5 < X/X_m < 0: \frac{\varepsilon_m}{\varepsilon_{max}} = -0.02 + \left(-0.138 * \exp\left(-0.5 * \left(\frac{\frac{x}{X_S} + 0.23}{0.12}\right)^2\right)\right) \tag{5}$$

$$0 < X/X_m < 1.8: \frac{\varepsilon_m}{\varepsilon_{max}} = -0.038 + \left(-0.9 * \exp\left(-0.5 * \left(\frac{\frac{x}{X_S} - 1.09}{0.449}\right)^2\right)\right) \tag{6}$$

$$1.8 < X/X_m < 3: \frac{\varepsilon_m}{\varepsilon_{max}} = 0.21 + \left(0.796 * \exp\left(-0.5 * \left(\frac{\frac{x}{X_S} - 2.5}{0.22}\right)^2\right)\right). \tag{7}$$

For the twin propeller, Equations (9)–(11) are proposed to predict the entire scour profile to establish a 2D twin propeller scour model.

$$-0.5 < X/X_{m,twin} < 0: \frac{\varepsilon_m}{\varepsilon_{max}} = -0.23 * \exp\left(-0.5 * \left(\frac{\frac{x}{X_S} + 0.14}{0.09}\right)^2\right) \tag{9}$$

$$0 < X/X_{m,twin} < 1.6: \frac{\varepsilon_m}{\varepsilon_{max}} = 0.04 + \left(-1.03 * \exp\left(-0.5 * \left(\frac{\frac{x}{X_S} - 0.87}{0.36}\right)^2\right)\right) \tag{10}$$

$$1.6 < X/X_{m,twin} < 2.6: \frac{\varepsilon_m}{\varepsilon_{max}} = -0.19 + \left(0.91 * \exp\left(-0.5 * \left(\frac{\frac{x}{X_S} - 2.08}{0.22}\right)^2\right)\right). \tag{11}$$

Notation:

C = clearance distance from propeller tip to bed;
C_t = thrust coefficient;
D_h = propeller hub diameter;
D_p = propeller diameter;
d_p = the distance between twin-propeller;
d_{50} = average sediment grain size;
F_0 = densimetric Froude number;
n = number of revolutions per second/minute;
N = blade number
β = blade area ratio;
t = time;
$\varepsilon_{m,\ twin}$ = depth of maximum scour of twin-propeller at time t;
$Z_{D,\ twin}$ = maximum deposition height of twin-propeller at time t;
Ω = experimental coefficient;
Γ = experimental coefficient;
X_m = distance to location of maximum scour for single-propeller;
$X_{m,twin}$ = distance to location of maximum scour for twin-propeller;

Author Contributions: W.-H.L.'s long term research series in ship propeller jet induced scour with former supervisor G.H.; Y.C. and W.-H.L. contributed to establish 2D scour model to predict twin-propeller scour; Y.C. wrote the manuscript with revisions, recommendations and validations from W.-H.L., T.Z., C.S. and G.H.

Funding: This research was funded by the Natural Science Foundation of Tianjin City: 18JCYBJC21900 and The APC was funded by the Natural Science Foundation of Tianjin City: 18JCYBJC21900.

Acknowledgments: The authors wish to extend their gratitude to the School of Civil Engineering at Tianjin University for laboratory space; Queen's University Belfast, University of Plymouth, University of Oxford, Dalian University of Technology, University of Malaya (HIR ENG47), Universiti Teknologi Malaysia, and Southern University College for their past support; and professional bodies EI, IEI, IET, BCS, IEM, IEAust, ASCE, ECUK, SCUK, BEM, FEANI, AFEO, MINDS and academy AAET for membership support and available resources.

Conflicts of Interest: The authors declare no conflict of interest.

References

1. Kim, Y.G.; Kim, S.Y.; Kim, H.T.; Lee, S.K.; Yu, B.S. Prediction of the manoeuvrability of a large container ship with twin-propeller and twin rudders. *Mar. Sci. Technol.* **2007**, *12*, 130–138. [CrossRef]
2. MarCom Working Group 180. *Guidelines for Protecting Berthing Structures from Scour Caused by Ships*; PIANC: Brussels, Belgium, 2015.
3. Hamill, G.A. Characteristics of the Screw Wash of a Manoeuvring Ship and the Resulting Bed Scour. Ph.D. Thesis, Queen's University of Belfast, Belfast, Northern Ireland, 1987.
4. Gaythwaite, J. *Design of Marine Facilities for the Berthing, Mooring, and Repair of Vessels*; ASCE: Reston, VA, USA, 2004. [CrossRef]
5. Albertson, M.L.; Dai, Y.B.; Jensen, R.A.; Rouse, H. Diffusion of submerged jets. *Trans. Am. Soc. Civ. Eng.* **1950**, *115*, 639–697.
6. Blaauw, H.G.; van de Kaa, E.J. *Erosion of Bottom and Sloping Banks Caused by the Screw Race of the Manoeuvring Ships*; Delft Hydraulics Laboratory: Delft, The Netherlands, 1978.
7. Verhey, H.J.; Blockland, T.; Bogaerts, M.P.; Volger, D.; Weyde, R.W. *Experiences in Netherlands with Quay Structures Subjected to Velocities Created by Bow Thrusters and Main Propellers of Mooring and Unmooring Ships*; PIANC: Brussels, Belgium, 1987.
8. Hamill, G.A. *The Scouring Action of the Propeller jet Produced by a Slowly Manoeuvring Ship*; PIANC: Brussels, Belgium, 1988.

9. Lam, W.H.; Hamill, G.A.; Robinson, D.; Raghunathan, S. Observations of the initial 3D flow from a ship's propeller. *Ocean Eng.* **2010**, *37*, 1380–1388. [CrossRef]
10. Hong, J.H.; Chiew, Y.M.; Cheng, N.S. Scour Caused by a Propeller Jet. *J. Hydraul. Eng.* **2013**, *139*, 1003–1012. [CrossRef]
11. Hamill, G.A.; Johnston, H.T.; Stewart, D.P. Propeller wash scour near quay walls. *J. Waterw. Port Coast. Ocean Eng.* **1999**, *4*, 170–175. [CrossRef]
12. Wang, S.; Lam, W.H.; Cui, Y.; Zhang, T.; Jiang, J.; Sun, C.; Guo, J.; Ma, Y.; Hamill, G. Novel energy coefficient used to predict efflux velocity of tidal current turbine. *Energy* **2017**, *158*, 730–745. [CrossRef]
13. Sun, C.; Lam, W.H.; Cui, Y.; Zhang, T.; Jiang, J.; Guo, J.; Ma, Y.; Wang, S.; Tan, T.H.; Chuah, J.H.; et al. Empirical model for Darrieus-type tidal current turbine induced seabed scour. *Energy Convers. Manag.* **2018**, *171*, 478–490. [CrossRef]
14. Ma, Y.; Lam, W.H.; Cui, Y.; Zhang, T.; Jiang, J.; Sun, C.; Guo, J.; Wang, S.; Lam, S.S.; Hamill, G. Theoretical vertical-axis tidal-current-turbine wake model using axial momentum theory with CFD corrections. *Appl. Ocean Res.* **2018**, *79*, 113–122. [CrossRef]
15. Jiang, J.; Lam, W.-H.; Cui, Y.; Zhang, T.; Sun, C.; Guo, J.; Ma, Y.; Wang, S.; Hamill, G. Ship Twin-propeller Jet Model used to Predict the Initial Velocity and Velocity Distribution within Diffusing Jet. *KSCE J. Civ. Eng.* **2019**, *23*, 1118–1131. [CrossRef]
16. Yew, W.T. Seabed Scour Induced by Twin-propeller Ships. Ph.D. Thesis, University of Malaya, Kuala Lumpur, Malaysia, 2017.
17. Hamill, G.A.; Kee, C. Predicting axial velocity profiles within a diffusing marine propeller jet. *Ocean Eng.* **2016**, *124*, 104–112. [CrossRef]
18. Chiew, Y.M.; Lim, S.Y. Local Scour by a Deeply Submerged Horizontal Circular Jet. *J. Hydraul. Eng.* **2016**, *122*, 529–532. [CrossRef]
19. Chen, L.; Lam, W.H. Methods for predicting seabed scour around marine current turbine. *Renew. Sustain. Energy Rev.* **2014**, *29*, 683–692. [CrossRef]
20. Mujal-Colilles, A.; Gironella, X.; Crespo, A.J.C.; Sanchez-Arcilla, A. Study of the Bed Velocity Induced by Twin Propellers. *J. Waterw. Port Coast. Ocean Eng.* **2017**, *145*, 04017013. [CrossRef]
21. Qurrain, R. Influence of the Sea Bed Geometry and Berth Geometry on the Hydrodynamics of the Wash from a Ship's Propeller. Ph.D. Thesis, Queen's University of Belfast, Belfast, Northern Ireland, 1994.

MDPI
St. Alban-Anlage 66
4052 Basel
Switzerland
Tel. +41 61 683 77 34
Fax +41 61 302 89 18
www.mdpi.com

Water Editorial Office
E-mail: water@mdpi.com
www.mdpi.com/journal/water